An OPUS book

Energy

A Guidebook

Janet Ramage

Energy

A Guidebook

Oxford New York
OXFORD UNIVERSITY PRESS
1983

Oxford University Press, Walton Street, Oxford OX2 6DP

London Glasgow New York Toronto
Delhi Bombay Calcutta Madras Karachi
Kuala Lumpur Singapore Hong Kong Tokyo
Nairobi Dar es Salaam Cape Town
Melbourne Auckland

and associated companies in
Beirut Berlin Ibadan Mexico City Nicosia

Oxford is a trade mark of Oxford University Press

First published 1983 as an Oxford University Press paperback
and simultaneously in a hardback edition

British Library Cataloguing in Publication Data

Ramage, Janet
Energy. — (OPUS)
1. Force and energy
I. Title II. Series
531'.6 QC73

ISBN 0–19–219169–1
ISBN 0–19–289157–X Pbk

Library of Congress Cataloging in Publication Data

Ramage, Janet, 1932–
Energy, a guidebook. (OPUS)
Bibliography: p. Includes index.
1. Power resources. 2. Power (Mechanics)
I. Title. II. Series.
TJ163.2.R345 1983 333.79 82–14207

ISBN 0–19–219169–1
ISBN 0–19–289157–X (pbk.)

Typeset by Wyvern Typesetting Ltd, Bristol
Printed in Great Britain by
Richard Clay (The Chaucer Press) Ltd
Bungay, Suffolk

For my parents, with love

Preface

This is a book about the ways we use energy, and some of the ways we might use it in the future. I have tried to make it reasonably comprehensive and balanced, paying equal attention to all the main energy sources, conventional or alternative. And I have tried to make it intelligible to people with no scientific or technical background. Indeed, its main aim is to overcome that first important barrier: the feeling that you can't even enter the debate because you don't understand the language. Chapter 1 starts with the claim that the vocabulary of this language, the technical terms, shouldn't be a problem. I can only hope that readers will agree, and also that they will not be too surprised to find space devoted to its grammatical structure, the fundamental scientific concepts and laws. We need these if we are not just to describe the energy technologies but to see why they are as they are, and I do not think that such basic concepts are necessarily out of place in a book like this. If nothing else, it will surely help in the debate on future power supplies if we start with a clear idea of the difference between a watt and a volt. More generally, technology changes all the time. Projects which are actively being developed as this is written will be abandoned before the words appear in print. And although power-stations are not designed overnight, details change continually. Perhaps any book for the interested layman should be judged less by what it tells its readers about the latest wind-power system or solar panel than by how well they can assess the next project they read about.

However, if the book is to serve its main purpose and help anyone who wants to understand – or join – the energy debate, it must be firmly based in the real world. I've tried to ensure this in several ways. The energy technologies are presented in the wider context of our changing patterns of production and consumption. There is discussion of the environmental effects of energy systems and of their safety, and an attempt to tackle the diverse problems of predicting demand and supplies. In the technical chapters I've often chosen to treat the particular rather than the general: the energy used during a specific car journey, or by a certain household. And there are very many facts and figures. I hope that the non-scientific reader doesn't find these too intimidating, but I think they are essential. There's not much point, after all, in arguing about the merits of wave-power without at least a rough idea of the kilowatts potentially available from a mile of Atlantic Ocean.

A list of relevant matters *not* included would be long. I've said little about the history of our energy systems, how we've come to be where we are. There are a few backward glances, but only to look at recent trends which might help us to forecast where we are going. There is nothing on the politics of energy. This is not because I think it unimportant, but because it is outside the scope of the book. My aim has been to give the facts where they are known, and to draw attention to some of the difficulties and uncertainties where they are not.

While writing the book I've enjoyed the great privilege of working in some of the world's more beautiful places, and this may account for the emphasis throughout on the tension between two opposing forces. It began within sight and sound of the rolling breakers on the coast of southern California, continued through a hot New England summer with a view across shady gardens in 'the other Cambridge', and was not quite finished during another summer in the sweet air of an Alpine village. Not wild, uninhabited regions but places where people have lived for centuries; and technology has undoubtedly improved their lives, bringing fresh water, protection against harsh climates and many other benefits. Yet each is now threatened – by more people, more building, more traffic, a new power-station, a new road. I'm sure that any reader will know of similar examples of this central dilemma.

There are many people to whom I owe thanks for making the book possible: the City of London Polytechnic for the grant of leave which enabled me to start it – and my colleagues in the Department of Physics who cheerfully took on my work-load during my absence; my hosts at the University of California, San Diego and at Harvard, especially Professor Penner of the Energy Center at UCSD (to whom credit is due for the nice term *energy arithmetic*). And I must thank many librarians: the staff of the Science and Engineering Library at UCSD whose interest and enthusiasm, being boundless, were made freely available to a temporary visitor, many other helpful people in Harvard's Widener, Cabot and Baker libraries, the Graubündner Kantonalbibliotek, and others; and of course the staff of the City of London Polytechnic libraries – Helen Dalton in particular, who treated with equal good humour requests that books be obtained yesterday and failure to return them weeks later. It has been a great pleasure to work with Nicola Hunter, Susan le Roux and Hilary Feldman of the Oxford University Press. In the nicest possible way, they have pointed out confusing explanations and incomprehensible tables, rearranged diagrams which didn't fit the page, corrected spelling mistakes and typing errors, and brought order to the book. I thank them all very much. Any remaining disorder is of course my responsibility.

Lastly, I would like to thank the person with whom I've discussed many parts of the book at some length. I would like to, but acknowledging a helpful comment on Chapter 7 seems inadequate when the whole endeavour must bear the print of more than twenty years' discussion, debate and argument. The needs of a working wife are quite modest: a husband who will tolerate frequent interruptions of his own work, offer interesting and valid comments but not mind if they are ignored, and maintain a warm enthusiasm for a project which means many months of domestic chaos. I have been very lucky.

J. R.

May 1982

Contents

List of figures

Figures by Illustra Design

List of tables

List of abbreviations*

A amp
a.c. alternating current
AGR advanced gas-cooled reactor
bbl barrel
boe barrel of oil equivalent
Bq becquerel
BTU British thermal unit
BWR boiling-water reactor
CANDU Canadian deuterium-uranium reactor
CEGB Central Electricity Generating Board
CHP combined heat and power
Ci curie
cm centimetre
cu cubic
cwt hundredweight
EJ exajoule
ft foot
gal gallon
GDP gross domestic product
GW gigawatt
HAWT horizontal-axis wind turbine
HWR heavy-water reactor
in inch
J joule
K Kelvin
kg kilogramme
kWh kilowatt-hour
lb pound
LFMBR liquid-metal fast breeder reactor
LOCA loss-of-coolant accident
LWR light-water reactor
m metre (except in mpg and mph)
Mbd million barrels a day
mi mile
MJ megajoule
mm millimetre
mpg miles per gallon
mph miles per hour
Mte million tonne
MW megawatt
μg microgramme
PJ petajoule
psi pounds per square inch
PWR pressurized-water reactor
s second
SCF standard cubic foot
SNG substitute natural gas
sq square
Sv sievert
tce tonne of coal equivalent
te tonne
TW terawatt
u atomic mass unit
V volt
VAWT vertical-axis wind turbine
W watt
yr year

* For definitions see Appendixes A and B.

Part I Energy Today

Prologue

Well-known expert: World Primary Energy Consumption at the present moment of time is ten terawatts. In accordance with our high-growth scenario . . .

Voice: Please.

Expert: . . . demand is projected to reach eighty terawatts in the year 2030. Question?

Voice: Please, what is a terawatt?

Expert: 1,000,000,000,000 watts. With a fossil fuel production potential of five hundred million barrels of oil equivalent per day . . .

Voice: Please, how many terawatts in a barrel of oil – er – equivalent?

Expert: . . . we obtain a high-probability prognosis of a non-negligible shortfall on the long-term time horizon. Your question is incorrect.

1 The Global Picture

The problem

We've all heard about the World Energy Crisis. It has featured in newspaper headlines and television programmes, and has been the subject of a great deal of public debate. A fairly typical account of the problem might run as follows.

> We are consuming energy at an ever-increasing pace, and reserves of oil and natural gas will be exhausted within a few decades. Our present way of life depends so critically on these two key fuels that unless we take urgent action the whole edifice will collapse, leaving our children shivering and starving in an over-populated, Stone Age world.

A little dramatic, but straightforward enough; and some solutions are equally straightforward:

> THE NUCLEAR FUTURE: World power needs could be met by 100,000 breeder reactors, while fuel for transportation would come from the liquefaction of coal, requiring the strip-mining of some 1,000 square miles a year. Problems of the environment and of hazards to health should not be exaggerated.
>
> THE SOLAR FUTURE: We must return to a low-technology way of life, rising and going to bed with the sun, which will provide power for our modest needs. We must grow our own food, build our own shelter and weave our own blue jeans.

Then, of course, there are those who deny the whole problem. It's all a plot by 'Them' and if we just ignore it, the energy crisis will go away, to be replaced by next year's fashion in disasters.

How do we know which expert to believe? We are offered a variety of unappealing solutions to a problem which may or may not exist and it's not surprising that many of us are uneasy. Opinion is confused with fact all too often in the energy debate while the religious fervour of the advocates of one or another solution to all our problems makes rational analysis difficult. If we do feel that there are important issues here, which we should all try to understand because they affect us all, then it seems that we have little choice. We must try to find our own way to the facts and assess them for ourselves.

Unfortunately the path is not made easy. At the first step we sink into a morass of jargon: terawatts and petajoules, therms and quads, barrels of oil or tonnes of coal equivalent. Then there are the technological blocks, hiding-places for certain experts who are so concerned to protect their own territory that they mislead when they should inform. But it is definitely not the moral of this book that we should give up in despair. Quite the contrary. If it has one, it is that the energy world is not some inaccessible place wrapped in impenetrable mystery, but an area open to any reasonable person. Anyone who has grasped the idea that lengths are sometimes measured in inches and sometimes in miles can come to terms with the various units of energy, which then cease to be jargon. And the way to deal with technological blocks is, naturally, to break them into little pieces and make a good smooth road. In plain words, if we make sure not to get lost at the beginning, and advance carefully, step by step, there is no reason why the science and technology should present much difficulty.

So much for the facts and figures and the technology, the essential background; but the energy debate is about more than these simple things. It involves economics and politics, and value-judgements about the sort of world we want to live in – much more difficult subjects. The size of world oil reserves, and whether coal is a clean fuel or not, are determined not only by technology but by how much we are prepared to pay for our energy. Physics tells us the temperature at which a meltdown will occur in a nuclear reactor, but not the probability that two valves are accidentally left closed, another fails to close when it should, and a maintenance record tag covers up a warning light, all at the same time. To ignore economic, social and political factors in energy planning would be as foolish as to ignore known facts about the physical world; but we do not have, as with the laws of physics, a firm basis for including these factors in our calculations.

The first two-thirds of this book are largely concerned with factual matters: with our present sources and uses of energy and with technology. Not that these topics are entirely free of uncertainty and conjecture, as we shall see when we look at estimates of world use of wood for fuel or the likelihood of large-scale power from nuclear fusion or ocean waves. Economic factors enter too of course (nothing could be harder fact than the behaviour of the price of heating-oil in the past ten years), but as long as we are concerned with the description of things as they are, these present only the same difficulties as some of the other data. When we come to the third part of the book, however, and look at attempts to predict the future, we are on much less sure ground and must proceed with extreme caution. Of three possible approaches to futurology, therefore, two will be adopted in the concluding chapters, and the third firmly rejected.

First, we can with reasonable certainty analyse proposed courses of action on an 'If . . . then . . .' basis. 'If all houses were insulated to the following standards then the energy saving would be equivalent to X million barrels of oil a day.' Or, 'If X tonnes of substitute natural gas were produced from coal then Y tonnes of additional carbon dioxide would also be produced.' How far people will choose to insulate their houses, or how far the carbon dioxide is harmful, are then separate questions to which we do not at present have clear-cut answers. The professional futurologist, who is required to offer forecasts for the year 2000, must of course attempt to answer all questions; so the second line of approach to the future is to look at some of these predictions and at the assumptions on which they are based. By trying to see how estimates of future demand and supply are obtained, and why some alternative courses are regarded as feasible and others not, we might learn to appreciate the problems and to look equally carefully at the solemn pronouncements of the experts and the fashionably easy answers.

And the rest is questions. The third option is rejected; this book offers no grand solution to all problems. It is a guidebook to the energy world, an attempt to point out and explain the most important features. Like many guidebooks, it includes warnings against hazards and doesn't entirely resist the temptation to award stars, but the choice of destination must be made by the traveller, not the guidebook. Only you can decide where you want to go.

Numbers of things

Any serious discussion of energy must be *quantitative*.

'My car uses very little oil.' In driving a thousand miles or standing in the garage? Compared with Saudi Arabian exports or with a horse?

The trivial example illustrates two requirements. We must know which type of quantity we are discussing (gallons per mile, per year, or just gallons) and we must have a suitable *unit* in which to measure the quantity.

The energy world is unfortunately plagued by a multiplicity of units. Every present-day schoolchild knows that the recommended unit for quantity of energy is the *joule*; but open a book on energy resources and what do we find? Barrels of oil, tons (short, long or metric) of coal, cubic feet of gas, British thermal units, kilowatt-hours, terawatt-years. Hardly a joule to be seen. If we want to understand the energy debate as it is carried on in the real world, we have no option but to come to terms with all of these units. Accordingly, one aim in this first chapter is to introduce the art of 'energy arithmetic' – of converting between the different ways of measuring quantities of energy.

The need to use extremely large numbers can also lead to problems. Most of us can visualize ten objects, perhaps 100; but who can picture 1,000,000,000,000? We cannot avoid using numbers of this magnitude, but we can try to make them more manageable. We can use shorter names, or special ways of writing them. (Appendix A gives an account of these methods and can be used as a glossary of terms such as terawatt or megajoule.) Then we can try wherever possible to convert very large numbers into quantities we can grasp. If the world rate of energy consumption is 10 million million watts and the world population is roughly 5,000 million people, then the *per caput* consumption is about 2,000 watts: the equivalent, as we shall see, of about one gallon of oil a day for each man, woman and child. The energy input to a modern power-station, 3,000 million joules per second, means more when we see that it corresponds to some seven tons of coal a minute. How we know these conversion factors is the subject of later parts of this chapter. For the moment, however, we shall return to the central problem of the book.

World energy consumption

'The world is consuming primary energy at a rate of about ten terawatts.'

The above statement raises at least four questions.

How is energy consumed?

One of the most fundamental laws of physics states that *energy is conserved*. The total quantity stays constant. You cannot create energy or destroy it. If you have ten units of energy at the start, you have ten units of energy – somewhere – at the finish. In this sense, we never consume energy.

It is, however, a matter of great practical importance that energy can take many different forms, and that what we *can* do – and have done at least since Man first used fire – is to devise means for converting from one form to another. When we talk of consuming energy this is what we mean: converting from the chemical energy of wood, coal, oil or gas, from nuclear energy, or from the gravitational energy of stored water in a reservoir, into heat or electrical energy or light or the so-called kinetic energy of a moving vehicle. *Consumption is conversion.*

What is primary energy?

The answer to this question is less clear-cut, because it depends on our present technology and may well change. The main current sources of primary energy are the fossil fuels: coal, oil and natural gas. To these we can add the energy of hydroelectric or nuclear power-stations, the energy of wood or other biological materials used as fuel, and contributions from a few additional sources such as geothermal energy. (The 'raw' energy content of the food we eat is not customarily included in the primary energy count.)

The rather arbitrary nature of the criterion becomes evident if we consider solar energy. It counts if special systems such as solar panels or photovoltaic cells are used, but not if we simply warm ourselves in the sunshine. Nevertheless, drawing demarcation lines need not be too serious a problem – and no one has yet passed a law forbidding common sense, even in this field.

What are ten terawatts?

Appendix A tells us the number of watts in 10 TW, but here the important point to notice is that a watt (or a terawatt) is a *rate* at which energy is used and not a unit of energy itself. Technically a watt is a unit of *power*, of energy per second. One watt is by definition one joule per second. Thus a 750-W electric heater is converting electrical energy into heat (and if it glows, some light) at a rate of 750 joules in each second. Conversely, a kilowatt-hour is a unit of *energy*: the amount converted in one hour at a rate of one kilowatt. As a kilowatt is 1,000 J/sec, it follows that one kilowatt-hour must be 3,600,000 J, or 3.6 MJ (Table 1.1).

To conclude, we might rewrite the statement at the start of this section:

> The world is converting the energy of its primary resources into other forms at a rate of about 10 million MJ in each second.

Table 1.1 Power and energy

Power		
1 watt	=	1 joule/second
1 kilowatt	=	1,000 joules/second
1 kilowatt	=	3,600,000 joules/hour
Energy		
1 kilowatt-hour	=	3,600,000 joules
or 1 kWh	=	3.6 MJ

Which would be more illuminating if we had a better picture of the amount of energy called one megajoule. We therefore postpone for a few pages a fourth, very important question in order to look further at quantities of energy.

Per caput energy consumption

10 TW divided evenly among all the inhabitants of the world gives each of us, as we have seen, 2 kW of primary power. What does this mean? Firstly, it does *not* mean that we can each run a 2-kW electric heater continuously, even if we decide to use our entire personal supply for that, because each kilowatt of electric power usually needs 3–4 kW of primary input. If you run your 2-kW electric heater for one-quarter of the time – six hours a day, year in, year out – you are probably just about consuming your fair share of world energy. Nothing left for production of food or clothing, for TV or travel, or to build a roof over your head.

No one will be surprised to learn that world-wide use of energy is *not* evenly distributed. Table 1.2 shows the production and consumption patterns for different parts of the world, data which surely provide food for thought. Comparison of per caput *energy consumption* and per caput *annual income* shows a not unexpected correlation. However, there is by no means a simple constant proportionality, and, as we shall see, some writers have argued that rising standards of living don't necessarily mean the consumption of ever more energy each year.

Energy equivalents

In this section we shall see how energies can be expressed as equivalent amounts of two important fuels, and shall also look at two further energy units: the BTU and the calorie.

Table 1.2 World comparisons

The figures do not include non-commercial 'biofuels' (see Chapter 2). The last two columns are approximate multiples only, and are discussed in more detail later.

Region or country	**Percentages of world totals**			**Primary energy consumed per caput (MJ/yr)**	**Per caput energy and income in terms of the world average**	
	Population (%)	**Primary energy produced (%)**	**Primary energy consumed (%)**		**Energy**	**Income**
Industrialized countries	20	35	60	190,000	three times	three times
Communist countries	30	35	30	60,000	equal	two-thirds
Developing countries	50	30	10	13,000	a fifth	a third
Western Europe	9	10	20	120,000	twice	three times
Africa	10	6	2	12,000	a fifth	a third
Middle East	3	20	2	36,000	a half	equal
Bangladesh	2	0.015	0.04	1,200	a fiftieth	a twentieth
Brazil	2.5	0.3	1	23,000	a third	equal
China	22	9	9	24,000	a third	a half
India	15	1.4	1.7	6,200	a tenth	a tenth
Kenya	0.35	0.001	0.02	4,000	a fifteenth	a sixth
Switzerland	0.15	0.05	0.3	120,000	twice	six times
Britain	1.3	3	3	150,000	twice	three times
USA	5	22	30	350,000	five times	five times
USSR	6	20	16	160,000	twice	twice

Coal

In burning coal we convert its chemical energy into heat energy. This process is discussed further in Chapter 7, where we also see that coal is a very complex material – or a wide range of complex materials. It is not surprising then that the heat energy released in burning it varies widely – from under 20 MJ to over 30 MJ per kilogram. The figure of 29,000 MJ per metric tonne (1,000 kg or about 2,200 lb.) is commonly used as an international average, but we'll simplify our arithmetic for the present by using 30,000 MJ/te.

Suppose all the primary energy in the world were to come from coal. How much would be needed? We note that

$$1 \text{ year} \simeq 30 \text{ million seconds}^*$$

and

$$1 \text{ tce} \simeq 30{,}000 \text{ million joules}$$

where 'tce' stands for *tonne of coal equivalent*: the amount of heat energy released in burning one tonne of coal.

So burning one tonne of coal a year releases 30,000 million joules in 30 million seconds, or 1,000 J/s:

$$1 \text{ tce/yr} \simeq 1{,}000 \text{ J/s}.$$

The world rate of primary energy consumption is about 10 TW, and a terawatt is a million million joules a second, so we can say that the total consumption (all fuels of course) is

10,000 million tce/yr, or *10,000 Mtce/yr.*

The average person thus uses the energy equivalent of a little over 2 te a year, or roughly 15 lb of coal a day. Of course the fortunate European receives about twice this, and the average American five times as much.

Oil

Successive crises have made us all familiar with at least one measure of quantity of oil: the barrel. This odd unit, alien in a world of pipelines and supertankers, comes from the size of the barrels used to carry oil from the world's first drilled well, in Pennsylvania in the 1860s. One barrel of oil is almost exactly 35 Imperial or British

* The symbol ≃ means 'is approximately equal to'. Note that, like any 'equals' sign, it can only relate two quantities of the same kind: a time to a time, an energy to an energy, etc.

gallons (42 US gallons) and there are usually a little over seven barrels to the tonne.

The heat energy released in burning one tonne of oil is more than that from a tonne of coal: about 42,000 MJ compared with 29,000 MJ, and we would therefore need fewer tonnes of oil to provide the entire world energy. However, oil production and use are frequently expressed in barrels a day rather than in tonnes per year. After a little arithmetic we find that the world total primary energy consumption is

150 Mboe/d (sometimes written Mbd).

Actual world crude oil production is over 60 million barrels a day, so we see that oil currently provides more than 40 per cent of the world's primary energy.

The BTU and the Calorie

One of the few happy consequences of the continuing energy debate is a growing appreciation of the advantages – evident to physicists for the past century or so – of having *one* unit for energy. Even the United Nations Expert Group on Energy Statistics, which 'focused on problems of terminology and definitions of boundaries that arose in the elaboration of the conceptual and methodological approaches to be adopted in the development of energy balances', suggested that 'international and national statistical offices should consider adopting the joule'. Despite this incisive recommendation the millennium is not yet with us, and many countries and official bodies continue to use the BTU or the calorie as their basic measure of energy.

The *British Thermal Unit* was originally introduced as a unit of heat, as its name suggests. One BTU is the heat energy needed to warm one pound of water by one degree Fahrenheit. It is equal to 1,055 J, or approximately,

$$1 \text{ BTU} \simeq 1{,}000 \text{ J}.$$

The *therm*, familiar to all who pay gas bills, is 100,000 BTU and thus equal to about 100 MJ. Remembering (Table 1.1) that one kilowatt-hour is 3.6 MJ, we can write

$$1 \text{ therm} \simeq 29 \text{ kWh}$$

which is to say that you buy the same amount of energy when you pay for one therm or for 29 kWh. How efficiently you subsequently use it in each case is another question.

The BTU is much too small a unit for expressing national or world energy figures, so the countries who base their statistics on it have introduced the *quad* (from the American quadrillion or thousand million million). One quad is 10^{15} BTU, and world energy consumption, 10 TW, is thus about 300 quads a year.

The *calorie*, now popularly known in relation to food intake, also first appeared as a unit of heat: the heat energy needed to warm one gramme of water through one degree centigrade. The 'food' unit is 1,000 times greater than this, and is written kcal, or sometimes Cal. The joule equivalent for this larger unit is

$$1 \text{ kcal} \simeq 4{,}200 \text{ J}.$$

This leads to a further way of looking at our energy consumption. As we have seen, the average European uses about 120,000 MJ a year. If we assume that the *food* energy intake to support one adult is 2,000 kcal a day, or about 3,000 MJ a year, we find that the European is using the energy equivalent of about forty people. A further look at Table 1.2 shows that the number of such energy servants you have at your command depends very much on where you live. It is in a sense a measure of the extent to which you use the muscle-power of machines.

On facts and figures

There remains a fourth question which we should ask about the statement that the world consumes 10 TW: How do we *know* that it does? In other words, before venturing further into the sources of energy, we should perhaps discuss the sources of *data*. Where do the figures come from?

A first answer is that we find them in official statistics, technical journals and similar publications. But one shouldn't believe everything one reads in books – or anything in newspapers – and a little comparison brings out the importance of knowing exactly what the figures mean, and whether they are reliable. Consider, for instance.

> 'The UK chemical industry consumes 14 per cent of energy used by industry in the UK.'
>
> 'the (chemical) industry used over 15 per cent of the energy required by all manufacturing'
>
> 'the gross energy input (to the chemical industry) being 3419 × 10^6 therms per annum or 20.5 per cent of the total industrial energy consumption'

Three quotations from the same issue of a technical journal. Are they disagreeing on the facts, or talking about different things? Notice also the precision implied by the third statement. The author, it seems, knows that the answer is 3,419, not 3,418 or 3,420 million therms – an accuracy of one part in 3,000! Or, on a broader scale, for world oil production in 1978, we have the following figures, all published in mid-1979:

3,099.792 million tonnes (Mte)
3,084 Mte
22,158,251 thousand barrels.

The third presents a slight problem unless we know the number of barrels in a tonne to one part in 22 million. Since this is a US figure we'll adopt the US value for the world average barrels per tonne, 7.33, which gives us

3022.9538 Mte.

Of course if the barrels per tonne are really known only to one part in 733, we should really write 3023 Mte; but the original authors do not seem to have been inhibited in their claimed precision by discrepancies of 10 or even 70 Mte.

In collecting international data there is the particular problem of governments who publish wishful thinking; or of those who decline to publish at all, and are thus responsible for the appearance of the strange geographical entity known as WOCA (or sometimes N-C W). Many energy analyses and projections are based on WOCA data, which are much more detailed than those for the world as a whole. (WOCA is of course the World Outside the Communist Areas.) Beyond such political problems, however, there remain a number of difficulties, which we can perhaps characterize under two headings: *definitions* and *conversions*.

Definitions

World data usually start as national statistics, and since there are almost 200 countries it is hardly surprising that the terminology doesn't always match at the seams. Does 'production' include fuel used by the producer? Does 'consumption' include energy used for transmission of energy? The problems are illustrated by the many pages of explanatory notes associated with any set of official statistics. For instance:

> 'Hard coal mines consumption: this heading covers only coal used as a direct source of energy within the coal industry: it excludes

> the coal burned in pit-head power stations (which is included in "consumption for transformation, thermoelectric plants"), or the free allocations to miners and their families (regarded as household consumption and included as such in "other sectors").'

Without this, how are we to interpret a statement that precisely 72.55 per cent of Britain's coal is used for electricity generation? Without the pages of explanatory notes it would undoubtedly be better – and certainly just as informative – to say, 'A little over 70 per cent . . .'.

A further mismatch appears in comparing figures for *production* and *consumption* of a resource. One would hope that any difference would be accounted for by a change in stocks; but when production figures come from producers and consumption figures from consumers this is by no means always the case. In recent UN data, for instance, total (world) exports of crude oil exceeded imports by 15 Mte. Some of it may be on the high seas – in ships, one hopes – but the figures do illustrate a problem.

Conversions

If we want to talk about energy totals or to discuss the replacement of one source by another – oil by wave power, say – we need some common unit: we must be able to compare a barrel of oil a day with a square mile of Atlantic Ocean. We have already used a number of examples of equivalent amounts of energy but have not bothered too much about the *nature* of these relationships. On inspection we find that we have used 'equivalent' in at least four senses, and it is important to distinguish these.

First there are cases where the conversion between units is *universal* – like the number of millimetres in a foot. The relationships between joules, BTUs and calories are based on precise definitions and careful measurement. 1 kWh is exactly 3.6 MJ. In these cases we can safely treat the conversion factors as known, fixed quantities.

When we come to the real world, however, and quantities such as the heat content of a fuel, matters are not so simple. The heat content of a particular piece of coal, for instance, *can* be measured reasonably accurately, provided the conditions are carefully specified:

> Take one kilogram of dry coal at room temperature, burn it completely to produce carbon dioxide, water and ash, cool these products back to room temperature and measure the heat energy extracted.

Even this leaves open a number of questions, but we can obtain in this way *particular* energy equivalents. Another example might be the measurement of the solar energy arriving on a particular surface in a specified time.

Unfortunately life is not long enough to calculate total energy inputs and outputs by adding contributions from every pound of coal or pint of oil, so we must work with *average* values. This introduces difficulties. The average heat content of British coal is some 26,000 MJ/te but the internationally used value is 29,000 MJ/te. The number of barrels of oil to a tonne varies from about six and a half to nearly eight. It is obviously important to know *which* average we are using.

Finally, we have the question of *attainable* conversion factors. How much electrical energy do we in fact get from a tonne of coal, from a million joules of solar energy, from a million gallons of water in some reservoir? The answers depend on current technology and will certainly change. For the present we must decide whether or not – or how – to include such factors in the data for production and consumption. As we shall see, policies differ.

When we add to all these difficulties the further problem of *time* – that data are subject to revisions for a year or more after their first appearance – we might well ask whether there is any point in the whole activity. Is it worth while collecting all these doubtful figures and then using them for even more dubious predictions?

The answer is probably a qualified yes. It is surely better to know something than to know nothing, and to make some attempt to plan for the future instead of rushing over the cliff with eyes firmly closed. Perhaps we do not need quite the present army of data-collectors trying to extract the sixth significant figure for coal production; and we should certainly not believe anyone who tells us that the world will use precisely 16,231 TWh of electrical energy in 1995. But would anyone seriously recommend that we simply ignore the finite nature of resources and continue on our blissful exponential rise until we reach the catastrophic drop? We return to these questions at the end of the book, but shall assume in the intervening chapters that it *is* worth keeping our eyes open for alternatives, just in case.

Units and conversions

Certain guidelines governing the use of units and conversion factors are followed throughout the book. We have argued that the adoption of a single energy unit (the joule, say), while having the advantage of simplicity, would not help the reader faced with the therms, kilowatt-hours and barrels of oil of the real world. On the other hand, to use 'floating' energy equivalents for fuels, with different values for different countries, would lead to an intolerable complexity. We therefore follow a middle route:

- We shall use which ever unit for energy (or length, area, power, mass, time etc.) seems most appropriate in the context, with a slight bias towards the joule and other metric units.
- We shall adopt a *single* set of conversion factors between the different units. (Appendix B is a summary of these.)
- In giving numerical values we shall use the number of digits which seems justified by the reliability of the data, or where there is no loss of essential detail, suitable approximate values.

2 Primary Energy

Introduction

In order to plan for the future we need to know about the present. Whether our subject is the whole world or a single household, we must ask how much energy is being used today, in what forms and for what purposes; and it also helps to know how the situation has been changing in recent years. These are the questions discussed in this chapter and the next.

We'll start with the world picture. Then, to see the detail within the overall pattern, we look at the changing scene in four countries: Britain, Switzerland, the USA and India. The choice is not entirely arbitrary. The two Western European countries provide some interesting contrasts with their very different geographical conditions and energy resources. The USA must be important in any international study, as the extreme towards which we may all be moving and as the major consumer of almost everything; and India, although not quite at the other extreme, illustrates well the particular problems of a developing country.

The world

The contributions from the main primary resources are shown in Fig. 2.1 and Table 2.1. These world data should, as we have seen, be

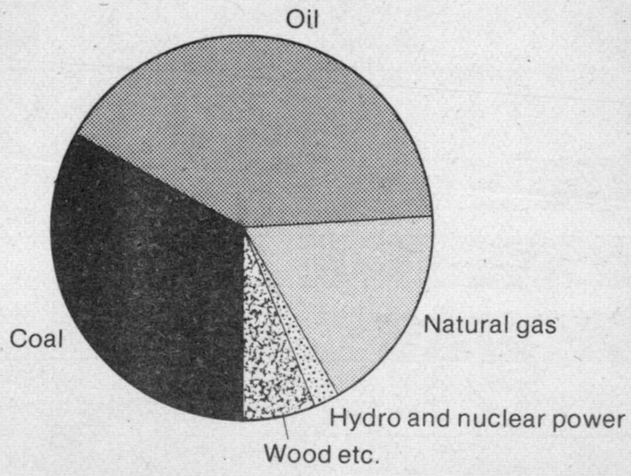

Fig. 2.1 World primary energy, 1980

accepted only with caution, but the general picture is clear enough. The striking features are that oil accounts for two-fifths of all our energy and the three fossil fuels together for over nine-tenths. Fig. 2.2 shows the 1980 data again, together with data for 1950 and 1965, and we see at once the increase in total consumption, doubling in the first fifteen-year period and almost doubling again in the second. The swing to oil and gas at the expense of solid fuels is evident, but it is worth noting that coal consumption has nevertheless more than doubled in thirty years.

Between 1950 and 1980 the population of the world rose from some 2,500 million to about 4,400 million people, so the *per caput*

Table 2.1 World primary energy contributions, 1980

Total primary energy: 320 EJ ≃ 300 quad ≃ 7600 Mtoe per year, or 150 Mbd or 10 TW

Source	**Quantity**	**Energy (EJ)**	**Percentage contribution**	
			To total energy (%)	**Commercial energy only (%)**
Coal	3,600 Mte	105	33	35
Oil	3,000 Mte (60 Mbd)	128	40	43
Natural gas	50 M M SCF	56	18	19
Power-station output				
Hydro	1.5 M M kWh	5.8	1.8	1.9
Nuclear	0.6 M M kWh	2.3	0.7	0.8
Biofuels				
Wood	50,000 M CF	12	4	—
Wastes	500 Mte	7	2	—

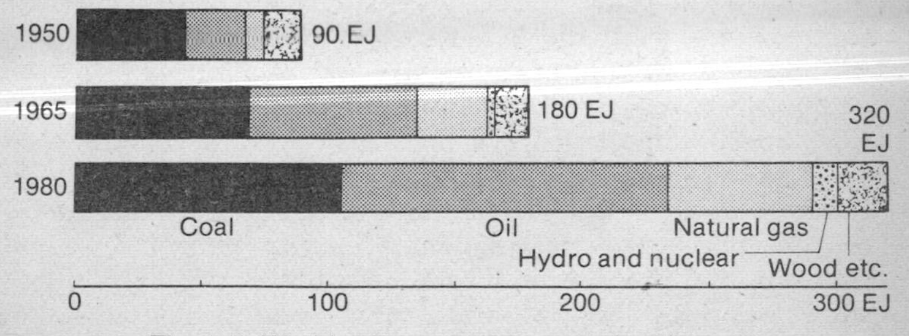

Fig. 2.2 World primary energy growth, 1950–80

energy has more than doubled: our galloping consumption owes more to our increasing appetites than to our growing numbers.

Interpretation of data

Before moving on to national figures we should perhaps look a little more closely at the three minor contributions to the world's energy. Hydroelectric power (equivalent to about twenty Niagara Falls) provides about a sixtieth of the total; two hundred and fifty nuclear reactors add half as much again; and the only other contribution large enough to appear on our diagrams is from 'wood etc.'. Each of these raises problems of data, and although they don't greatly affect the world total they can be important for individual countries and significant in international comparisons.

The category 'wood etc.' includes all fuels not otherwise mentioned: wood, plant wastes, dried dung, urban refuse – anything which is burned to produce heat or power. Determining the quantity and energy content of these is no simple matter. Much of the consumption is outside the developed nations, and methods for collecting and analysing data are accordingly less standardized. Then the fuel is often 'non-commercial' – nobody pays money for it – so the economists' methods of keeping track of quantities cannot be used. It might seem unimportant whether or not we have accurate figures for these fuels, and they do account for only a small fraction of the total. But for much of the developing world they are essential for life. Without some means of cooking, people starve. And there is increasing concern because growing populations are denuding ever-larger areas of all wood. It may well be very important indeed to know how much energy these fuels provide, and to start looking for some substitute in order to prevent a *real* fuel famine – a very different thing from our little local petrol shortages.

Hydroelectricity and nuclear power also raise problems, but these have more to do with the presentation of the data. At the simplest, we must be careful to distinguish between *generating capacity* and *output*. No power-station delivers its full rated power all the time. It may be out of commission, or demand may be low. A 500-MW plant could deliver in one year 4.4 million megawatt-hours of electrical energy if it ran continuously at full power, but the actual energy delivered may be only three million: a *capacity factor* of 68 per cent (Fig. 2.3). And then we need to know whether the given output allows for losses due to electricity used in the plant itself and in transmission to the consumer.

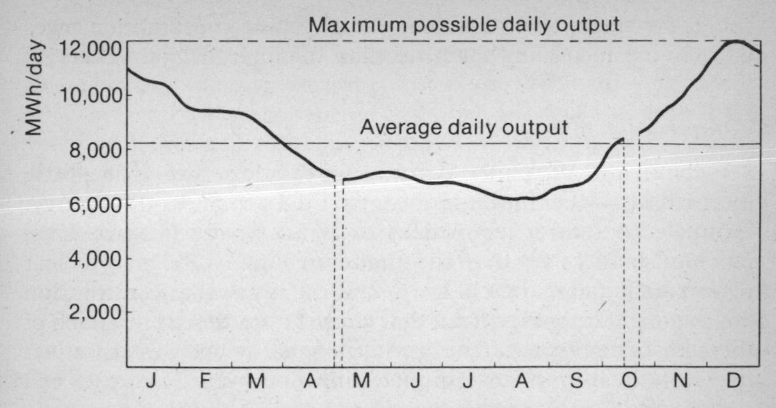

Fig. 2.3 Output from a large power-station during one year

1. The maximum possible yearly output is 4,400,000 MWh.
2. The actual output for the year is 3,000,000 MWh, so the CAPACITY FACTOR is 68%.
3. The large variation in demand in the course of each day, and at weekends, is not shown by this curve.
4. The use of air-conditioning can mean that the annual variation is reversed – in parts of the USA, for instance.

In any case, you may well ask why the *output* from power-stations is counted as primary energy. Shouldn't it be the *input*? Unfortunately there are difficulties with this. We'll look at the details in later chapters, but a brief summary of the facts will bring out the problem. There are three main types of power-station. A *hydroelectric* plant converts nearly all the energy of its falling water into electrical energy. In *fossil fuel* plant, nearly all the energy of coal, oil or gas is converted into heat, and a little over a third of this is in turn converted into useful electrical energy. The electrical output is thus about a third of the primary energy input. *Nuclear reactors* also produce heat, and a little under a third of this is typically converted into useful electrical energy. However, as we'll see in Chapter 8, it is less simple to decide on a figure for the 'primary input' to the reactor. Given this situation, a country must choose how to present its figures, and different countries choose differently.

Some countries, Switzerland included, enter the actual input energy for hydroelectric plant and a 'nominal input' equal to about three times the known output for nuclear plant. Britain adopts a more radical approach, with a nominal input for *both*. The true hydroelectric input is multiplied by about three, the argument being that this gives a better measure of its value as primary energy compared with, say, a certain number of tonnes of coal. Another method is to treat hydroelectric and nuclear plant as special cases, saying nothing about the input and presenting the electrical output as a form of primary energy. This has the advantage of simplicity and is the method we adopt, but it does have disadvantages. It does not reveal, for instance, whether the fuel for nuclear reactors is produced within the country or imported, and the waste heat from nuclear plant won't appear anywhere in the figures.

The main purpose of the above discussion is to warn the reader. In comparing the data given here with official statistics, or comparing official statistics for different countries, it is obviously essential to know how the various authors have done their sums. To repeat, the method used here – in this chapter and later – shows the power-station *output* as primary energy contribution from both hydroelectric and nuclear sources, for all countries.

Britain

The United Kingdom is now almost self-sufficient in energy. The jubilation which has greeted this news is a little surprising since, except for one short period, Britain has been more than self-sufficient in energy throughout recorded history. At the start of the

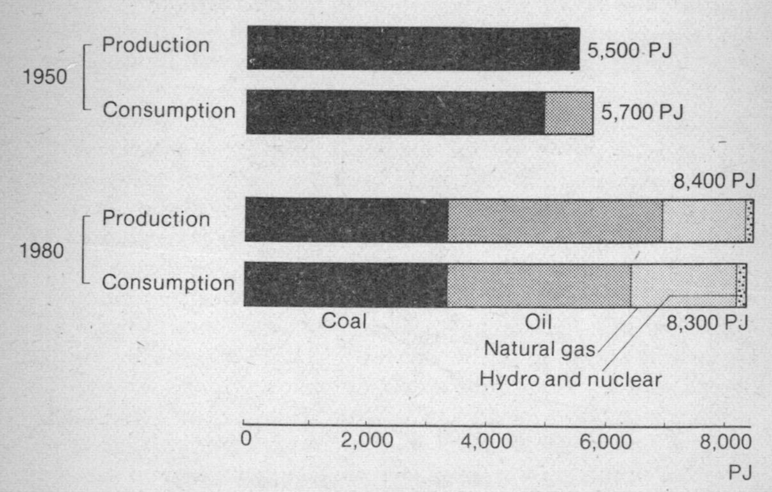

Fig. 2.4 Production and consumption of primary energy, Britain 1950 and 1980

Second World War the energy of coal exports was still more than twice that of oil imports. But of course the internal combustion engine does not run conveniently on coal, and oil and natural gas are preferred fuels for many other purposes, so self-sufficiency in these is without doubt important.

The patterns of production and consumption of primary energy are shown in Fig. 2.4, with the 1950 picture for comparison. (Fig. 2.5 shows in graphical form the changes over a longer period for the two main fuels.) We see that the relative contributions of the main resources are much the same as for the world as a whole. Hydroelectricity plays a smaller role and nuclear power a somewhat greater one, while non-commercial fuels do not contribute significantly. Given that the world pattern for commercial fuels is heavily dominated by the developed countries, the broad similarities are not surprising.

The rise and fall of British coal production is an extreme variant of the world pattern. World-wide, the contribution fell from two-thirds of commercial energy in 1950 to one third now, despite a doubling in the quantity of coal produced. For Britain, the fall in consumption was from nearly 90 per cent to 40 per cent, and production almost halved. (As elsewhere, it has risen somewhat again in the past few years.)

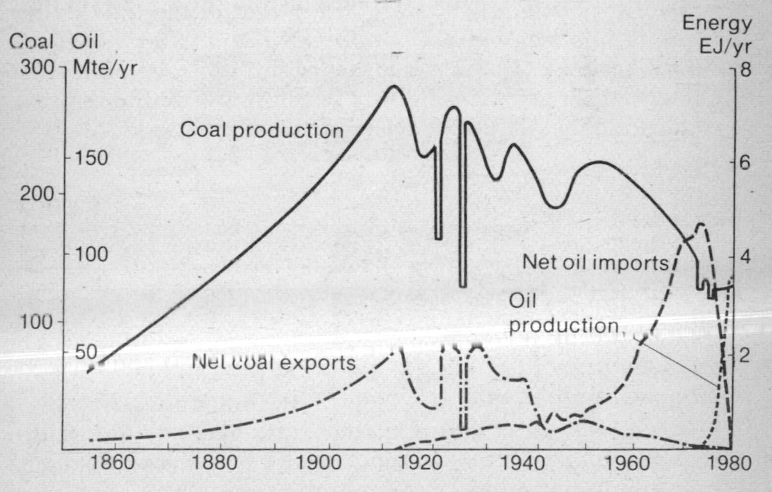

Fig. 2.5 Coal and oil, Britain 1860–1980

1. Year-to-year fluctuations of a few per cent are not shown in these curves.
2. The British value for the energy content of coal (26 GJ/te) is used.

Table 2.2 National primary energy contributions, 1980

	Britain	**Switzerland**	**USA**	**India**
National total primary energy (all in PJ)				
Production	8,400	220	67,000	8,200* (4,200)
Consumption	8,300	760	77,000	9,000* (5,000)
Per caput daily consumption				
Oil	1 gal	1½ gal	3 gal	⅓ pt
Coal	12 lb	½ lb	15 lb	1 lb
Natural gas	80 SCF	15 SCF	250 SCF	¼ SCF
Hydro and nuclear electricity	2 kWh	20 kWh	6½ kWh	⅙ kWh
Biofuels	—	1 lb	—	2 lb
All primary energy	410 MJ	320 MJ	960 MJ	36 MJ* (20 MJ)
Percentage increase in per caput energy 1950–80	**30%**	**200%**	**65%**	**65%* (150%)**

* Including estimated biofuels consumption. Figures in brackets show commercial fuels contribution only.

The secondary forms into which primary energy is converted, and the final uses, are discussed in the next chapter. We might, however, note here that only a little over 70 per cent of the total energy reaches the *final* consumer: less than 6,000 of the 8,300 petajoules shown in Table 2.2. 'Losses' in conversion and transmission account for nearly 30 per cent, a high figure which reflects Britain's heavy dependence on fossil fuels for electric power generation.

Switzerland

It is probable that most people would, if asked, agree with both the following statements:

> The possession of adequate energy resources, and in particular self-sufficiency in oil and natural gas, is a critical factor for the economic well-being of any country today.
>
> Switzerland is one of the world's more fortunate countries, with a sound economy, a strong currency and a higher standard of living than almost any other.

Yet Switzerland has virtually no fossil fuel resources: no coal, no oil, no gas.

There are of course some who will deny the second statement – Swiss commentators in particular, who deplore inflation (still in single figures) or rising unemployment (about a third of a per cent); but it remains the case that by any reasonable measure of well-being – per caput income, proportion of the population above the poverty level, nutrition, housing – Switzerland stands at present as a counter-example to the easy generalization that economic survival *requires* energy self-sufficiency.

Switzerland does possess one major source of energy. About a quarter of her land surface lies 1,000 m or more above the remaining three-quarters, and with a mean annual precipitation of the order of 1,000 mm (30–40 in) the hydroelectric potential is considerable. Since the first installation 100 years ago (to light a St. Moritz hotel) the waters of the Rhine, Rhône, Inn and Ticino and their tributaries have provided increased capacity each year, and now supply about a sixth of the energy consumed. (Fig. 2.6.)

A second 'local' energy source comes from a technology in whose use Switzerland seems to be ahead of many other countries: the burning of wastes or refuse to produce heat and electric power. It is now generally appreciated that with each tonne of refuse dumped we throw away some 10,000 MJ of usable heat energy. Our rubbish production is estimated to be about one kilogram per person per day (two in the USA), and by burning two-fifths of this total Switzerland adds some 5 per cent to her national energy production.

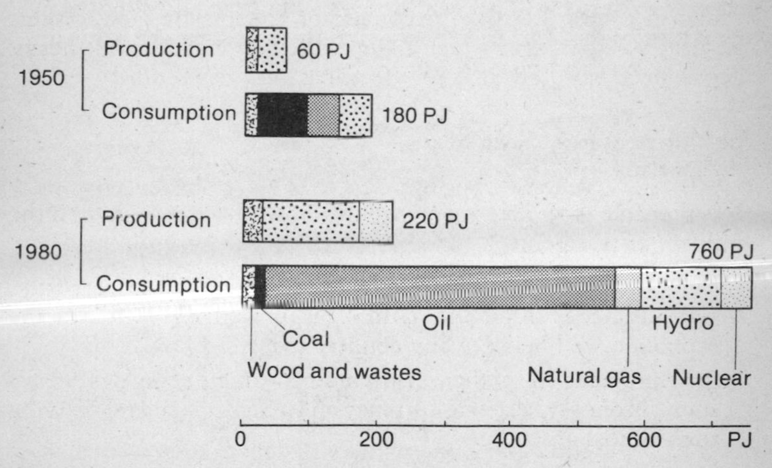

Fig. 2.6 Production and consumption of primary energy, Switzerland 1950 and 1980

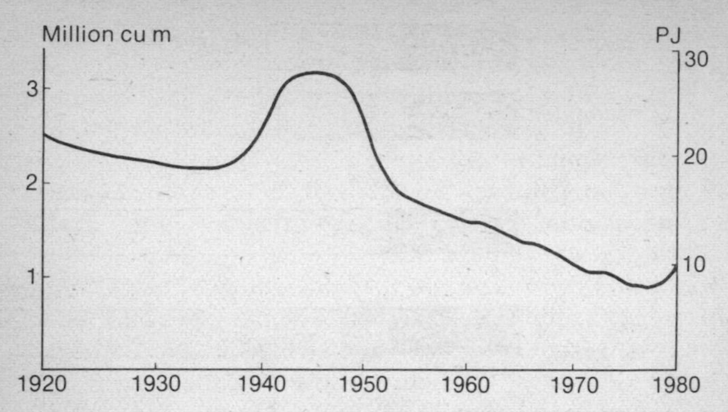

Fig. 2.7 Annual consumption of wood as fuel, Switzerland 1920–80

The remaining indigenous contribution is also worth noting. Wood, which has fallen to a negligible fraction of the energy total in most industrial countries, still provides a per cent or so of fuel energy consumed in Switzerland. Fig. 2.7 shows the general decline over recent years but there are indications that with the rising price of fossil fuel, consumption is again increasing. Nearly half an acre of forest per head of population, three-quarters of it in common ownership and publicly accessible, makes estimation of the amount of wood used for domestic heating difficult, and figures from different authors can vary by 50 per cent or more. The graph does, however, reveal clearly that this is an energy reserve which can be drawn on quickly in a period of shortage.

The United States of America

The US is everyone's favourite example of conspicuous consumption, in energy as in material goods. Per caput consumption is five times the world average and over twice that of Western Europe, while an apparently insatiable thirst for petrol has led to a rake's progress from self-sufficiency to extreme dependence in a couple of decades. The world's leading oil producer until 1974 became its leading importer in 1976.

Yet to turn from this spectacle with a pious shudder would be a mistake. In the first place, all the evidence suggests that given the chance, people throughout the world will adopt rather than reject the American way of life: four-car families, fifteen-channel TV, central-heating, air-conditioning, electric toothbrushes and all; so

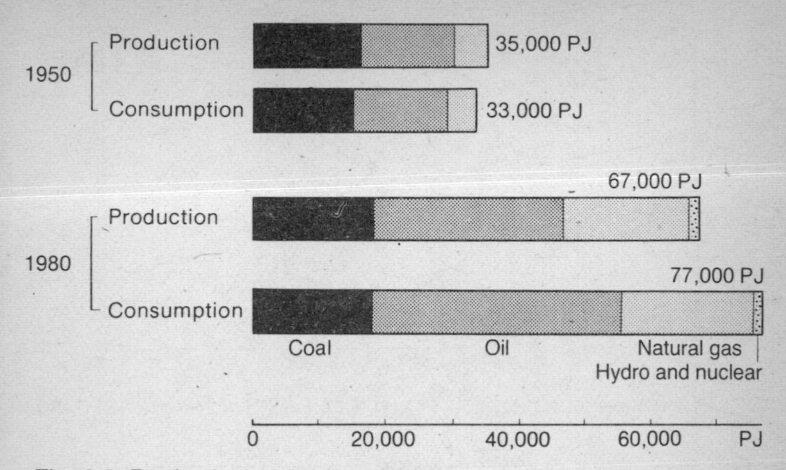

Fig. 2.8 Production and consumption of primary energy, USA 1950 and 1980

we would do better to study carefully this present example of our future. Furthermore, the people of the US are aware of their problems and have in the past decade unleashed their inventive talents in the fields of energy use and resources. They already have much useful experience to offer in the technological and social problems of large-scale introduction of conservation measures, or alternative sources of energy.

Comparison of the present situation with that 30 years ago shows the now-familiar increases (Fig. 2.8). The graphs for the major fuels over a slightly longer period reveal, however, that there have been a number of shifts within this trend (Fig. 2.9). Until the middle 1950s coal production had been falling; but then came growing awareness of the finite nature of reserves. Predictions that oil and gas would start to run out in a decade or so undoubtedly played a role in reversing the trend. The declines in oil and gas production did indeed occur, but were somewhat later and much less catastrophic than had been predicted, for several reasons. First, how much of a resource is 'available' depends on how much someone is prepared to pay to extract it, so rising international prices mean increasing local reserves. Then the completion of the Alaskan pipeline brought an extra one and a half million barrels a day of oil – a 15 per cent bonus for US production. Conservation also helped; and overall, the US leaves the 1970s with energy consumption very little above that at the start of the decade, despite a 10 per cent increase in population.

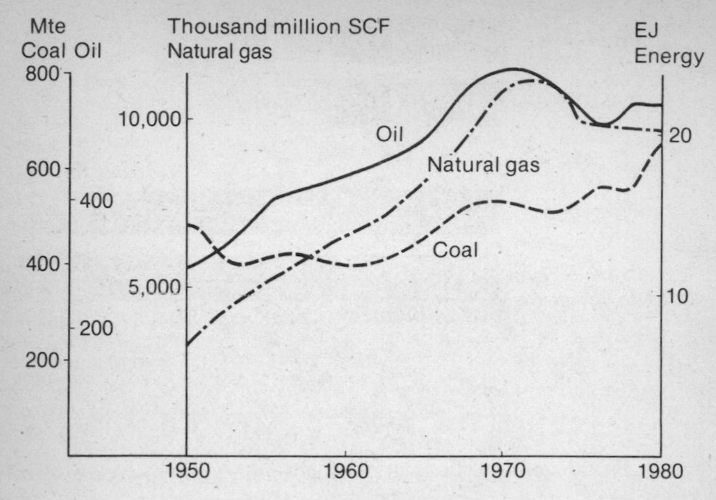

Fig. 2.9 Annual production of fossil fuels, USA 1950–80

The problem which remains, however, is the dependence on imported oil: a critical issue not only in the global use of energy but in the possible consequences for us all when a major power relies on the stability of a not very stable region of the world.

India

A per caput consumption which is little more than a thirtieth of that in the US. A hundred times the Swiss population with just over ten times the total energy. The contrast could hardly be more striking.

However, before starting any detailed analysis of India's uses of energy we need to look carefully at the data (Fig. 2.10). Not so much at the commercial sources, where the figures show that coal production has trebled, that power from hydroelectric and nuclear plants has increased by a factor of 15, and that indigenous oil now supplies nearly half India's needs. These are probably fair indicators. The problem lies in the large part played by non-commercial 'biofuels'. (Some writers also believe that official population figures are an underestimate, which would mean that per caput consumption is even lower than that shown in Table 2.2.)

The biofuels' energy is the sum of three contributions: from wood,

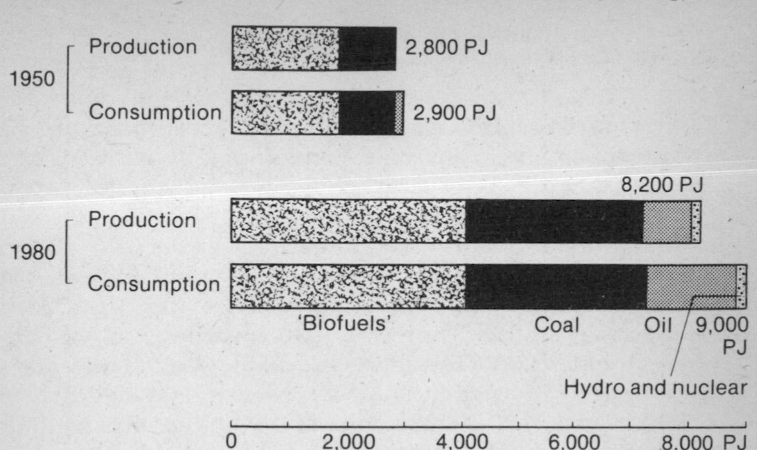

Fig. 2.10 Production and consumption of primary energy, India 1950 and 1980

from the dried dung used as cooking fuel throughout much of the country and from plant wastes which are also burned. Data for all these are uncertain. One survey estimated the amount of wood used as domestic fuel to be over 100 Mte. The official figure, based on forestry returns, was 9 Mte! We probably have here a larger-scale version of the problem discussed in the Swiss context: wood used as fuel is often 'non-commercial' and obtaining reliable data is therefore difficult. Estimates for the quantity of dung cakes used and the corresponding energy consumption also vary considerably, as might be expected. The figures come from two sources: the number of households using the fuel (and their consumption), and the number of cattle (and their production). The same problems, of how much is produced, what fraction is used for fuel, and how much energy this provides, occur again in estimating the contribution from agricultural wastes. So we are left with the difficulty of deciding on a figure to represent the total of three very uncertain contributions. All authorities agree that the energy from biofuels is important. The figure which we use makes its contribution about equal to that of the commercial energy. It is difficult even to give upper or lower limits, but perhaps we can say, somewhat tentatively, that the total is unlikely to be less than half our figure and is probably not twice as much.

Comparisons

A study of Table 2.2 or of the 1980 figures in the diagrams shows that our four chosen countries undoubtedly have very different patterns of consumption. There are some similarities. The average Briton and Swiss use much the same energy, and indeed if we regard natural gas and oil as interchangeable heating fuels, and further note that nearly all Britain's coal is used in electricity production, with an energy output equal to about a third of the input, the differences in consumption per person and in the general pattern largely disappear.

Not so of course with the other two countries, and one very disappointing feature of Table 2.2 is the almost identical percentage increase over thirty years in per caput consumption in India and the USA. Not much sign of a reduction in that thirty-to-one ratio. A twentieth of the world's population is still using a quarter of its primary energy, and the absolute gap between the wealthiest country and the developing nations is wider now than thirty years ago. Then we have Switzerland, not the first to spring to mind as a developing country, starting from a 1950 figure which would today be regarded as typical for Latin America and increasing consumption by a startling 200 per cent, mainly through a twelve-fold rise in the use of oil. It is important of course to remember when comparing all these figures that 'per caput consumption' is no more than total consumption divided by population, and that the ways in which the fuels are distributed – between for instance factories, households and motorists – can be just as significant as the overall data.

From the standpoint of resources then, we have one country in the happy position of producing almost all its present needs. The others, like much of the rest of the world, are all heavily dependent on imported oil. Switzerland has the advantage of a large input from a resource with no problem of diminishing reserves, and the USA has of course the option of becoming self-sufficient by reducing consumption to merely twice the European value. India has little option for reducing anything, as most of her population live at a level of energy consumption which is regarded as barely sufficient to support life.

3 Patterns of Consumption

Energy balances

To the best of our knowledge, energy is conserved; so the total annual consumption of a nation, a factory or a household *must* be equal to its total annual supply. If we draw up a balance sheet with energy income on one side and expenditure on the other, and if we are careful to include entries for stock changes, losses and so on, then physics tells us that the books must balance.

Unfortunately the procedure turns out to be less than straightforward. The principal categories of primary energy ('income') are of course known, but when we come to expenditure we are faced with a range of options. We may want to know how much energy is used for each of a variety of *purposes*: heating, lighting, driving machines, etc. Or in national statistics the interest may be in the use by different *consumers*: industry, transport, households and others. The two cases obviously require quite different categories on the consumption side of the balance sheet. Then we may want to trace the use of each fuel separately from production to consumer, or alternatively just add all the input contributions and see how the total energy is distributed.

As an example of one approach, let us look at an energy balance sheet for Britain for 1980 (Table 3.1). We start with a figure for home production of each fuel, expressed in petajoules. Imports are added and exports subtracted to give a grand total of 9,000 PJ, from which we must further subtract allowances for non-energy uses of oil (lubricants, raw materials for the plastics and other petrochemical industries, etc.) and for fuel used by foreign-going ships (not normally counted as home consumption). After taking into account stock changes, we are left with a gross consumption of 8,300 PJ.

Turning to the consumers, we calculate the total energy used by adding the energies of petrol, diesel oil, fuel oil, natural gas, coal, electricity, etc. purchased or produced in each sector. We thus obtain a total consumption figure – only 5,900 PJ. The discrepancy is no great surprise, of course, because we know in general terms the fate of the missing 2,400 PJ. They were consumed in mining and

Table 3.1 Energy balance sheet, Britain 1980

Supply	Coal	Oil	Natural gas	Hydro-electricity	Nuclear	**Total**
Production	3,300	3,600	1,400	20	120	8,400
Imports (+)	300	2,500	400	—	—	+3,200
Exports (−)	—	2,600	—	—	—	−2,600
						9,000
Non-energy	—	300	—	—	—	− 300
Bunkers	—	100	—	—	—	− 100
To stock	300	—	—	—	—	− 300
Gross consumption	**3,300**	**3,100**	**1,800**	**20**	**120**	**8,300**

Consumption	Coal etc.	Oil products	Natural gas	Electricity	**Total**
Industry	400	700	650	300	2,000
Household	300	100	900	300	1,600
Transport	—	1,500	—	10	1,500
Other	100	300	200	200	800
Net consumption	**800**	**2,600**	**1,750**	**800**	**5,900**

All data in petajoules

refining, became waste heat in power-stations, were used in distributing energy to the final consumers, and so on. There is no scientific mystery; just a problem of accounting procedure. Is this type of balance sheet satisfactory then? Does it tell us what we want to know, or does it – like many balance sheets – conceal more than it reveals?

It certainly allows us to compare one country with another in rather broad terms, but if our concern is to improve the efficiency of energy use we undoubtedly need to know more about the gap in the middle. Where *are* these losses? How much energy is used up in producing one barrel of oil or one tonne of coal? How much in transporting the coal to the power-station and in generating electricity? More fundamentally, what is meant in the balance sheet by a consumer, or by consumption, or useful energy? If houses were perfectly insulated no energy would be needed to keep them warm; so shouldn't we enter all heating fuel under 'losses'? Similarly, if you drive at constant speed along a level road the useful energy consumption is 100 per cent, 30 per cent, 15 per cent or zero per cent of the input fuel energy, depending on your definition of 'useful' (see Chapter 6).

Evidently we must first decide what we want to know and then design the balance sheet. In national balance sheets, as in the example above, it is customary to take as 'consumer' the last individual or institution to pay for the energy *as energy*. You pay for 10 kWh of electrical energy, and whether you use it to run a motor or to heat the air outside your windows, it is entered as 'energy used' and not as 'losses'. When you buy a book, you are of course paying for the energy which was used to make the paper, but that is already accounted for in the fuel purchased by the paper industry, and doesn't appear again. The line is clearly arbitrary, but it is convenient because it matches energy accounting with financial accounting. When we want the answers to other questions – how the paper industry or a household uses its energy, for instance – then we need further information arranged in different categories.

The electrical generating industries appear on national balance sheets in a dual role. On the one hand they are major consumers of primary energy (mainly coal) in most countries; but they are also energy producers, selling their product to the final consumers. It seems that they should appear on both sides of the balance sheet – or perhaps in the middle. A relatively simple procedure, if we are studying broad patterns of consumption, is to start by subtracting from the primary fuels the amount which each contributes to electricity generation. We can then ask how the remainder of the primary fuels, and also the generated electric power, are shared by the consumers.

In the following sections we look at this overall picture for the world as a whole and for our selected countries. The remainder of the chapter is then devoted to two particular aspects of energy consumption: the energy costs of transportation, and the ways in which households use energy.

World uses of energy

World total annual energy consumption is, as we have seen, just about 300 EJ for commercial fuels alone, a little over if non-commercial fuels are included. In Table 3.2 we see how the contributions from the main sources are used. The second column gives the input energy to electrical generating plant and the last four show the distribution between the four classes of consumer. The energy contributions to the sectors thus fall into five categories:

- *Coal*: burned directly as fuel, or after conversion into coke, town's gas or other fuel products.
- *Oil*: all petroleum products, including petrol, diesel oil, different types of fuel oil, etc.
- *Natural gas*: mainly burned directly as fuel.
- *Biofuels*: wood, plant and animal wastes burned as fuel.
- *Electric power*: energy supplied in this form, from all primary sources.

The same information is presented in a different form in Fig. 3.1, and certain main features are obvious at once. Nearly half the final energy from commercial fuels is used in industry, with the household and 'other' sectors sharing rather less than a third in roughly equal parts. Transportation uses nearly half the world's oil, or over a fifth of the total final energy: enough to transport the entire world population about 10,000 miles by bus or 1,000 miles by Concorde.

In view of the known inaccuracies in the data we make no attempt in these world statistics to deal with the losses which occur in production, transportation and distribution of energy, except that – as the observant reader will have noticed – the fossil fuel input for electrical power is about three times the electrical energy output. When we turn to individual countries the statistics are in general

Table 3.2 World patterns of consumption, 1980

Source	Total contribution	To generate electric power	Available for distribution	Distribution to final sectors			
				Industry	Transport	Household	Others
Fossil fuels							
Coal	105	45	60	35	5	8 to 12	8 to 12
Oil	130	13	117	45	50	8 to 10	10 to 12
Natural gas	56	8	48	25	—	10 to 15	8 to 12
Biofuels	20	—	20	—	—	20	—
Hydroelectric power	5	5	—	—	—	—	—
Nuclear power	3	3	—	—	—	—	—
	Distributed electric power from all sources		29	12	1	about 10	about 6
Total energy to sectors			**270**	**120**	**55**	**about 60**	**35 to 40**
Percentages to sectors			**100%**	**45%**	**20%**	**20%**	**15%**

All data in exajoules

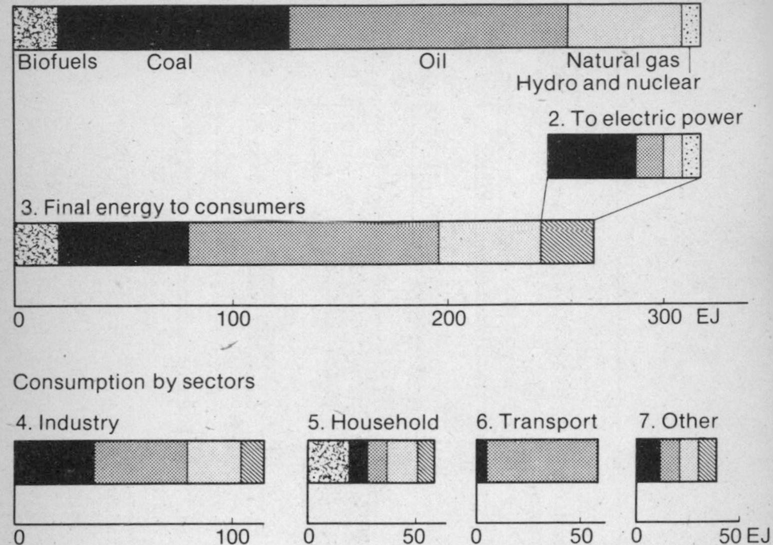

Fig. 3.1 World patterns of consumption

more reliable (or at least more consistent) and we can hope to trace the losses in greater detail.

Uses of coal – a detailed example

When we confront the energy statistics for the industrial nations of the world the problem is not so much a dearth of data as an *embarras de richesses*. Fig. 3.2, showing an energy flow for just one primary fuel in one country, is a simplified diagram omitting much available detail. We shall not attempt to deal with every source in each country in even this degree of detail, but shall use this example to see a little of what lies behind the tables of statistics and to discuss some of the questions which we can ask about national data. The example is of general interest because, as we shall see, three of our four countries use much of their coal for generation of electric power – as do many others in the world – and this use is particularly important for at least three reasons. First, coal is the most plentiful of the fossil fuels; secondly, it is the dirtiest; and thirdly, the heat

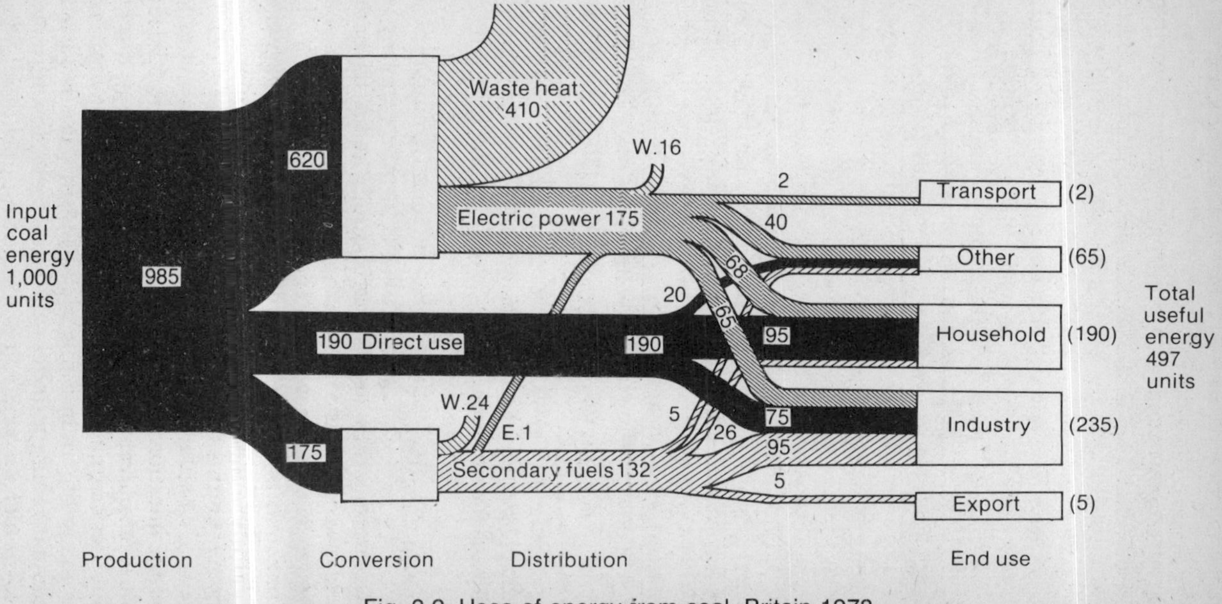

Fig. 3.2 Uses of energy from coal, Britain 1978

losses in power generation are a major contribution to energy waste.

We start at the left-hand edge of the diagram, where we find an input of 1,000 units of coal energy. *Which* unit does not matter, as the proportions will be the same regardless of the unit chosen. This input, the mined coal, is distributed for three main purposes: generation of electricity (62 per cent), direct use as coal (19 per cent) and conversion to secondary fuels: coke and other solid fuels, gas, etc. (17.5 per cent). Of the missing fifteen units, ten are accounted for by coal used in the mining process, and the remaining five are a 'statistical difference' – an unkind interpretation of which is that only 995 units could be accounted for when preparing the national statistics, and this entry is to balance the books.

Following first the *direct use* line, we see that half the coal which is directly burned is used in households and most of the remainder by industry. (This direct use has been falling in all industrial countries.) With the replacement of 'town's gas' by natural gas, almost all the *secondary fuel* goes, as we see along the bottom line, to industry. We note that 18 coal units are lost in the conversion to secondary fuels and a further 24 disappear as waste heat. One unit of surplus electricity is sold to the electrical supply industry. An omission from both the 'direct' and 'secondary' lines is any allowance of energy for *transporting* the fuels. About five units of equivalent coal energy are needed to carry 1,000 units of coal 100 miles by rail, and roughly five times this for road transport. In Britain, where many major industrial users are near the coal-fields, the secondary fuels deduction is too small to appear on our diagram, but the overall transportation energy for coal is probably about five units (mainly as diesel oil) and we should really reduce the 'useful energy' figure by this amount. (For much larger countries such as the US or India, where coal may be transported thousands of miles, the energy costs can of course be considerably greater.)

Coming finally to the major user of coal energy, we also meet the major culprit in production of waste heat: the generation of *electric power*. Of 620 units entering the power-stations, only 174 reach the final consumer as useful electrical energy, an efficiency of 28 per cent. Nearly three-quarters of the coal input is wasted. Why this is, we shall see when we deal with the technical aspects of generation and distribution, but it is unfortunately the case that over a third of the total energy of coal mined in countries such as Britain, the US and India disappears out of the chimneys and into the cooling systems of power-stations.

Four countries

To reduce the large body of available data to manageable size, the basic information for the four countries is presented as a set of diagrams (Figs. 3.3–3.6). These follow the pattern used in Fig. 3.1 with the addition of entries for energy losses. As before, all quantities are expressed as equivalent amounts of energy with the individual contributions from different fuels displayed for each sector. One or two comments and reservations may be worth bearing in mind.

- The first bar in each diagram is simply the 'consumption' bar of the primary energy diagram in Chapter 2.
- Bar (3) shows the losses associated with the production, processing and distribution of the fossil fuels and the losses in generating and distributing electric power. In the interest of simplicity, amounts such as the electric power used in oil refineries or the diesel oil needed to transport coal are omitted. This means that the entries in bars (3) and (4) do not add up exactly to bar (1).

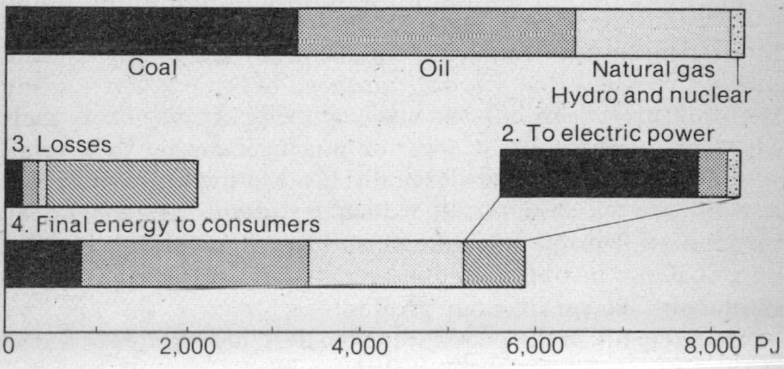

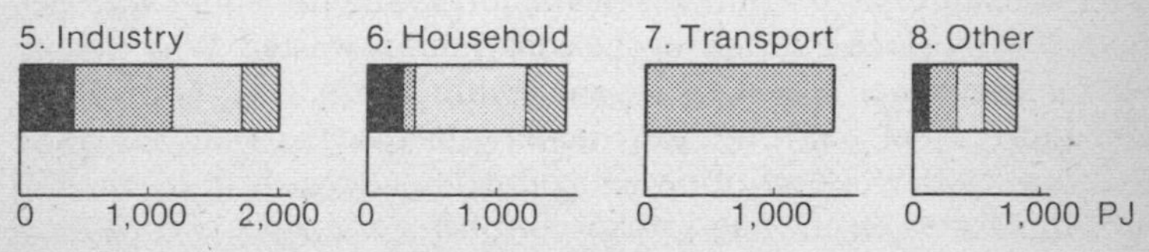

Fig. 3.3 Patterns of consumption, Britain

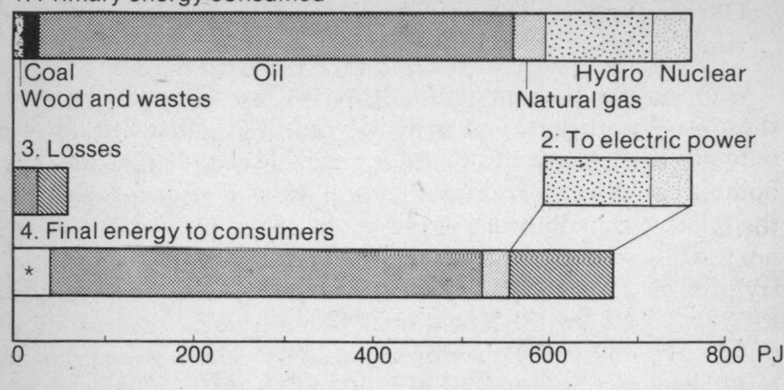

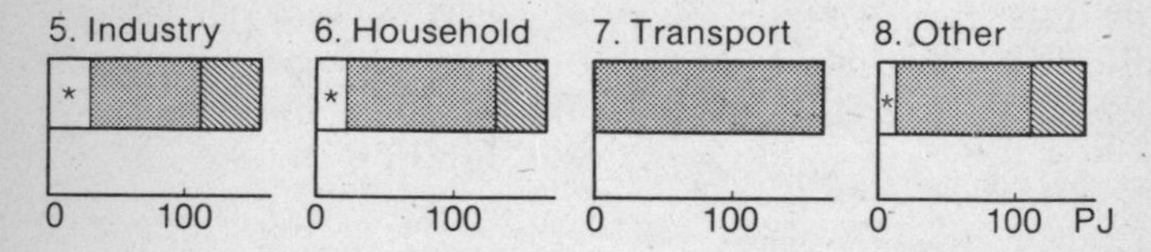

* Includes contributions from wood and wastes, coal, distributed heat, and in the individual sectors, natural gas.

Fig. 3.4 Patterns of consumption, Switzerland

- Bars (5) to (8) show how the final energy is shared by the consuming sectors. Note that, except for the electrical energy, and a small amount of distributed heat ('district heating') in Switzerland, these still reflect the original sources. Gas or liquid fuel derived from coal still appears in the 'coal' segment; gases produced from oil in refineries are entered as 'oil', etc. Very small contributions, such as electric power used for transport in all the countries, natural gas in India or wood in the USA, are omitted.
- The allocation of energy to sectors involves many detailed decisions, which may differ from country to country. Is agricultural machinery 'transport' or 'other'? How about street lighting? Is an office block on a factory site 'industry' or 'other'? Can we always separate dwellings and 'other' buildings? (Swiss data, for instance, often give just one figure for heating oil for these together.) In general, the figures for energy used in industrial *production*, which accounts for most of (5), and for transportation are probably the most reliable.

- Finally, we must note that the diagrams are not all to the same scale. To show all four national totals on one scale the bars of the USA diagram would have to be ten times longer and the Swiss ten times smaller.

What do the diagrams tell us then? A few obvious points: that these days transportation is heavily oil-based; that Britain is the only one of the countries to use appreciable coal-derived fuel in the home; that, if we treat oil and natural gas as interchangeable fuels, the relative contributions of these jointly and of electric power are not too different for the three industrial countries; that Swiss industry, unlike the other three, uses virtually no coal. The very small total losses for Switzerland – under a tenth of the primary input – follow in part from the major use of hydroelectric power but also reflect the fact that unlike Britain and the USA, whose primary input is largely crude oil, Switzerland imports two-thirds of her oil in final-use form, so the refinery losses are on the balance sheet of some other country.

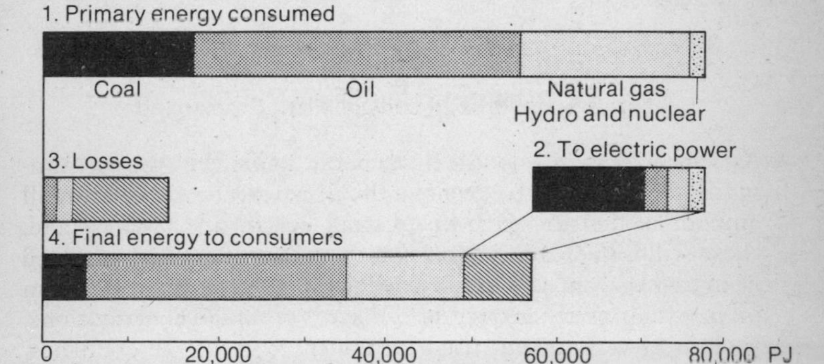

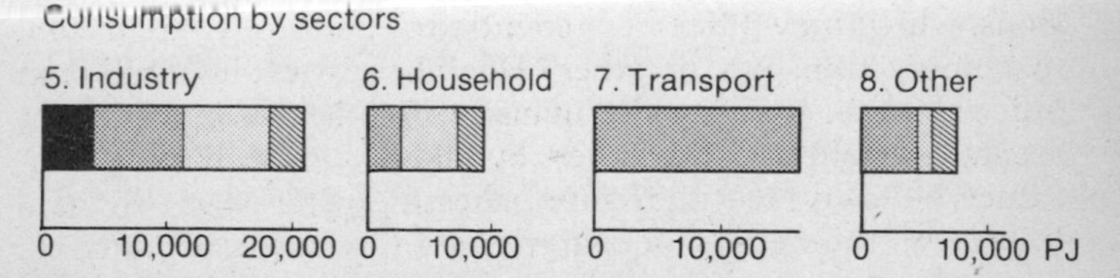

Fig. 3.5 Patterns of consumption, USA

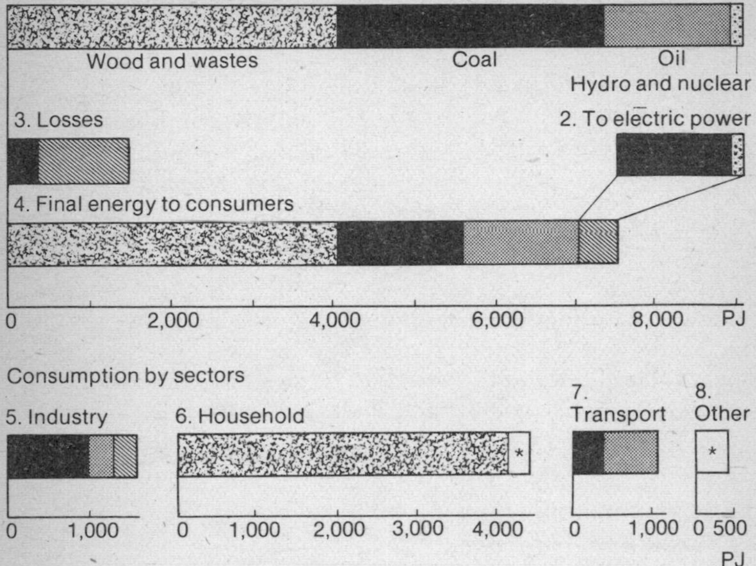

Fig. 3.6 Patterns of consumption, India

What the diagrams do *not* tell us, of course, is the final use to which the energy is put. We can reasonably deduce that most of the oil going to (7) is in the form of petrol and diesel oil, and that to (6) and (8) is fuel oil for heating; but to know how much electric power is used for motors and how much for heat we need more data. Nor do we learn anything here about the efficiency of final use, an important subject to which we must return later.

Converting the sector totals into per caput energies, we obtain the figures in Table 3.3, which show at once that the average Briton and the average Swiss use much the same energy at home and about the same for transportation whilst the American uses half as much again in the home and three times as much in moving goods and people. We shall return to these two sectors, but in many ways the most interesting difference between the four countries is in industrial consumption. As we see, there are very wide variations in the energy used per caput for industrial production, and one might suggest three general reasons for this:

- Differences in *quantity* of industrial output per caput.
- Differences in *type* of industrial output.
- Differences in *energy efficiency* in industry.

In comparing Britain with the USA there is evidence to suggest that the first of these factors is the main one. If people use more (and larger) cars, and more (and larger) refrigerators then they need not only more petrol and electric power to run them but more industrial output to provide them. American per caput quantities are in many instances about twice those of Britain: steel production, agricultural production, motor vehicle production, cars per hundred people, and so on. In a very broad sense the USA is Britain scaled up by a little under ten times: four times the population and a bit over twice the per caput use of everything. (An obvious exception is land area, with the average Briton occupying two-thirds of an acre while the American enjoys ten times as much.)

Switzerland is a rather different matter. Industrial energy consumption is little more than a quarter of the American figure yet per caput production is if anything slightly *greater*. One measure of national output is the gross domestic product, and Table 3.4 allows us to compare this, and energy consumption, for our four countries: we can find an *energy/GDP ratio* for each. The actual value of this ratio will of course depend on current prices. (If inflation puts up the GDP next year, the ratio will fall, even if we produce exactly the same goods and use exactly the same energy.) Similarly, international comparisons depend on how we convert from pounds, francs or rupees to dollars. Do we use official exchange rates or some other measure, such as national or international purchasing power? We must obviously treat the figures with some caution, but nevertheless the differences in national ratios are startling. How do the Swiss produce their money with so little energy?

Table 3.3 Final energy: national consumption by sectors, 1980

Country	**Sector**				
	Industry	**Household**	**Commercial and other**	**Transport**	**Total**
Britain	36	29	14	27	**105**
Switzerland	25	28	26	28	**110**
USA	95	45	41	77	**260**
India	2½	6½	½	1½	**11**

All data in gigajoules per caput

Table 3.4 Energy and gross domestic product

	Britain	**Switzerland**	**USA**	**India**
Per caput final energy consumption (MJ)				
Total	105,000	110,000	260,000	5,000*
Industry	36,000	25,000	95,000	2,300
Gross domestic product†				
1980 GDP (billions)	£200	SF160	$2,600	Rp1,100
Per caput GDP	£3,600	SF25,000	$12,000	Rp1,600
Per caput GDP, dollars‡	$7,000–$8,000	$14,000–$16,000	$12,000	$150–$200
Percentage contributed to GDP by industry	35%	50%	35%	20%
National energy/GDP ratio (MJ per 1980$)	14	7	22	28
Energy consumed by industry per dollar of industrial output (MJ/$)	14	3½	23	60

* Excluding biofuels.
† The GDP is the total inland production of the national economic system.
‡ The ranges correspond to different choices of exchange rate. Mid-range values are used in energy ratios.

Reflection soon produces an explanation. Switzerland, as is well-known, earns her money from milk, butter and cheese produced by cows grazing on alpine meadows, from tourists doing likewise, and from rich Arabs grazing in Zürich banks. None of these activities is very energy-intensive; so the reason that Swiss industry uses little energy is that, like India, she doesn't have much industry. A simple explanation. But false. Industry in Switzerland contributes a *greater* fraction of GDP than in Britain or the USA: nearly half compared to about a third. The fact is that the *industrial average energy intensity* (the final line in Table 3.4) is very low compared with the other countries.

This does seem to be a genuine phenomenon, not to be spirited away by a little fiddling with exchange rates. In later chapters we shall see why there is good reason to expect high energy intensities in developing countries such as India, but the explanation of the Swiss figure probably lies in a combination of the second and third factors mentioned on page 43. There *are* differences in the types of industrial activity: mining, and iron and steel production play a smaller role in Switzerland, and these are all energy intensive. Then more of the energy input to Swiss industry is in the 'efficient' form of electric power. However, to find whether these factors alone can account for the difference, or whether there is a genuinely more efficient use of the *same* types of energy in making the *same* goods in Switzerland, we need to compare the energy intensities of, say, Swiss and American manufacturers of turbine engines, or saucepans, and this is not simple. In Chapter 13 we discuss the question of energy intensities of different activities, a matter of some importance both in comparing one industry with another and in planning for the future. As we shall see there, the analysis needs care. There have been some studies comparing different countries, but sorting out the many factors involved is a complicated process – and drawing lessons from the comparison even more difficult.

People travelling

> You consume over 100 gallons of petrol a year if you drive five miles to work every day in city traffic. Making the journey by jumbo jet, your share would be not much over fifty gallons. Aircraft are more energy-efficient than cars.

True or false?

One could equally well prove the reverse. With four occupants and a reasonably clear road the energy cost per passenger-mile in a car need be only one-hundredth of a gallon, while the average for all commercial flights is about five times this; so flying is *less* efficient than driving.

Much has been written about the energy costs of transportation, and probably the only statement which is true under all circumstances is that the further you travel (in the same vehicle) the more energy you use. It is approximately the case that the *faster* you travel the more energy per mile you use, but if you drive a car – or ride a bicycle – very slowly indeed, even this rule breaks down.

If we ignore bicycles and bullock-carts the principal means of transporting people and goods are at present private cars and commercial road vehicles, trains (steam, diesel or electric), aircraft and ships. All except steam and some electric trains use oil for their power, which explains why about 90 per cent of transportation energy comes from this source. A fifth of world energy consumption is used in transportation, and the saving from a 10 per cent reduction would equal Africa's entire consumption. Limited reserves and the costs and uncertainties of imported oil are of course the factors which have brought the term 'energy crisis' to the headlines, so any reduction in the use of this fuel – by substitution, increased efficiency or reduced mileage – must be of immediate interest. Would it solve America's fuel problems to require car-pooling by all commuters, with four passengers to each vehicle instead of one? Could Britain become a major oil exporter by pricing half her cars off the road and extending instead of closing down her railway system?

A whole range of economic, social and political issues enter into questions of this sort, but at least we can make a start by looking at the present situation. How much *do* we travel, and by what means, and how much energy do we use on different types of journey? Table 3.5 shows a few current data. Until we are all required to report our odometer readings to the authorities every week, any passenger mileage figure for private cars is bound to be an estimate. In Europe, the present average seems to be about 16,000 passenger-miles a year for a private car: 9,000–10,000 miles with average occupancy of a little under two people. (At 30 mpg this means just under 60 passenger-miles per gallon.) In the USA the annual miles per car are probably rather over 10,000 and occupancy again about two people, but the number of cars is much greater and the mpg much lower. In fact the so-called 'fuel economy' (an odd term, considering the figures) *fell*, in the US, from an average 17 mpg in 1945 to 15 mpg by 1974 (15 and 13 miles per US gallon). The fitting of anti-pollution

Table 3.5 How we travel

	UK	Switzerland	USA	India
Annual distance travelled by the average person (miles)				
Private car	4,400	6,200*	11,000	15
Bus or coach	600	300	700	150
Rail	400	900	200	190
Air	—	—	1,000	—
Total	**5,300**	**7,400***	**13,000**	**360**
Private cars per 100 people	28	30	56	$\frac{1}{8}$th
Average passenger-miles per gallon (oil equivalent) for all modes of travel	60	60	40	120

* The Swiss figures include appreciable contributions from foreign tourists.

devices was blamed for some of the deterioration but there are doubts about this, and the fall had been continuous before they became mandatory. With rising fuel prices, fuel economy also rose – to about 20 mpg by the end of the decade – largely because more people bought imported cars. Legislation now requires the 'fleet averages' of the American manufacturers to reach about 30 miles per (Imperial) gallon in the 1980s.

The *national average* figures for passenger-miles per gallon depend of course on the distribution of traffic between different modes. India, with about three times the US value and twice that of the other two countries, shows what happens if instead of travelling nine-tenths of the way by private car and the rest by public transport, you reverse these proportions. One-third of a mile per gallon may not sound very good, but when it refers to a train carrying 900 people the passenger-mile cost is about half a megajoule – equivalent to about one three-hundredth of a gallon of oil. As we saw at the start, the question of energy consumption per passenger-mile for different transport modes is a vexed one, and the passenger occupancy is obviously relevant. Fig. 3.7 shows the consumption per passenger-mile at full normal capacity, but the range can be very wide in practice. The average for all London Transport Underground journeys is 100 passenger-miles for each train-mile, but a particular train could carry as few as five or six people – or more than 2,000. (A non-scientific estimate based on rush-hour experience of the Northern, or Misery Line in the days when rolling-stock and drivers were in short supply.)

As an example of saving oil by substitution let us look at the

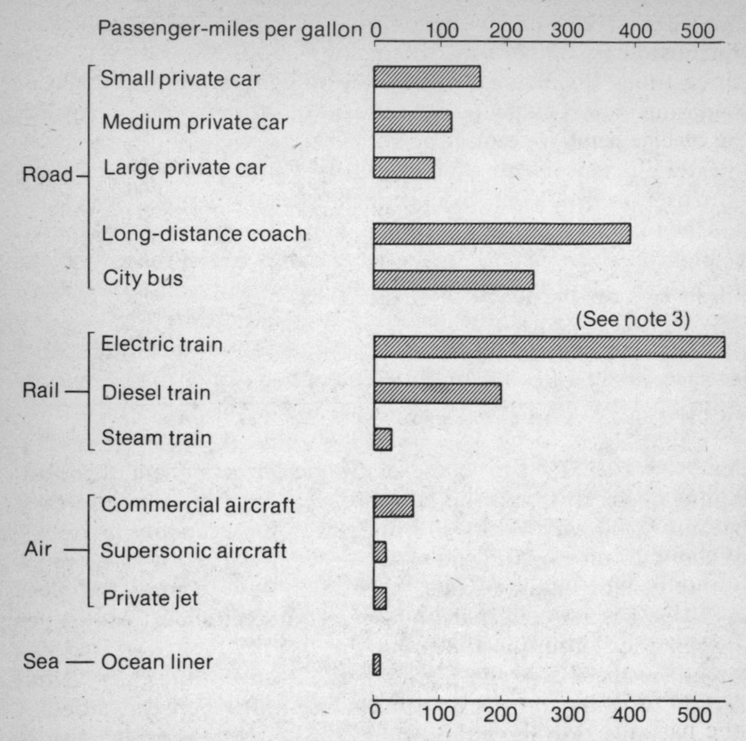

Fig. 3.7 Passenger-miles for an energy consumption equivalent to one gallon of oil, at full normal passenger capacity

1. The passenger-miles will be proportionately less for part-full vehicles.
2. There can be wide variations even at full capacity. The data here are based on averages for the four countries studied.
3. If the electric power for the train comes from thermal (fossil or nuclear) power-stations, the passenger-miles per gallon equivalent of *primary* fuel will be about a third of the figure shown.

suggestion that electric trains replace commuter cars. The energy could then come from sources such as hydroelectricity or coal-fired or nuclear power-stations. (British Rail understandably but unfortunately changed from coal to oil for its electric power in the early 1970s, just before oil prices rose. In the later seventies it changed to gas, whose price is now starting to rise.) If we assume that a commuter's car carries on average one and a half occupants and achieves twenty mpg (in Europe) we can calculate from Fig. 3.7 that a replacement train would be as efficient with one seat in eighteen occupied –

but we must be careful in the comparison. If the power is produced at the customary 30 per cent efficiency, the train would need about three times this average occupancy to compete. An interplay of opposing factors characterizes any analysis of proposals to replace oil by electricity for motive power. There are economies due to the greater efficiency with which an electric motor converts its input into useful work – over 75 per cent compared with under 20 per cent for the internal combustion engine; but unless hydroelectric power is plentiful there is a balancing diseconomy in the generating station. Then the savings which accompany mass transport are in a sense offset by the fact that electric power can *only* be used in public systems. (Devices for private vehicles to draw power from the road surface are still in the land of dreams; and although the battery-driven car may be on its way, millions of travellers are hardly going to wait at the roadside for two decades until it reaches them.)

The ancillary costs – in energy and money – of any proposed new system must enter, and it has been estimated that the true energy consumption of a transportation system is about twice its annual fuel input, the extra being needed to manufacture vehicles or rolling-stock, maintain roads and ways, and provide all the other inputs to the system. (Is anyone surprised, these days, to learn that *paper* accounts for about 2 per cent?) Unless we know that users of a new system will give up their cars altogether there is a danger that the energy-saving measure will add another energy-*using* element to the national total, with a net increase in consumption. Then of course the capital costs of new mass-transit systems are high, and there may be cheaper ways to achieve the shift from oil.

We could of course travel less. Or improve the fuel economy of our vehicles. (Some technical possibilities are discussed in Chapter 6.) The savings can easily be calculated using the data in this section, and you can find the result of, say, travelling 10 per cent fewer miles, or increasing by a third the mpg of private cars. Unfortunately it seems that we are not yet doing enough about either. Despite the massive price rises and all the concern with fuel economy, Britain was using about a third *more* petrol in 1980 than in 1970. (Total oil consumption fell, but mainly through the replacement of oil by gas for heating.)

People at home

Clearly we aren't doing too well when we travel, so how about staying at home? Table 3.3 showed that each European consumes about 30 GJ of energy every year for 'household' purposes, and an American half as much again. What we do with it is another subject where the number of investigations is rising about as fast as the price of oil. The results show, to no one's surprise, that 'we' have very varied habits; and the differences can be as great within one country as between countries. You have only to consider the heating needs in Chicago (absolutely essential for survival) and in San Diego (hardly necessary), or to note that much of rural Switzerland has no piped gas but still plenty of wood, to see that the variations are not just matters of taste either.

Rather than attempting extensive international comparisons, it seems more useful here to concentrate on a few questions which are important to almost everyone, and to use one set of figures to give an idea of quantities. The data used are for 'the average British household'. The Swiss, as we've seen, use about the same per caput household energy, but with a very different distribution between fuels. Given the sort of variations mentioned above, it hardly makes sense to talk of an 'average' American household, but where consumption depends on the sizes of houses, a rough working average is a 50 per cent increase over the European figure, while for many appliances, doubling is about right. The whole analysis is virtually irrelevant to India, as the power consumption of a dishwasher or colour TV does not feature very largely when your problem is how to find enough wood to cook the next meal.

There are about 20 million households in Britain, with an average of some two and three-quarter people per household. Table 3.6 shows the pattern of consumption in 1980. (The customary domestic energy units are used.) It is of course improbable that any actual household uses exactly the average mix of fuels, and one aim of the analysis will be to see how the same needs can be met by different patterns of consumption.

What are these needs? We can list four categories:

- essentially electrical: lighting and many appliances
- high temperature heat for cooking
- medium temperature heat for hot water
- lower temperature heat for space heating.

We'll see in later chapters the importance of the distinction between different types of heat, but the main reason for separating them here is that it's what we do in practice. Many of us use different fuels for cooking, producing hot water and heating our houses. Let us look at

Table 3.6 Energy consumption in an average British household, 1980

Annual consumption		**Therms**
Coal	600 kg or 12 cwt	150
Oil	about 30 gal	50
Natural gas		390
Electricity	4,100 kWh ('units')	140
Total		**730**
		(equivalent to 21,000 kWh)

each in turn then, and try to see where the energy goes. Although this part of the book is mainly concerned with the present, we might also keep our eyes open for possible future improvements.

Essential electricity

The average British household uses about 250 kWh a year for lighting; effectively two 60-W bulbs running for just under six hours a day. For the future it is hardly likely that we'll change to a different fuel for this purpose, or that household demand for light will fall dramatically. (Public buildings may offer more scope for reduction.) However, a very important point to notice is that only about *twenty* of the 250 kWh appear as light. All the rest immediately becomes heat – which of course helps to warm the house but hardly in an ideal way and rather expensively, with full-cost electricity.

Why not just produce the light, without the heat? For reasons discussed in Chapter 10 this is not possible with ordinary filament lamps, so we must look to other types. A large sodium discharge tube is about the most efficient, giving some 15 times the light energy output of a 60-W bulb for three times the electrical energy input. But there are a couple of problems. First we don't like living in a yellow glare, and secondly we don't *want* 15 times the output of a 60-W bulb. We want nice, small, intimate lights. The problem of colour has largely been solved in the fluorescent tubes – with some loss in efficiency – but reducing the size has been more difficult. Small 'domestic' discharge lamps, designed to replace light bulbs, are now coming on the market, however, and it may be that before long we'll be saving perhaps half of those 250 kWh.

Coming to larger appliances which depend essentially on electricity (like TV sets) or are usually electrical (like refrigerators) it helps to divide these into three groups, in terms of their *annual energy consumption*. For an individual appliance this is calculated by multiplying the power rating, the wattage in kilowatts, by the number of

Table 3.7 Annual energy consumption by domestic appliances, British households

All figures are for a household of average size (2.7 people) for one year.

Group	Appliances	Annual consumption each appliance	Consumption for group*
low	vacuum cleaner, iron, spin-drier, toaster, radio, record player, electric blanket	less than 100 kWh	500 kWh
intermediate	refrigerator, black-and-white TV, washing-machine,† electric kettle	200–300 kWh	1,000 kWh
high	freezer, colour TV, clothes drier, dishwasher†	500–1,000 kWh	2,500 kWh

* By a household owning most of the appliances

† Consumption includes full energy 'cost' of hot water

hours a year for which it runs. (A 1-kW appliance running continuously consumes 8,760 kWh a year.) Household consumption naturally depends on the number of appliances and their use, so the totals in Table 3.7 are averages for households possessing (and using) *most* of the appliances in each group.

Adding the 250 kWh for lighting we find then that a 'minimum consumption' household of average size uses about 750 kWh a year of electrical energy and a 'maximum consumption' household over five times as much. All this, of course, before adding contributions for cooking or heating.

Cooking

With present cookers and present habits about 60 therms a year are used for cooking with gas or about 1,200 kWh with electricity. Gas cookers thus use about 50 per cent more energy than electric cookers (unless the people using gas cook 50 per cent more meals!) But we must as usual recall that three times as much primary energy is needed to produce the electrical energy in the first place.

Poor insulation of cookers means heat losses, and it is claimed that consumption could be halved by improving this. Microwave ovens, with very little heat going anywhere except into the food, need only a fifth as much energy; but the uses of these efficient devices are limited and it is unlikely that they will totally replace conventional cookers – except perhaps in the households of wealthy batchelors who live on heated-up pies. (The *costs* of improved uses of energy are not much discussed here, but will appear many times in later chapters.)

Hot water

We like our hot water supply at between 130 and 150°F (55 and 65°C) and we use from 10–15 gallons a day per person. Recalling the definition of the BTU and the therm (page 12) and knowing that a gallon of water weighs 10 lb, we deduce that if the incoming water is at 50°F we need between 80 and 150 therms a year to heat the water for the average household. Let us take 120 therms, or 3,500 kWh, as our figure.

But we must remember the losses. Even if you draw no water at all, an energy input is needed to keep the water hot, because where ever there is a temperature difference there will be a heat flow: heat energy flowing from the hot water, through the tank and any material around it, to the cooler air outside. The *rate* of this energy flow depends on three things:

- the *temperature difference* between the hot water and the surrounding air
- the *total surface area* through which heat flows: top, bottom and sides of the tank
- the *thickness and insulating qualities* of the material covering these surfaces.

The first fact tells us at once that the hotter the water the greater the loss rate. If the surrounding air is at 65°F, the loss from water at 150°F will be about a third faster than from water at 130°F. It also follows that we waste energy by leaving the immersion heater on all night. Keeping the water hot means continually replacing lost heat. With the heater off, the water cools (because of the lost heat) but the loss rate decreases as it becomes cooler. So the total loss during the night – the heat energy we must replace – is less than if the water is kept hot all the time.

Insulating the tank and hot water pipes makes an enormous difference. With poor insulation, heat losses can add 50 per cent or more to the energy consumption, and if the tank is in the loft the house doesn't even benefit much from the lost heat. Very good lagging can reduce the loss to under 20 per cent a year: an input of 140 therms to produce 120 therms' worth of hot water (Table 3.8(A)).

Table 3.8 Producing hot water

The figures show the fuel input needed to produce 120 therms of useful heat – enough for hot water for an average household for a year. The three examples are (A) a very efficient well-lagged system, (B) an average system and (C) a very inefficient system with little insulation.

	Energy needed for hot water (therms)	**Tank and pipes**			**Required input to tank (therms)**	**Boiler**		**Required input to boiler (therms)**	**Annual fuel consumption**
		Temperature (°C, °F)	**Insulation**	**Losses (therms)**		**Efficiency (%)**	**Losses (therms)**		
A	120	55°C, 130°F	good	20	140	70	60	200	4,100 kWh electricity *or* 16 cwt coal *or* 200 therms gas *or* 120 gal oil
B	120	60°C, 140°F	average	40	160	55	130	290	4,700 kWh electricity *or* 23 cwt coal *or* 290 therms gas *or* 180 gal oil
C	120	65°C, 150°F	poor	60	180	40	270	450	5,300 kWh electricity *or* 35 cwt coal *or* 450 therms gas *or* 280 gal oil

If the water is heated electrically then this (about 4,100 kWh) is the total energy needed, because an immersion heater converts pretty well all the input electrical energy into useful heat. But if a fossil fuel boiler is used there will be further losses – up the chimney or out of the sides. We'll discuss combustion in Chapter 7, but all we need to know here is that only if the boiler is exceptionally well maintained is the efficiency likely to reach 70 per cent: 7 therms of useful heat for each 10 therms of fuel energy. At best, we then need about 200 therms of input to obtain 140 therms for the water tank. The worst can be very, very bad, as Table 3.8 shows.

Hot air

Finally, we like to keep warm. Again, the only reason for continuous heating is that the house continuously gives up energy to the colder air outside. Like the water tank, it loses heat by conduction – mainly through its walls, roof and windows. Furthermore, if we are not to suffocate we need a change of air and the new air, like the replacement water in the tank, must be heated.

How good – or bad – your walls are as insulators is measured by their 'U' value. In the metric version now used, this is equal to the rate of heat loss in *watts* through each *square metre*, for a temperature difference of *1°C* between inside and outside. So the loss to be replaced by the heating system can be calculated:

watts = U value × area in square metres × temperature difference in °C.

(The metric version is not too difficult to use, because 1 sq m is roughly 10 sq ft.)

Windows are quite important because a pane of glass is a good conductor of heat. And there is the roof, of course. To get an idea of the quantities we note that a metric U value greater than two for a wall or roof is dreadful (though some older houses approach it), that the normal cavity-wall value of about one and a half can be at least halved by injecting insulating foam, and that ordinary windows have U values of between five and six, reduced to about three by double glazing.

Air may seem rather light stuff – airy, in fact – but nevertheless there is about a quarter of a tonne of it in an average house; so each complete air change means warming this mass. One change an hour is enough for comfort and this needs an average heat input rate of about 80 W for each degree (°C) difference between indoor and outdoor air. With our typically draughty British houses we could

Table 3.9 Heat losses from houses

The figures are for a semi-detached house with 900 sq ft (90 sq m) floor area on two floors. Its total external wall and roof area is about 140 sq m.

Description of house	**Effective U value of walls, roof & windows**	**Heat loss rate (W) per °C temperature difference**		**Annual heating hours**	**Average temperature during heating period (°C)**			**Total annual heat loss***	
	(W/sq m/°C)	**walls etc. + air change**	**total**		**indoor (a)**	**outdoor (b)**	**(a) − (b)**	**kWh**	**therms**
The energy-saving house Well insulated walls and roof, double-glazed windows. Few draughts, so one air change per hour.	1	140 + 80	220	4,380†	20	7	13	12,500	430
The leaky house Normal cavity walls, some roof insulation, ordinary windows. Draughty, so two air changes per hour.	2	280 + 160	440	4,380†	20	7	13	25,000	860
The hot house As the energy-saving house.	1	140 + 80	220	8,760††	21	10	11	21,000	720

* Loss rate times hours times temperature difference. † 18 hours a day for 8 months. †† 24 hours a day for 12 months.

easily be heating the upper atmosphere at three times this rate.

Once we have the basic data for the house there are two more things we need to know in order to work out the heat required: how cold it is outside and how warm we want it inside. Strictly we would need to know the temperature difference for every hour of every day throughout the year. Published tables of 'degree days' for different places give this information, but we'll shorten the calculation by taking the annual average outside temperature to be 50°F (10°C) and the 'winter average' (eight months!) as 45°F (7°C). We must then decide

- what temperature do we want indoors?
- for how many hours a day is the house heated?
- for how many months a year?

A possible decision might be 68°F (20°C), 18 hours and 8 months, which would look like hardship to many Americans, and self-indulgence to a lot of Britons. Table 3.9 shows the annual heat loss from a very well-insulated house with these requirements, and for comparison the losses under the same conditions for an average present-day British house. A third example shows the losses for the well-insulated house but with more luxurious heating.

Conclusions

We can now do a few sums to see how the figures for the different types of use might add up to give the national average data. We take (Table 3.10) a family of average size, who have more than the bare minimum of electrical appliances and use an immersion heater for summer hot water and a gas boiler in the winter. For cooking we allow the option of either electricity or gas, so when the contributions are added there are two possible sub-totals. Subtracting these from the total average consumption tells us how much energy is left for heating the house. Of course our family doesn't have to use either of these particular divisions between fossil fuel and electricity: the available energy is about 400 therms, to be used in any form.

Unfortunately, as we see from Table 3.9, this is less than even the lowest of the calculated annual heat losses. However, we have omitted one important item. All the energy consumed by electrical appliances eventually turns into heat energy, so if they are all inside the house their 1,300 kWh provide an extra 45 therms of 'free' heat annually. The heat from cookers is about the same, and if the house is sensibly designed, losses from tank and pipes, and to some extent

Table 3.10 How one household might use the annual average energy

Annual consumption		**kWh**		**Therms**
Lighting		250		—
Electrical appliances (Table 3.7)				
Basic	500			
Refrigerator	300			
Colour TV	500			
Total, appliances		1,300		—
Cooking	*either*	1,000	*or*	50
Hot water (Table 3.8)				
Immersion heater, $\frac{1}{3}$ of year		1,550		—
Gas boiler, $\frac{2}{3}$ of year		—		190
Sub-total without space heating				
	either	4,100	+	190
	or	3,100	+	240
Total annual consumption, average household		4,100	+	590
Available for space heating	*either*			400 therms
	or	1,000 kWh	+	350 therms

the boiler, also contribute. So do the occupants. At about 75 W each, or about one therm for every 400 person-hours. Of course not all this heat appears when and where it's needed, but if only a half is useful the gain is over 100 therms a year (Table 3.11).

Our sample family, if they consume no more than the national average energy, thus have effectively some 500 therms a year for space-heating: just about enough to replace the heat losses from a centrally-heated semi-detached house if it has good insulation and an efficient boiler. Does this mean, then, that the average British family lives in a well-insulated, well-heated house? Unfortunately

Table 3.11 'Free heat'

Annual heat production	**Therms**
Electrical appliances, 1,300 kWh	45
Cooking	50
Tank and pipes	40
Boiler, 60% of loss, $\frac{2}{3}$ of year	50
People, 14,000 person-hours	35
Total	**220**
Useful contribution, 50% of total	**110 therms**

not. It means that our average leaky, draughty house almost certainly doesn't achieve even a modest level of whole-house heating. But the main lesson we learn is that it would be *possible* for everyone to live comfortably without any dramatic increase in national energy consumption – and without even considering contributions from such exotic new sources as the sun. Whether we can afford to do it, or afford not to do it, are subjects we return to in later chapters.

Part II Technologies

4 Energy Conversion

Introduction

In the past three chapters we have built up a fairly detailed picture of the way things are. Now we move on from the energy of the world to the world of energy, and the next eight chapters are concerned with *why* things are the way they are. Not of course why people behave the way they do. If the answer to that difficult question is to be found in any book, it is not in this one. Our concern is with the simpler matters of why *things* behave as they do. What picture of the physical world and its laws allows the engineer to predict that high-pressure steam hitting a turbine which is in turn linked to massive coils of wire will produce electric power; or that a suitable wafer of silicon, lying in the sun, will do the same?

To understand such matters is important not only for engineers. We all need to know about the energy systems we use, in order to use them as efficiently as possible, if for no other reason. Whether you are considering the merits of combined heat and power for the industry which you happen to control, or the advantages (or otherwise) of leaving the immersion heater switched on all night, an appreciation of the basic science at least clarifies the issues, and may even lead to a clear-cut answer. Then there is the future. We are not all of us directly involved in planning national power systems for the 1990s, but most of us are concerned to reduce next winter's fuel bills, a subject not unconnected with thermal conductivity or the efficiency of the combustion process. We might like to know how much useful hot water we are likely to get from the solar energy which reaches us; or on a wider scale, whether windmills or the waves of the ocean could possibly lead to electric power in useful amounts. Scientific laws do not prescribe what we ought to do, but they can say what is possible – or more important, impossible – and may often tell us what will be the consequences of some proposed course of action.

This is not intended to be a textbook of science, and no single work *could* be a textbook of all the technologies from turbogenerators to the internal combustion engine. The plan, therefore, is to be selective in order to avoid being superficial. We start with the basics, the picture of a physical world made up of atoms. Heat is very important for all energy systems (either we are trying to produce it or we are trying to get rid of it) and when we understand the difference at the atomic level between a hot object and a cold one, we begin to see why we can or cannot convert heat energy into other forms. If we look at electric currents on a similarly detailed scale, we see why wires get hot and (a natural extension of this) why the beauty of the countryside is marred by enormous pylons. The burning of fossil fuels uses oxygen and produces carbon dioxide; a nuclear reactor, switched off, continues to generate heat; a normal 100-W light bulb produces about 7 W of light and 93 of heat. All these are fairly direct consequences of basic science.

The fact that the average car engine is used most of the time at well under its already low optimum efficiency, on the other hand, is a matter of technology. It is not an inevitable consequence of any law of science but a result of particular requirements we put on the engine for other reasons. Further questions, such as whether we can get useful quantities of electric power from the sun or the tides, or of liquid fuel from coal to replace petrol, are also technological – in the wider sense that cost is an essential factor. Science, as always, sets limits to the possible, while current technology sets limits to the attainable; and it is our task to look at the costs – financial and environmental – and at the potentialities of future alternatives such as these.

The programme is clear then. First the basics, and then the technology, selecting for detailed study some systems which are important because they play a major role and some where the possibilities for improved efficiency are particularly evident. Of the future technologies, we shall concentrate on those which, because they could satisfy a major need, seem most likely to be developed; but we shall also look at a few which are attractive for other reasons: because they use renewable rather than finite resources, or because they are clean.

Energy systems

We use the term *energy system* for the sequence of processes which leads from primary resource to end use, starting when we acquire a certain number of joules and finishing when we have – literally – no further use for them. All energy systems, past, present or in the foreseeable future, have certain common features. Every system involves processes of energy conversion and of transportation or distribution, and these processes share the important characteristic of producing *wastes*; waste energy in particular. This is an unavoidable consequence of the basic laws, and our main concern in this chapter is to discuss these laws and to find out what they say about limits to efficiency. In this way we can develop some criteria for later use when we come to assessing different processes and technologies.

Of course the details will be very different for different energy systems. One which we all know rather well leads from the oil well to the petrol tank of a car. Here the energy enjoys a rather uneventful life. Apart from a fairly dramatic episode in the refinery, the fuel remains a liquid throughout, with the energy stored chemically in its molecules from primary resource to end use. This is by no means always the case. In a nuclear power plant at least three major conversions take place: from nuclear energy to the thermal energy of a hot fluid, then to the energy of a rotating turbine and finally to electrical energy. In contrast, chopping down a tree and burning the wood involves rather few processes (which does not *necessarily* make it more efficient).

Another everyday sequence of processes starts in the coal mine and leads eventually to the electric kettle, and in this case we have provided some data (Table 4.1). An important fact (one that is no

Table 4.1 How to make a pot of tea, using about 1 oz of best coal

	Energy (joules)
1 Heat content of the coal: 25 grammes at 29,000 MJ per tonne	725,000
2 Subtract energy needed to mine and transport the coal: 8% of the above	58,000
	667,000
3 Generation of electrical energy, at 34% efficiency	227,000
4 Subtract transformer and distribution losses, 18% of electrical energy	41,000
5 Energy delivered	**186,000**
Energy to run a 1-kW kettle for 3 minutes: 1,000W × 180 sec	180,000
or	
Energy to heat 1 lb of water from 62°F to 212°F: 150 BTU at 1,055 joules per BTU	158,000

(Leaving a few joules for warming the pot.)

surprise by now) is that about three-quarters of the original energy is lost between primary resource and end use; and if we look carefully at the loss processes we find that in every case the missing energy has become heat. Of course, if we ask how else we could make a pot of tea using just one ounce of coal, we realize that electricity generation is not by any means the only energy-wasting process.

The inevitability with which waste heat occurs is closely related to another feature of all energy technologies: they are once-through processes. None of them is cyclic. In no case do we replace all the energy in the form in which we found it. This is just as true for the so-called renewable resources (solar energy, wind and wavc power, etc.) as for the fossil fuels. *We* never renew our energy resources. The sun may, however, do it for us; and the chief difference between 'renewable' and 'non-renewable' resources is the rate at which this happens. Plants grow and reservoirs refill in months or years but the formation of coal or oil took millions of centuries.

A matter of convenience

Why do we need all these inefficient processes, changing the Earth's resources in irreversible ways and wasting valuable energy? Why do we transport oil and gas thousands of miles instead of using the coal or wood on our doorstep? One answer becomes evident if we rephrase the question and ask why more pots of tea are made with water heated by electricity or gas than with water heated on coal fires. The answer is *convenience.* Convenience in use and convenience in distribution to a very large extent govern present choices of energy, at least in the industrial countries. It could be argued that the criterion of convenience even outweighs that of cost in many cases.

Plans for future energy sources must surely fail if they do not take into account the superb qualities in this respect of three present energy carriers: oil, gas and electricity. Superb, that is, for the uses to which we put them. So before continuing our study of waste and inefficiency, let us look briefly at what these qualities are.

Petrol or diesel oil for transportation

Given that we want to travel separately at 50 mph or in groups at 500 mph, any alternative energy source would have to match the following properties:

- energy content about 160 MJ per gallon

- can be pumped through pipes for distribution and during use
- storage in tanks, which can be refilled many times
- final user can load fuel at about 1,000 MJ a minute (for a car, or much more than this for an aircraft).

Consider some possible alternatives. First the electric vehicle. A lead-acid rechargeable battery (normal car battery) stores about 2 MJ when fully charged. So we would need about eighty of these to replace the stored energy of each gallon of gasoline. If all eighty were simultaneously on charge we could 'load' about a quarter of a million joules a minute, and after a few hundred rechargings we would need a new set of batteries.

How else might we propel a vehicle? You may care to consider a solid fuel (coke and charcoal have both been used). Natural gas? Its heat content per kilogram is very good, 50 MJ or so, but gases are much less dense than liquids and to match one gallon of petrol we would need about 160 cu ft – or to use high-pressure tanks.

Of course these are rather unfair pictures, as we shall see later, and we cannot rule out other possibilities; but the fact remains that, at present, petrol is the most compact form of portable energy that we have.

Electricity for industrial, commercial and domestic use

Provided that we don't need to move too far from the nearest outlet socket, there can hardly be a more versatile or convenient energy carrier than electricity. With normal household wiring we can draw on about a fifth of a million joules a minute continuously at each socket; and ten times this rate presents no particular difficulties. There is no problem of storage (for the user) and the energy is totally clean (for the user). Conducting wires take up very little space. We can drive motors, heat water, light lights, all at the turn of a switch. Any proponent of a 'low electricity' future would have to face, at the least, an army of indignant housewives.

Natural gas for heating

The requirement may be process heat for industrial purposes, space heating in buildings, domestic hot water or cooking. Whenever we need hot anything, gas will provide. The rate of supply can be varied almost instantaneously from zero to several million joules a minute in a domestic system, or to some orders of magnitude more than this in an industrial installation, a flexibility which is hard to beat.

One conclusion which we might draw from these observations is that people are not entirely irrational in their choices. The millions of Britons who shifted from coal to oil, gas or even electricity to heat their houses, or the thousands of Californians who changed their hot water systems from solar energy to gas when it became available, were not doing so simply to ensure that there would be future energy crises. Unfortunately no one told them, back in the halcyon days, that a time would come when availability of resources and efficiency of use would join convenience and cost as determining factors.

Inevitable heat

The question is *why* there must be heat losses in every process. The answer, unfortunately, is more in our stars than in ourselves. There may be (undoubtedly are) losses which are due to our incompetence, but it is important to understand that there is always a *necessary* inefficiency. Necessary in the sense that the laws of physics tell us that it is bound to occur.

We start with four observations:

- All energy tends to become heat energy.
- All energy conversion processes are once-through. None is completely reversible.
- A generator whose input is *mechanical* energy (hydroelectric power-station, car generator, bicycle dynamo) can convert from this to electrical energy with almost no losses. The same is true for the reverse process in the electric motor.
- Any system whose input is *heat* energy (fossil fuel or nuclear power-station, internal combustion engine) converts much less than 100 per cent of its input into useful output.

These statements obviously lack precision, but they are sufficiently definite for us to be able to add two comments: that the word 'almost' is essential in the third, for otherwise it would contradict the second; and that the first observation tells us the fate of the 'missing' energy in all processes. Putting all these thoughts together we notice that they have a 'one-way' aspect in common, and that this arises in connection with heat energy. It seems that it is only too easy to go from any other form of energy to heat but that there are severe constraints on the reverse process.

We are of course surrounded by examples of this truth. Heat is produced whenever an electric current flows, but putting hot water in the electric kettle does not generate a useful voltage in the element. We bring the car to a skidding halt, producing quantities of heat in brakes and tyres. Making use of this warmth to set the car in motion again is unfortunately not possible.

To see what is so special about heat energy, we need to look at heating and cooling processes at a very detailed level. When we heat something, we add energy. Where does this energy go? The physicists' answer is that it goes into increasing the motion of the atoms of the material. It is very important that when we think of materials as consisting of these tiny particles, we have a *dynamic* picture in mind, not a static one. The air around us consists mainly of molecules of nitrogen and oxygen – and a lot of empty space. The distance between molecules is about ten times their size (it is their consequent freedom to move which makes air a gas and not a liquid or solid), and we can visualize the molecules of the air at normal temperatures as like a swarm of gnats, moving incessantly. Their average speed is high: about 1,000 mph, and when the air is heated this speed increases, the change between ice-cold air and air on a very hot day being about 5 per cent.

For solid materials the detail is different but the effect much the same. The atoms of a solid cling together fairly closely (one reason why you don't fall through your chair), but not so closely that they cannot move. Each can oscillate in the available space between all its neighbours, and the hotter the material the more vigorous the oscillations. Our picture of a solid is thus of a quivering mass of particles.

For any material, then, we have a way of looking at two important quantities: *temperature* is simply a measure of the average speed of the atoms or molecules ('hotter is faster'), and *heat* is the energy added or removed in order to change the temperature. With this in mind, we see what it is that distinguishes heat from all other forms of energy: the characteristic quality of thermal motion is that it is *disorganized* motion. The gnats in our analogy may or may not be moving at random but the molecules of the air certainly are. They move with a variety of speeds in all directions, bouncing off each other and the walls of the room, changing both speed and direction all the time. (The pressure which gas exerts on any surface is the result of these continual impacts.) The same picture of random motion holds for the vibrating atoms of the solid material.

If this is heat, then the principle that all energy tends to become heat energy is no more than a law which we all know: *things tend to become disorganized.* Which is one, rather informal, way of expressing the Second Law of Thermodynamics. (We already know the First Law of Thermodynamics: it is the Law of Conservation of Energy.)

The above 'statement' of the Second Law is hardly going to tell us how much waste heat to expect from a power-station. It is far too imprecise to lead to quantitative results. However, it does throw light on some of the other processes which we have discussed. Consider, for instance, a stationary car as an assembly of atoms moving at random. In another car, moving north-east at 30 mph, all the atoms have a *directed* motion superimposed on their random movement, and this is the extra energy of the moving vehicle. As the car stops, this extra energy must go somewhere. It becomes heat – further *disordered* motion of the atoms. We now have a warmer, stationary vehicle. To reverse the process and recoup the heat energy requires part of the random motion to become ordered again, something which is very, very unlikely to occur spontaneously. It is easy to produce chaos. Order presents more of a problem.

Refrigerators, heat pumps and a very important rule

One of the one-way processes which we meet every day is the flow of heat from a hot object to a cooler one. You pour hot coffee into a cold cup and the almost instantaneous result is slightly cooler coffee and a much hotter cup – assuming that there is much more coffee than cup. A little hot coffee in a large cold mug leads to a slightly different result; but in either case there is a flow of heat energy from hot coffee to cold container. We would be very surprised indeed if the reverse were to occur, with the coffee becoming even hotter and the cup ice-cold. Yet, provided the heat energy gained by the coffee is the same as that lost by the cup, energy is still conserved in this bizarre case. Once again, we must look to the *Second* Law of Thermodynamics, not the First, to tell us the direction in which things happen.

'Hotter means faster.' The atoms of a hot object have higher average speed than those of a cold object. To take another domestic example, suppose you were to observe with astonishment one day that one end of a metal fork was becoming red-hot while ice crystals were forming on the other end, as it lay on the table. This would mean that, from having the same average speed everywhere, the faster atoms were now on the whole at one end and the slower at the other. Rather as though 1,000 people entered a theatre, sat down at random, and then found that nearly all the men were on one side and nearly all the women on the other. A very improbable occurrence, with long odds against its happening, and longer the greater the number of people. Now the fork contains a very large number of atoms indeed (about 10^{24}, or 1,000,000,000,000,000,000,000,000) so your astonishment would be entirely justified. We see that the Second Law has something to do with all this when we reflect that, once again, the issue is order versus chaos. The people in the theatre and the atoms in the fork will not *spontaneously* adopt the more ordered arrangement: 'Heat will not of itself flow from a colder to a hotter body.' The natural way is the reverse, the direction of increasing chaos.

Yet heat *can* be transferred from a colder to a hotter body. Every refrigerator or air-conditioner does it. Take a jug of milk at room temperature and put it in the refrigerator. It spontaneously loses heat to the (colder) surroundings, heat which must be removed if the interior is not to warm up. So the refrigerator switches on and 'pumps' heat out to the (warmer) room. As we know from its power consumption this uses energy and the situation is as shown in Fig. 4.1(a). Notice that all the energy has to go somewhere, so the heat output to the room must be greater than that removed from the refrigerator.

In comparing the performances of two refrigerating units, we might ask how much heat each removes in an hour, at a given electric power input, the one which 'pumps' more heat for the same input being the better unit. Fig. 4.1 includes data for a refrigerator of modest size where the rate of removal of heat energy is 2,100 BTU an hour for a power consumption of 250 W. Is this a good or bad performance? We can't really tell unless the heat and the electrical energy are in the same units, so we convert both to megajoules a day (b), revealing that the heat removed is two and a half times the energy input. The *coefficient of performance*, or COP, is 2.5. Is *this* good? How do we know what should be possible?

The answer comes from the Second Law of Thermodynamics. With its aid we can calculate a value for the coefficient of performance even if we know nothing about types of refrigeration unit or how they work. The inside and outside *temperatures* are all we need to know. The value obtained is of course for an 'ideal' refrigerator, and represents a best possible performance which could never quite be reached in practice. Nevertheless the result is very important, because we learn that no designer, however brilliant, could increase the COP above this for the given temperatures. The input energy to pump a certain amount of heat will *never* be lower. We can see that

(a) Power input and heat removed

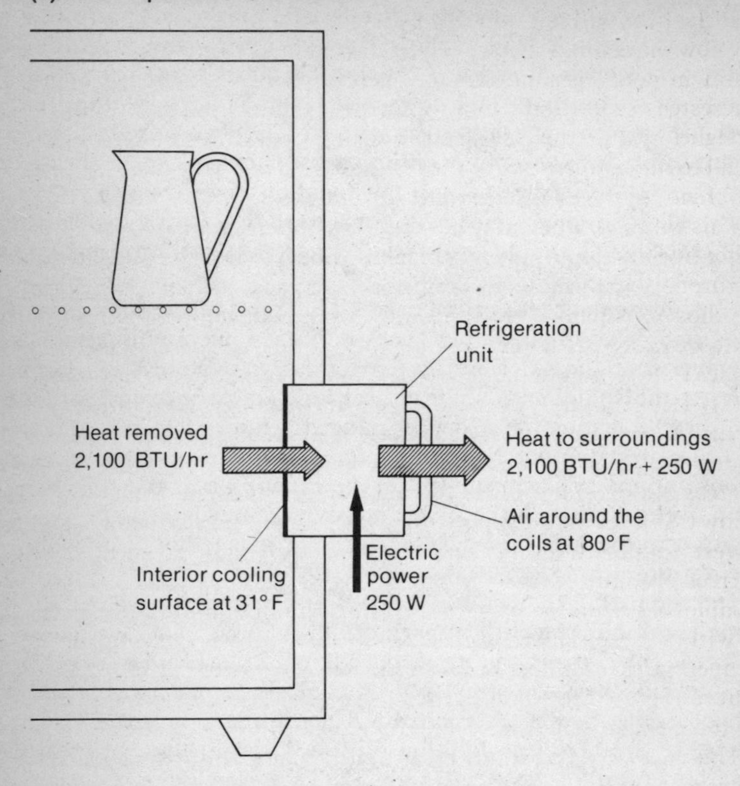

(b) Energy balance and coefficient of performance

Electrical energy input:
= 250 W × 4 hr
= 1 kWh
= 3.6 MJ

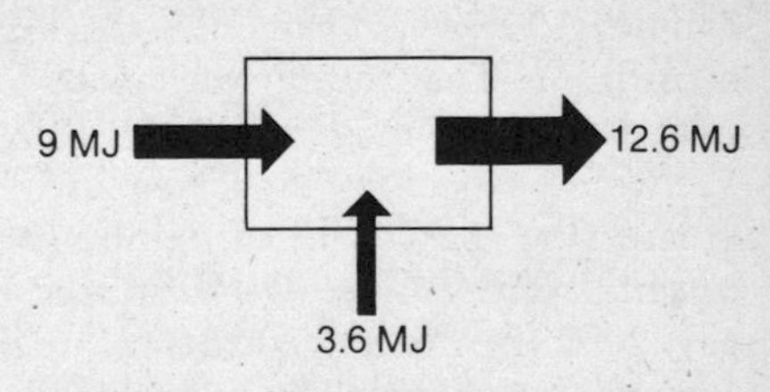

Heat removed:
= 2,100 BTU/hr × 4 hr
= 8,400 BTU
= 8.9 MJ

Coefficient of performance = 8.9 ÷ 3.6 = **2.5**

Fig. 4.1 Refrigerator performance

the Second Law might impose a limit of this sort, because it says that the heat will not flow out by itself. So *some* input is needed – but now the claim is that we can tell how much. To substantiate this claim in detail is unfortunately beyond our present scope. (The interested reader is referred to Further reading.) But the result is straightforward: the 'ideal' coefficient of performance is simply equal to *the temperature of the interior divided by the temperature difference between interior and surroundings*.

This simple statement needs one very important qualification: it is *not* true for just any old way of measuring temperature. We should be surprised if it were, considering how our usual temperature scales are chosen. When we measure on the centigrade (properly, Celsius) scale we choose as zero the temperature of freezing water, while 0°F comes from, amongst other things, winter weather in Danzig in the 1700s. It would surely be most odd if a universal law were to depend on either of these arbitrary choices. What we need is a zero point which itself has universal significance. There is one, called *zero degrees absolute*, and it is 273°C below 0°C or (the same thing) 460°F below 0°F; a very low temperature whose importance is that it is the lowest possible. Nothing can ever be cooled right down to zero degrees absolute (that's the *Third* Law of Thermodynamics), but complex refrigeration systems can bring tiny amounts of substances to within a thousandth of a degree or so, at which extremely low temperatures a number of interesting changes occur. The random heat motion of the atoms becomes less and less. Things freeze up, even on the atomic scale. So no more chaos.

This is a very important result, with much wider applications than keeping the ice-cream cold. As we shall see, one immediate consequence is that a power-station or internal combustion engine can never be 100 per cent efficient. What does it say, then, that is so significant? That the energy needed to transfer a certain amount of heat depends on the temperature difference. This seems reasonable. We would expect the input to a refrigerator or air-conditioner to be greater the hotter the surroundings. But it also depends on the *absolute* temperature of the interior. Even if the difference stays at, say, 20°C the input needed is greater if the whole system is run at lower temperatures. To take this to the extreme, as the interior temperature approaches zero degrees (absolute) the rate at which heat can be pumped out also approaches zero, no matter how much power you supply. This really is, as far as we know, a universal law and it is enormously important for all energy discussions, not because we want ice-cream at $-460°$ but because of the restrictions it imposes at *any* temperature.

Returning to the domestic refrigerator, we can now compare the actual performance with the ideal. Using the temperatures shown in Fig. 4.1, we calculate the 'ideal' COP:

$$(80 + 460) \div (80 - 31) = 11.$$

We see that the real refrigerator achieves only a quarter of this (Fig. 4.1(b)). It needs four times the power input for a given rate of heat extraction. This is not unusual, and a system which reached 50 per cent of the ideal would be considered good.

There is the further question why we need to remove heat for four hours a day. We all know that a refrigerator switches on occasionally even if the door is never opened. It must be pumping something, and that something is heat which leaks in all the time from the surroundings. Thus the average daily energy consumption depends on how well insulated the appliance is, as well as how many times you open the door and what you put in to be cooled. And the inside and outside temperatures will affect this consumption in two ways: through the COP and through the rate of heat gain. If you stand your refrigerator in a warm place you are making a double contribution to its energy consumption. It has been estimated that improved insulation – which would of course increase the cost of the appliance – might reduce power consumption by about a third; but if you restrict the heat flow so that the coils are in a pocket of warm air, you might *increase* it by 50 per cent: an extra half a kilowatt-hour a day for the refrigerator in our example.

Finally, mention should be made of another device which is basically the same as a refrigerator: the heat pump. Here (Fig. 4.2)

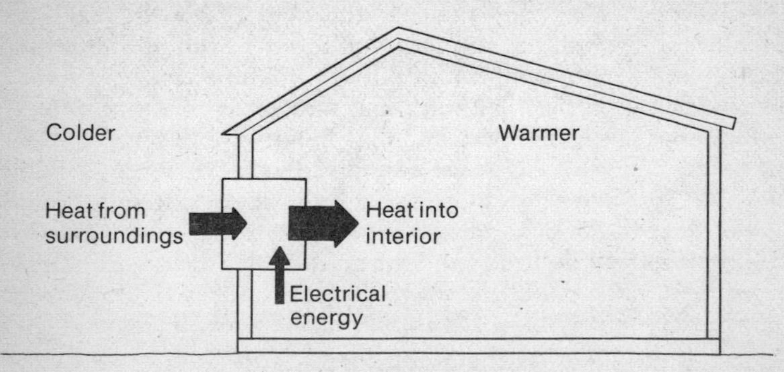

Fig. 4.2 Heat pump for a house

heat is taken in from the cold surroundings and transferred to the building – together of course with the heat from the electrical energy input. With a COP of about three, the heat supplied to the building is thus *four* times the electrical consumption. This should be a very good way of heating a house, but unfortunately the capital cost and the poor reliability of some early heat pumps have limited its more widespread adoption. The gain of course deteriorates (with the COP) as the temperature of the air, ground or water from which the heat is pumped falls, and there comes a point where it is so little that the system might as well switch to direct electrical heating. This is expensive in many countries, so heat pumps are not ideal where there is extreme cold. One suggestion which sounds attractive – if the technology is reliable – is to combine a heat pump with solar panels. Taking in heat from water warmed in the panels instead of the cold ground outside, the pump should work at a much better COP.

Heat engines – and another rule

Now that we've solved the case of the domestic refrigerator, a power-station presents no problem: it is simply a refrigerator run backwards. Consider Fig. 4.3. In (a) we have the refrigerator, using electrical energy, removing heat from the cold region and delivering to the warmer surroundings an amount equal to the heat removed plus the input energy. In (b) there's the power-station, taking in heat from hot steam, turning some of this into electrical energy and rejecting the remainder into the cooling water. The heat *pump* (refrigerator) becomes a heat *engine* (power-station) if you just reverse the arrows. It's not surprising, then, that the Second Law once again determines the performance, and it is in fact quite easy to obtain from the 'pump' rule the rule for the ideal heat engine: *the maximum possible ratio of useful output energy to input heat energy is equal to the temperature difference divided by the input temperature.* Here we have it: the reason for a world-wide energy waste equivalent to 24 million barrels of oil a day. Fig. 4.4 shows the numbers. Using high temperature steam and with plenty of cooling water to carry away the waste heat without itself becoming hot, the 'ideal' efficiency is about 60 per cent. This ideal value is sometimes called the *Carnot efficiency*, in honour of the French engineer who, much ahead of his time, thought his way to the general rule – though he probably never built a 'heat engine' of any sort himself. As we'll see in the next chapter, even the best modern turbines can only

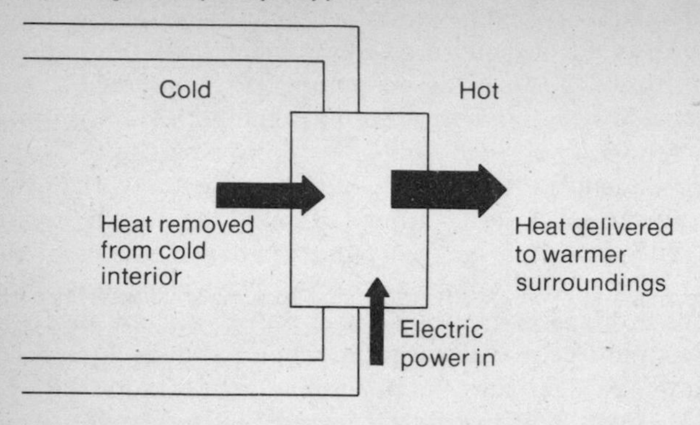

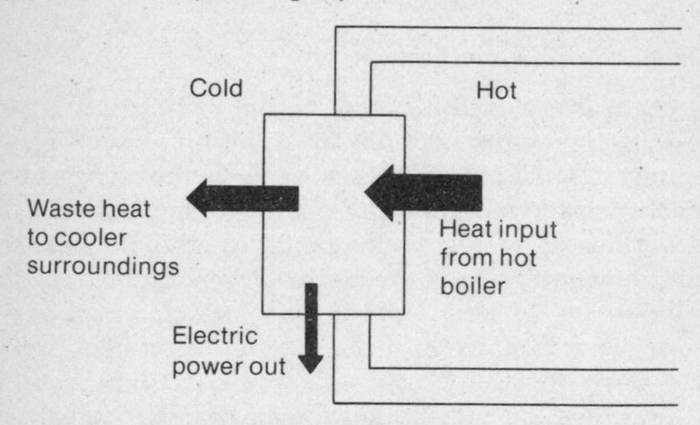

Fig. 4.3 Refrigerator into power-station?

achieve about 80 per cent of the Carnot efficiency, and that is before all sorts of other losses are taken into account. The complete power station, from fuel input to electricity output, is doing well if it manages 35 per cent efficiency overall.

To conclude, the important message of this and the preceding section is that there are rules which govern the performance of *any* system which comes under the heading 'heat pump' and of *any* system which comes under the heading 'heat engine'. These rules are universal. They are not consequences of our choice of materials or of the particular design, but apply whenever we transfer heat from a lower to a higher temperature, or produce useful work from heat

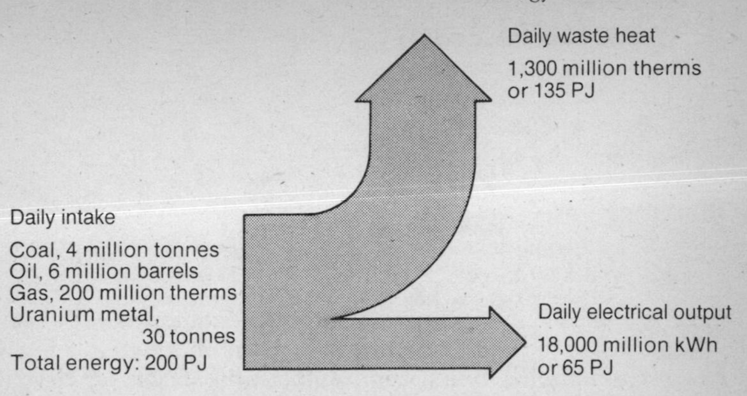

Fig. 4.4 All the world's thermal power-stations

energy. Of course, the whole or any part of our present physics can at some future time be proved false, but as far as our present understanding goes the message is clear: *you cannot beat the Second Law.*

Footnote

In case anyone who has been listening to scientific gossip is wondering where that ineffable thing called *entropy* comes into all this, the answer is simple: we've been talking about it throughout the chapter. Entropy is the physical scientists' way of quantifying chaos.

5 Electrical Energy

Electricity: a new force

Electrotechnology is a relative newcomer – compared with coal-mining, say, or the working of metals, or paper-making. Only in the late 1870s did generators become sufficiently reliable for electric power to compete successfully with coal or gas, and the world's first electric street lighting was switched on just a hundred years ago. The growth of the electrical industry in its first fifty years (Fig. 5.1) was probably faster than for any other newly-introduced technology over a comparable period. The rate of increase then declined a little, but it is worth noting that despite the recent world recession the rise in consumption in the past fifteen years has been greater than during the entire preceding three-quarters of a century. And as we've seen, electric power production now accounts for a quarter of the world's consumption of primary energy.

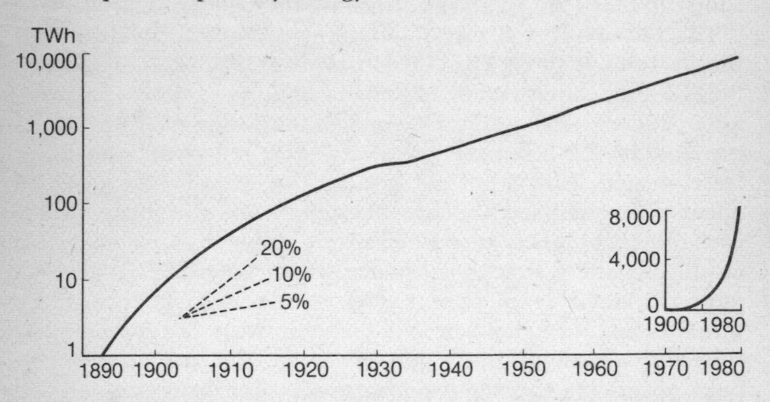

Fig. 5.1 World annual electric energy consumption, 1890–1980

1. Notice the *scale* of the main curve. Each step corresponds to *multiplying* the consumption by ten. The small graph shows how the growth looks if plotted with a straightforward linear scale.
2. The dotted lines show the slopes for different annual growth rates. Sliding a rule across from these to the main curve allows us to see the approximate rate of increase at any time.

The efficient production, distribution and use of electric power are accordingly important matters, and we shall look at some aspects of present methods in this chapter. With as wide and varied a field as electrical technology, we can of course only touch on a few topics, but there is one merit in this area of energy studies: the field may be wide but the central core of basic science is very compact. We shall need to extend the atomic picture of materials, and to introduce a few new laws dealing with the specifically electrical behaviour of things; but this will be enough to allow us to discuss in some detail several important technological matters.

Let us begin by returning to the atom, which we saw in the last chapter as the basic constituent of everything – solid, liquid or gaseous. Closer investigation reveals that atoms themselves are composed of smaller entities. Almost all the mass of any atom is in its nucleus, a concentration of dense matter roughly at the centre. We shall say more about nuclei in Chapter 8, but here we are concerned with the outer part of the atom: a swarm of electrons, very light particles which contribute well under one thousandth of the total mass. One might visualize an atom as a miniature solar system; but when we try to explain how the fast-moving electrons are held in place, calculations show at once that gravity – the pull that holds the planets as they swing around the sun – is far too weak a force. If that were all, the electrons would fly off in all directions.

The force which holds an atom together and gives it its stability is the *electrical force* between the electrons and the nucleus, and this type of force differs in one extremely important respect from the force of gravity. *Everything* experiences the pull of gravity, not only towards the Earth but towards every other object to a greater or lesser degree. This is not the case for the electrical force, which affects different objects in different ways. Some are pulled towards each other, but others are pushed apart – and many are not affected at all. Imagine a strange substance which is pushed rather than pulled by gravity. A piece of it when released will shoot vertically upwards, and if you hold on to it, you will rise or fall depending on the relative strengths of the pull on you and the push on it. With a piece of suitable size, the two forces will just balance and you will float. There is of course no such 'negative-gravity' material, as far as we know, and the parallel with electrical forces is in any case not perfect, but it is the idea that is important: that we can have forces of repulsion as well as attraction and that these can cancel each other.

An atomic nucleus and an electron are electrical opposites. The nucleus is pulled one way by any outside electrical force and an electron the opposite way. The nucleus is said to have 'positive electric charge' and the electron 'negative electric charge', names which are of course quite arbitrary. (In Chapter 8 we meet particles called neutrons, which have no charge.) There is a strong electrical force of attraction between the nucleus and the electrons in an atom, and as long as the electrons stay near the nucleus – which they do with some tenacity – the atom as a whole is neither pushed nor pulled by outside electrical forces. It is electrically neutral because the positive charge of the nucleus is *exactly* counter-balanced by the negative charge of all the electrons. If we do manage to remove an electron from an atom, it can then move off under the influence of any outside force, leaving the atom (now properly called an *ion*) with a net positive electric charge.

All electrons are completely identical with each other – 'as alike as the same pea in a pod' – and the number needed to make up the complete atom is the single most important characteristic of a substance. Atoms of the element copper, for instance, have 29 electrons (Table 5.1). They may lose one or two under some circumstances, but if the number in the complete, neutral atom is not 29, it is not copper but some other element. A carbon atom has six electrons, oxygen eight, and hydrogen only one. The whole of chemistry is a consequence of these electron numbers: they determine whether the element is a metal or a gas, what compound molecules the atoms will form with atoms of other elements, and – an important question, as we'll see in the next chapter – how much energy is needed to take apart these compounds.

Amps

That some materials are much better conductors of electric currents than others is common experience. When a wire breaks, the current in the circuit stops because the air between the broken ends is a good insulator. The current in a household circuit flows in the copper wires and not out through the surrounding walls because copper is a much better conductor than the plastic insulation around the wire. When we ask why this is, we are really asking what it is that makes a metal metallic. The answer, in terms of the atomic picture, is that a metal has large numbers of *free electrons*. Every atom has released one or more of its electrons, and these move around freely within the metal at very high speeds. When we introduce an electric force along a wire, as we do in connecting a battery to, say, a torch bulb (Fig. 5.2), the whole assembly of free electrons is moved along by the force, and

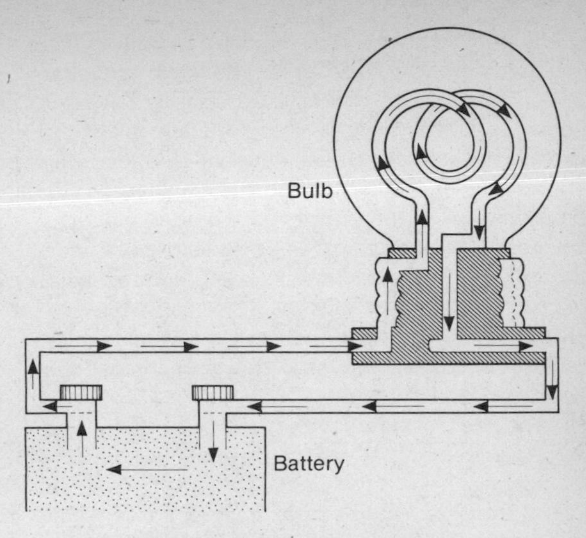

Fig. 5.2 An electrical circuit

it is this drift of the cloud of electrons which constitutes the current. To allow a current to flow *continuously* there must obviously be a complete circuit: a loop. (Otherwise electrons would pile up at some point, producing a massive electric charge there, and stopping the flow.) The same current must thus be flowing at every point along the wire, and also through the supply. If the bulb filament is broken, no current flows at all, because none can pass across the gap. Air, like other insulators, does not normally have many free electrons and therefore will not carry much current. (Of course, if something happens which breaks up the atoms, producing free electrons, then a current can flow: we've all seen sparks, and streaks of lightning.)

We'll return to the supply, the cause of the current, shortly; but for the moment let us look at an important *effect* of the current. As the cloud of high-speed electrons drifts down the wire, they collide continually with the ions of the metal, and these collisions have two consequences. First, they tend to slow down the electrons. In other words, they provide *resistance* to the current flow; and secondly, they increase the motion of the ions, the vibrations which, as we saw in the last chapter, are related to the temperature. They *heat* the metal.

The amount of current which is flowing in a wire obviously depends on how many moving electrons there are and how fast they

Table 5.1 The elements

The upper left-hand figure is the number of electrons in the atom. The figure beneath the symbol represents the mass of the atom, and is further discussed in Chapter 8. One of the most interesting discoveries in chemistry is the repeating *periodic* pattern shown here, in which elements with similiar chemical properties fall in the same vertical columns.

^{1}H 1.008																	^{2}He 4.003
^{3}Li 6.939	^{4}Be 9.012											^{5}B 10.81	^{6}C 12.01	^{7}N 14.01	^{8}O 15.999	^{9}F 19.00	^{10}Ne 20.18
^{11}Na 22.99	^{12}Mg 24.31											^{13}Al 26.98	^{14}Si 28.09	^{15}P 30.97	^{16}S 32.06	^{17}Cl 35.45	^{18}Ar 39.95
^{19}K 39.102	^{20}Ca 40.08	^{21}Sc 44.96	^{22}Ti 47.90	^{23}V 50.94	^{24}Cr 52.00	^{25}Mn 54.94	^{26}Fe 55.85	^{27}Co 58.93	^{28}Ni 58.71	^{29}Cu 63.54	^{30}Zn 65.37	^{31}Ga 69.72	^{32}Ge 72.59	^{33}As 74.92	^{34}Se 78.96	^{35}Br 79.91	^{36}Kr 83.80
^{37}Rb 85.47	^{38}Sr 87.62	^{39}Y 88.91	^{40}Zr 91.22	^{41}Nb 92.91	^{42}Mo 95.94	^{43}Tc (97)	^{44}Ru 101.1	^{45}Rh 102.91	^{46}Pd 106.4	^{47}Ag 107.87	^{48}Cd 112.4	^{49}In 114.8	^{50}Sn 118.7	^{51}Sb 121.8	^{52}Te 127.6	^{53}I 126.9	^{54}Xe 131.3
^{55}Cs 132.91	^{56}Ba 137.34	57–71 (see below)	^{72}Hf 178.5	^{73}Ta 180.95	^{74}W 183.85	^{75}Re 186.2	^{76}Os 190.2	^{77}Ir 192.2	^{78}Pt 195.1	^{79}Au 196.97	^{80}Hg 200.6	^{81}Tl 204.4	^{82}Pb 207.2	^{83}Bi 209.0	^{84}Po (209)	^{85}At (210)	^{86}Rn (222)
^{87}Fr (223)	^{88}Ra (226)	89–103 (see below)															

Lanthanides	^{57}La 138.9	^{58}Ce 140.1	^{59}Pr 140.9	^{60}Nd 144.2	^{61}Pm (145)	^{62}Sm 150.4	^{63}Eu 152.0	^{64}Gd 157.3	^{65}Tb 158.9	^{66}Dy 162.5	^{67}Ho 164.9	^{68}Er 167.3	^{69}Tm 168.9	^{70}Yb 173.0	^{71}Lu 175.0
Actinides	^{89}Ac (227)	^{90}Th 232.0	^{91}Pa (231)	^{92}U 238.0	^{93}Np (237)	^{94}Pu (244)	^{95}Am (243)	^{96}Cm (247)	^{97}Bk (247)	^{98}Cf (251)	^{99}Es (254)	^{100}Fm (257)	^{101}Md (258)	^{102}No (255)	^{103}Lr (256)

	Element	Atomic number		Element	Atomic number		Element	Atomic number		Element	Atomic number
Ac	actinium	89	Er	erbium	68	Hg	mercury	80	Sm	samarium	62
Al	aluminium	13	Eu	europium	63	Mo	molybdenum	42	Sc	scandium	21
Am	americium	95	Fm	fermium	100	Nd	neodymium	60	Se	selenium	34
Sb	antimony	51	F	fluorine	9	Ne	neon	10	Si	silicon	14
Ar	argon	18	Fr	francium	87	Np	neptunium	93	Ag	silver	47
As	arsenic	33	Gd	gadolinium	64	Ni	nickel	28	Na	sodium	11
At	astatine	85	Ga	gallium	31	Nb	niobium	41	Sr	strontium	38
Ba	barium	56	Ge	germanium	32	N	nitrogen	7	S	sulphur	16
Bk	berkelium	97	Au	gold	79	No	nobelium	102	Ta	tantalum	73
Be	beryllium	4	Hf	hafnium	72	Os	osmium	76	Tc	technetium	43
Bi	bismuth	83	He	helium	2	O	oxygen	8	Te	tellurium	52
B	boron	5	Ho	holmium	67	Pd	palladium	46	Tb	terbium	65
Br	bromine	35	H	hydrogen	1	P	phosphorus	15	Tl	thallium	81
Cd	cadmium	48	In	indium	49	Pt	platinum	78	Th	thorium	90
Cs	caesium	55	I	iodine	53	Pu	plutonium	94	Tm	thulium	69
Ca	calcium	20	Ir	iridium	77	Po	polonium	84	Sn	tin	50
Cf	californium	98	Fe	iron	26	K	potassium	19	Ti	titanium	22
C	carbon	6	Kr	krypton	36	Pr	praseodymium	59	W	tungsten	74
Ce	cerium	58	La	lanthanum	57	Pm	promethium	61	U	uranium	92
Cl	chlorine	17	Lr	lawrencium	103	Pa	protoactinium	91	V	vanadium	23
Cr	chromium	24	Pb	lead	82	Ra	radium	88	Xe	xenon	54
Co	cobalt	27	Li	lithium	3	Rn	radon	86	Yb	ytterbium	70
Cu	copper	29	Lu	lutetium	71	Re	rhenium	75	Y	yttrium	39
Cm	curium	96	Mg	magnesium	12	Rh	rhodium	45	Zn	zinc	30
Dy	dysprosium	66	Mn	manganese	25	Rb	rubidium	37	Zr	zirconium	40
Es	einsteinium	99	Md	medelevium	101	Ru	ruthenium	44			

move. We could measure electric currents as we measure traffic flow: the number passing per second. However, this would not be very convenient, owing to the very small amount of electric charge carried by each electron, so we use a more practical unit: the ampere, usually abbreviated to *amp*. (One ampere corresponds to about 6 million, million, million electrons passing per second.) Apart from the supply, two main things determine the current flow. It depends on the particular metal, not only because different metals have different numbers of free electrons, but because the electrons in different metals flow at different rates, for the same supply conditions. There are good and bad conductors, even among metals, and a copper wire will carry about six times as many amps as an identical iron wire with the same supply: the resistance of the iron wire is six times that of the copper. Then of course the *area* available for the current flow matters, so that if we compare two otherwise identical wires, one with twice the area for the electrons to flow through, the latter will carry twice as much current, all other things being equal – a very important point, as we'll see, for the transmission of electric power.

Volts

Basically it is an electric force which causes a current to flow, and what a battery or the generator in a power-station does is to provide this force. A useful way of visualizing the process is through an often-used analogy: the flow of water along a pipe. Water flows along a pipe because there is a pressure difference between the ends. How this is produced – by a reservoir or by a pump – doesn't matter. In either case, the larger the pressure difference the greater the flow rate. The analogous 'pressure difference' between the two ends of a wire carrying a current is the *voltage difference*. (It is customary to drop 'difference' and speak of the 'voltage across' a wire, or any component in a circuit, the word 'across' meaning in this case 'between one end and the other'.) It is the voltage maintained across the wire which causes the current to flow, and the greater the voltage, the greater the current. Notice that this voltage is needed even to keep the electron cloud moving along at constant speed – just like the water in the pipe. This is because the electrons continually give up energy to the ions of the metal; and the supply is therefore doing two things. It acts as a 'pump', bringing electrons 'up' from the low voltage end of the wire so that they can flow back in at the high voltage end; and as it does this, it is continually replacing the energy lost in collisions: the energy which is heating the metal. *A voltage supply is an energy supply.*

Voltages are of course measured in volts. For ordinary metals, we find in practice that the current is proportional to the voltage: doubling the voltage causes the current to double, and so on. If the voltage measured in volts is divided by the current in amps, the figure obtained is the resistance (measured in ohms) of the wire. With a little rearrangement, this statement can be written as

$$\text{voltage} = \text{current} \times \text{resistance}$$

or more briefly, if less elegantly,

volts equal amps times ohms.

Watts

Power is the *rate* at which energy is converted from one form to another (joules per second, calories per day, or barrels of oil equivalent per year). To see what this is for our simple electrical circuit consisting of a supply and a wire, we follow a sample electron around the loop. It is given energy as it is 'lifted up' by the supply, and this energy is then dissipated as heat during the journey along the wire. So energy is converted from the electrical energy of the supply into thermal energy of the metal at a rate which must depend on the energy gained by each electron and the number of electrons flowing per second. In other words, the power is proportional to the *voltage* and to the *current*.

To make life simple, the unit we call one volt has been chosen so that the power in watts (joules per second) is just *equal* to the voltage in volts times the current in amps. Another simple rule:

watts equal volts times amps.

A slight complication in practice is that the normal mains supply does not provide a steady voltage, like a battery, but an *alternating voltage* which changes direction every hundredth of a second (in Europe, or every hundred-and-twentieth in the USA); so the picture must be of an electron flow which switches to and fro at this rate. However, if we use suitable *average* values for the varying voltage and current, the two rules (above and in the previous section) still apply, at least where heating effects are involved. These averages are what we mean when we talk about the 230 volt a.c. mains, or 13-A wiring – wiring which can safely carry a current of 13 amps without becoming too hot.

Transmission and distribution

The cost, in both energy and money, of transporting any energy from the place where we find it to the place where we want it is an important factor in decisions about the systems we use. There are energy costs in carrying coal or oil and in pumping liquids and gases along pipes, but here we are concerned with the costs of transmitting electric power. These can be very high: in financial terms, up to half the cost we pay as consumers, and in energy terms a waste of from 10 per cent to as much as 40 per cent of the generated power. Moreover, there can be few people who think that a series of large metal structures supporting a network of cables actually improves the landscape. To see why (or perhaps whether) these are really necessary, we need only carry out a simple calculation based on the results of the past three sections.

Suppose that electric power is to be supplied to a village fifty miles from the power-station, and that the peak requirement is just one megawatt. The total current flowing, into all the houses in the village, must then be about 4,000 amps (Table 5.2); and this current must also flow along the connecting lines from the power-station, generating heat as it does so. We decide that we do not want to lose more than a tenth of the power in heating the countryside, so the total voltage drop along the 100 miles of cable must not exceed 24 V. This means in turn that the entire resistance of the cables must be less than about six-thousandths of an ohm. We have a table showing the ohms per 100 miles for different thicknesses of copper wire – about the best conductor there is. Inspection shows that we need cables about *two-and-a-half feet* in diameter!

This is evidently not feasible; yet we know that electric power *is* transmitted, and over longer distances and at higher rates than in the example. How is it done? To see the answer we work backwards. We need thinner cable, and thinner cable means greater resistance. So the only way to keep the heating effect low is to reduce the *current* in the cables. How can one megawatt be delivered with less than 4,000 amps? By having a higher supply voltage. If this were 100 times greater – 24 kV instead of 240 V – then the current, for the *same power*, need only be forty amps, which could be carried without much loss in quite a modest cable. The answer, then, is to use much higher voltages; but this solution brings with it two immediate problems. Both are related to the fact that the greater the voltage the better the insulation needed to prevent currents flowing where they should not. In the first place, a household supply at 24 kV

Table 5.2 More power to the people

The aim To deliver power to a village without losing most of it as heat in the transmission cables.

The data The power needed by the village is 1 MW, the normal supply voltage is 240 V and the distance from the power-station is 50 miles.

The specific requirement The energy wasted as heat in the cables should not exceed 10% of that supplied:

10% of 1 MW = 100,000 W

Step 1 Find the current flowing to the village to provide 1 MW at 240 V:

watts = volts times amps
1,000,000 = 240 × amps

The required current is thus about 4,200 A, and this must flow along the 100 miles of wire.

Step 2 Find the voltage drop along the cable, for 100,000 W heat loss with 4,200 A current:

100,000 = voltage × 4,200

The voltage drop along the entire cable must therefore not exceed 24 V.

Step 3 Find the maximum permitted cable resistance:

volts = amps times ohms
24 = 4,200 × ohms

The resistance must therefore not exceed six-thousandths of an ohm, or roughly 1/170 ohm.

Step 4 Using the data given for copper wire, estimate the required diameter.
(The figures are approximate, but adequate for our rough calculation.)

Wire diameter		**Resistance of 100 miles of wire (ohms)**
in.	**mm**	
$\frac{1}{4}$	6	100
1	25	5
2	50	$1\frac{1}{2}$
10	250	$\frac{1}{20}$
20	500	$\frac{1}{80}$
40	1,000	$\frac{1}{300}$

would be very unsafe – or, at the least, would require impossibly massive insulation. So the 24-kV transmission voltage must be reduced to an acceptable 240 V, and this requires transformers, which of course cost money and which themselves lead to some additional energy dissipation.

The second problem comes from the fact that high voltages mean high pylons. The 240-V supply to a house can use cables slung on wooden poles or buried in a simple conduit, but transmission at tens or even hundreds of kilovolts requires a very different order of insulation. (For long distances, 400 kV is quite widely used, and 800-kV lines are being introduced.) As in so many cases, we have again a 'trade-off' situation. We, the consumers, want ever more electric power, and we'd like it as cheaply as possible. It is probably the case that *generating* costs show economies of scale, at least up to power-station outputs of several hundred megawatts: enough for a population of a few hundred thousand people. If we then add that power sharing makes it desirable to inter-connect these large units, we begin to see the need to transmit many megawatts over many hundreds of miles. Burying high voltage cables is more expensive than carrying them on pylons overhead; but no one likes the visual pollution of the latter method.

There doesn't seem to be any simple 'technical fix' to get us out of this difficulty. Better and cheaper methods of insulating underground cables are being developed all the time, but slowly. Superconducting cables (for some, the answer to all problems) do not seem likely to be with us for a while yet. Superconductivity, the disappearance of all resistance and with it all heating effect, sounds like the perfect solution, but unfortunately the superconducting metals turn out to be rather expensive alloys, often difficult to handle, and needing to be kept at about −250°C (or −420°F) in order to stay superconducting. The cost of reaching and maintaining these extremely low temperatures makes this, for the present, a rather expensive way of economizing, and we seem to be left with the usual three-way balancing problem, weighing against each other cost, energy efficiency and the environment. One thing is certain: whatever the marvels of modern technology, it won't resolve these issues for us.

Generators

The generation of electric power is such an important part of our use of energy (and as we have seen many times now, involves so much waste) that an awareness of how a power-station works is necessary for any serious discussion of either the present situation or the future. We shall therefore devote the rest of the chapter to these matters, starting with accounts of the two main components of any thermal power-station, the generator and the turbine, and then putting these together so that we can analyse exactly what goes into and comes out of a large modern power-station.

The input to almost all present generators is *mechanical power*. There are other possibilities: a battery uses chemical energy and a photocell uses light, but these contribute only tiny fractions of the world total power. In all other cases, from a 10-kW diesel generator to a 500-MW nuclear power-station, some mechanical force causes a machine to rotate and thus generates a voltage. The mechanical force itself can be produced by almost anything (even a person cranking a handle), but by far the most common means is a turbine driven by water or hot steam or other gases.

The principle of the generator is very simple, and like many simple principles it took a genius to discover it. In 1831 Michael Faraday found the link which is the basis of almost the whole of electro-technology. It is that *a voltage is generated whenever there is a magnetic effect which is changing with time*. Magnetic effects have of course been known for centuries. The compass needle, a small magnet aligning itself with the magnetic field produced by the Earth, was probably known to the Chinese 3,000 years ago, and was certainly used by European navigators in the Middle Ages; and the Romans already knew that a piece of iron brought near a magnet becomes 'magnetic' itself. The understanding that the two separate fields, electric and magnetic, were very closely related came, however, only in the nineteenth century, culminating in a series of great discoveries which have led not only to the electrical industry as we know it, but to television and the transistor radio and other delights of modern civilization.

A voltage is produced if the magnetic field through a loop of wire changes. *Change* is of the essence, because only as long as there is change is the voltage maintained; and the faster the change the greater the voltage. This is why continuous motion is required. A generator must therefore have three essential features: a *coil*, a *magnet*, and a method of *moving* the one with respect to the other. Fig. 5.3 shows a very simple possibility. If we were to connect a light bulb to the two leads A and B and rotate the coil, a current would flow in the filament of the bulb, just as if a battery were connected – except that this particular generator would produce an alternating

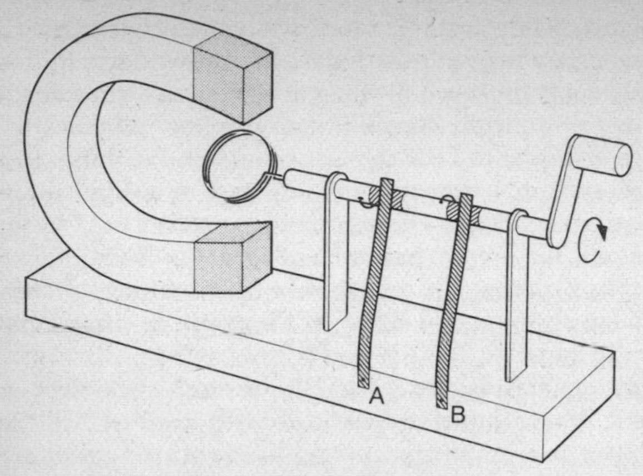

Fig. 5.3 A very simple generator

As the coil of wire is rotated the changing magnetic effect produces a voltage. Two springy metal strips rubbing against metal bands connected to the ends of the wire provide a continuous path for the current.

current, changing direction every time the coil flipped over in the magnetic field. The construction is on a rather grander scale, but the principle is exactly the same, for a power-station generator. The main differences apart from size are that the latter has a fairly complex system of coils wound on an iron core to give the greatest magnetic effect, and the magnet itself is an *electromagnet*: a separate set of coils also wound on a suitable shape made of iron or a magnetic alloy. This again increases the magnetic field, and also gives better control.

For an energy analysis of the generator we must start with the mechanical input power. In our hand-cranked example, this is the energy used per second in pushing the handle round and round. The *output* power is of course the voltage across the lamp multiplied by the current flowing in it, and the difference between input and output must be the losses due to friction and to the heating effect of the current in the generator wires. (In a large generator there are further losses as energy is dissipated in all the iron, both in magnetizing and demagnetizing it continuously and because some electric currents are produced in it.) If there is no load – no light bulb and hence no current and no output power – then *all* the input energy becomes losses, a very inefficient way of running a generator. At the other extreme, if the losses are very small, we can see what must happen as the light bulb is connected. The *voltage* produced in the coil depends on how fast it is rotated (one reason why speed control is very important in power-stations). Suppose we crank the handle at absolutely constant speed. With no load, very little force is needed, but as soon as someone connects the bulb, we find that it is necessary to *push harder* to keep the crank turning. We must change our rate of energy input to match the new rate of energy output – the power used in the lamp. The argument is precisely the same on the large scale. If the TV programme ends and 100,000 people plug in their electric kettles, the power-station must find a great deal of additional push – and find it fast, or there will be complaints about the 'drop in supply'.

Modern generators are probably the most efficient energy-conversion devices in large-scale use, with losses which can be as little as 1 per cent of the input power if they are run under optimum conditions. Unfortunately, the efficiency falls off if they are run at much less than their full rated power, a problem for the engineers who have to deal with the widely varying demand in the course of a day or a year. Under the best conditions, however, a hydroelectric plant, whose direct input is the mechanical energy of the water, can have an overall efficiency of about 85 per cent, from water energy input to final electrical output.

Turbines

Turbine (noun): Any motor in which a shaft is steadily rotated by the impact or reaction of a current of steam, air, water or other fluid upon the blades of a wheel.

In other words, anything from a windmill to a water-wheel – both of which are of course used to generate electric power. We'll leave these, however, to later chapters and confine ourselves here to the steam turbines of modern large-scale power plants. The reason that we deal mainly with these is not just that they are more common, but that thermal operation of a turbine is the prime case of the Second Law of Thermodynamics in action, with all its unfortunate consequences.

Any steam-engine is a heat engine in the general sense in which we used the term in Chapter 4. The heat energy of the burning fuel is used to turn water into steam in the boiler, the steam drives pistons or a turbine, producing mechanical energy, and the waste heat is removed in the condenser (Fig. 5.4). In an 'open cycle' system, fresh water is fed continuously to the boiler, while the 'closed cycle' arrangement recirculates the condensed steam (now water, of

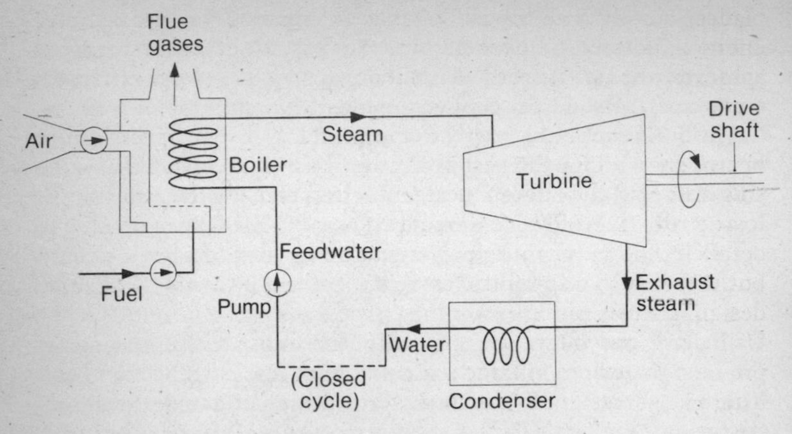

Fig. 5.4 The steam turbine

course) back to the boiler. Whatever the detail, the *heat engine rule* (p. 75) governs the performance, and to see what this means we'll look first at an early steam-engine and then at a modern turbine.

The old engine uses steam at its normal boiling point, 212°F, and we'll suppose that the condenser is cooled to perhaps 120°F (a little below household hot water temperature). The theoretical ratio of mechanical output to heat input is, for an ideal heat engine at these temperatures,

$$(212 - 120) \div 672 = 0.14$$

or 14 per cent efficiency. Already poor, but the actual achievement of those elegant old machines is unfortunately even worse: a miserable 5 per cent or so. Part of the responsibility for the difference lies in inefficient boilers, using much more coal energy than should be necessary to produce the steam, but the main cause is that the steam cycle is very far indeed from the 'ideal heat engine' studied by Carnot.

Things have improved, however. Present turbine systems use steam at temperatures up to 1,000°F, while a good cooling-water supply can hold the condenser temperature down to perhaps 80°F, giving a theoretical ratio of

$$(1{,}000 - 80) \div 1{,}460 = 0.63.$$

How nearly an actual system approaches this ideal depends on the turbine design. A modern turbine consists of a complex array of blades, closely interleaved so that the steam gives up as much energy as possible to the rotor. Nevertheless, the best that steam can achieve with a *single* turbine is not much over 30 per cent efficiency – about half the 63 per cent of the ideal engine.

The difference is due partly to the usual losses – friction, escaping heat, etc – which we expect in any system. But the major reason for this discrepancy between ideal and actual is that even a *perfect* (no losses) turbine would not be an ideal engine in the Carnot sense. The detailed thermodynamic analysis is too complex to reproduce here, but it proves to be a fruitful exercise, because it shows that certain design changes can bring up the efficiency by 15 per cent or more. Using not one but a series of turbines and extracting steam to pre-heat the water entering the boiler both improve the efficiency. There are practical limits to these measures of course: increased cost, and the increased likelihood of breakdown in a more complex system.

And unfortunately there's a very practical limit to higher steam temperatures, the obvious way to increase the efficiency. Normal steel will not survive steam at over 1,000°F for very long, and although special alloys have been tried, the return in efficiency is not generally considered worth the extra cost.

What is the best we can do, then? With steam at about 1,000°F and a pressure of some 200 atmospheres, with several turbine stages, and pre-heating, the 'before losses' efficiency can be brought up to perhaps 85 per cent of the ideal heat engine value:

85 per cent of 63 per cent = 53.5 per cent.

Losses will reduce this by a little under 10 per cent in the best case, giving a final efficiency of 45 per cent. Over half the input is still wasted – but at least it is a tenfold improvement on the early engine.

The search for yet further improvement must be a search for yet higher temperatures, and this means finding an alternative to steam. An obvious candidate is the hot gas from the furnace itself. Why use this to make steam to drive a turbine when it could do it directly? With gases at almost 2,000°F the thermodynamic gain should be considerable. In practice the *combined-cycle* system, using a gas turbine whose exhaust gas raises steam (at perhaps 900°F) for steam turbines, is probably the most efficient method available for producing mechanical energy from heat.

However, although some 15 GW of the world's 2,000-GW generating capacity uses combined-cycle systems, there are still problems to be solved. Not the least is the need for the gases to be very *clean* if they are not to foul the turbine blades. This means both physically clean (free of particles) and chemically clean (free of corrosive impurities), and in consequence most present systems use either natural gas or high quality oil – the premium fuels which everyone wants for everything. It's easy to see why there is great interest in the development of *coal-fired* combined-cycle power generation, and we return to this in Chapter 7.

A 660-MW power-station

The preferred size for generators has risen steeply in the past few decades: from a few megawatts – enough at that time for a small town – to tens of megawatts and now hundreds. Large turbo-generator systems probably *are* more efficient – as long as they are working. But the increased chances of a breakdown, and the more serious effect when it occurs, together with the extra distribution costs incurred by having a few, very large power-stations, gave rise to doubts about the wisdom of the 'larger must be better' philosophy, and it now looks as though the present systems which produce 500–700 MW will be regarded as optimum for some time to come. In regions where demand is high it is customary to combine a number of such units: five times 600 MW rather than three times 1,000; and we shall therefore take as our example for a study of inputs and outputs a 660-MW system.

The two diagrams in Fig. 5.5 show flows of energy and materials for an oil-fired, water-cooled installation. The figures are representative of any modern plant of this capacity, though there will be some differences and specific problems for coal or nuclear-powered systems. (These are discussed in later chapters.) We see at once that, for 660-MW output, the fuel energy input rate of 60,000 therms an hour (equivalent to 1,750 MW) means a plant efficiency of about 38 per cent and waste power output of over 1,000 MW. To see where it all goes, we'll work through each stage from oil input to electrical output.

The boiler

The boiler takes in about 1,500 tonnes an hour of water from the condenser and converts it to steam at 1,000°F (540°C) and a pressure of 180 atmospheres. It uses 150 te an hour of fuel oil: a flow rate of about ten gallons a second. To provide enough oxygen for full combustion (discussed further in Chapter 6) about 2,500 te an hour of air is needed. This is 2 million cu m, or equivalent to using all the air in

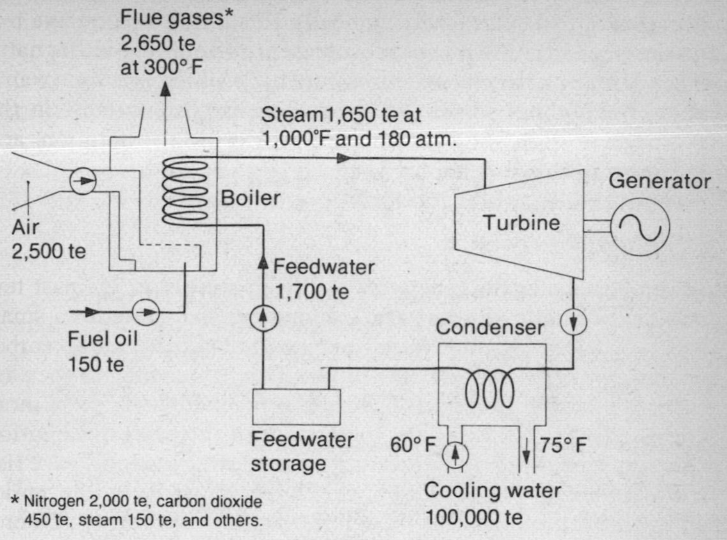

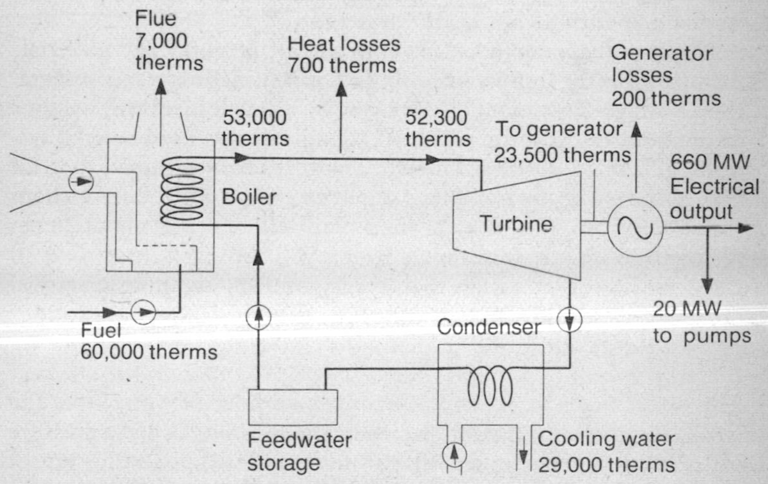

Fig. 5.5 Hourly flows in a 660-MW power-station

an average house twice every second, and the pumps to draw it in consume about 5 MW of electrical power. Oil is a 'clean' fuel, in the sense that there should not be a great deal of solid matter in the 2,650 te of hot flue gases which pass into the atmosphere every hour. Their temperature is about 300°F (150°C) and they carry with them some 7,000 therms of heat energy. That some heat is lost in hot flue gases is necessary (if they are not hot they won't go up the chimney) but the overall efficiency of the boiler is nevertheless nearly 90 per cent.

The turbine

The energy input to the turbine is about 52,300 therms an hour, in the form of roughly 1,500 te of superheated steam. At an efficiency of 45 per cent (see p. 93) the mechanical power output will be

45 per cent of 52,300 therms = 23,500 therms an hour,

which is a little under 690 MW.

The remainder of the input, about 29,000 therms an hour, becomes waste heat at this stage and must be removed to maintain the required low temperature and pressure at the turbine outlet.

The condenser

In our system the cooling is by flowing water from the sea or a river, which must carry away a little less than 30,000 therms an hour without becoming too hot. Now a therm is 100,000 BTU, and one BTU will raise the temperature of one pound of water by one degree Fahrenheit, so if we allow a 15°F rise in water temperature, each pound of water carries away 15 BTU; and the total water requirement is

$$30{,}000 \times 100{,}000 \div 15 = 200 \text{ million lb}$$

or about 100,000 te of water an hour. (We can visualize this as a six-foot depth of water flowing in a ten-foot wide channel at a speed of 10 mph.) Assuming that the water needs to be pumped up twenty or thirty feet, about 5 MW of pumping power will be required.

The generator

Running under ideal conditions, the generator can achieve 99 per cent efficiency, so the losses will be 'only' 5 MW or so. We then need to subtract the 20 MW required to run all the pumps and other equipment, leaving the specified 660 MW. Notice that transformer losses, and heat generated in distribution have yet to be deducted. As these will be at least 10 per cent, we should perhaps regard 600 MW as representing the maximum useful output – an overall efficiency of about 34 per cent.

Conclusions

The conclusions of this study are all there in the figures. Unless we have the resources for hydroelectric generation to provide the electric power we want, there is no present alternative to the thermal power-station, with its enormous output of waste heat and combustion products. (All power-stations except the hydroelectric are 'thermal' in this sense, though the combustion products are rather different in the nuclear case, as we shall see later.)

If we don't like the figures there are at least three courses of action open to us:

- To look for other ways of providing electric power, less wasteful and less harmful than present methods.
- To make use of the 'waste' heat.
- To use less electrical power.

These are not novel ideas, and much effort in many countries is being invested in each of them. All will recur at different points in the remainder of the book.

6 Oil and Natural Gas

Noble fuels

The name is by analogy with the noble metals (silver and gold, etc.) and there is some justification for it. In Chapter 4, we looked at some of the advantages of these two fuels; and it is important to bear in mind that it is not only in the over-developed countries that oil has become indispensable for the present way of life. When we deplore the Californians in their gas-guzzlers consuming megajoules and generating smog, we should also remember that oil drives public transport in Birmingham, Benares and Buenos Aires, and generators and irrigation pumps in even more remote places. Concern about the undesirable effects of coal or nuclear power makes oil and natural gas in many ways the preferred fuels for electric power production; and then there is the petrochemicals industry, a non-energy use of these versatile materials. Whatever the plans for the future, the present situation is that there is indeed no substitute for oil.

Whether the wells run dry in twenty years or fifty it is undoubtedly the case that supplies are finite and that growth in world-wide per caput consumption to the present US level could not be tolerated. The problem of oil (and natural gas) is therefore really the problem of finding suitable alternatives – either new technologies to use other energy sources directly, or substitute liquid and gaseous fuels derived from these more plentiful resources. We discuss these issues later, but in this chapter devoted to the noble fuels the question we must ask is how best *not* to use them – or how to use as little as possible where there is no present alternative. The prime case is transportation, because we do have the technologies to replace these fuels for heating or power generation, even if we don't much like them. So we look first at the nature of the hydrocarbons and at their function as energy carriers, to provide a background for analysis of the internal combustion engine (and also for later discussion of synthetic fuels). Then we consider all the energy-consuming processes in a moving vehicle, and ask how efficiency might be improved by changes in vehicle or engine design.

Hydrocarbons

That oil and natural gas, fluids which more or less propel themselves out of the ground and which burn with virtually no solid wastes, have advantages is obvious. But why are hydrocarbons – as opposed to, say, sand or seawater – good fuels at all? To approach an answer, and to see a little how they work, we need to look at their atomic composition.

A hydrocarbon is, as one might guess, a compound with hydrogen and carbon as two of its main constituents. Crude oil, or even the part which we separate and call petrol or gasoline, is not a single hydrocarbon but a mixture of many. Hundreds of different hydrocarbons have been identified in crudes from various parts of the world, though the principal constituents are far fewer in number. In general, crude oils contain 80–90 per cent by weight of carbon and 10–15 per cent of hydrogen, with up to 4 per cent of sulphur, some oxygen and nitrogen and traces of other elements. In looking at the carbon–hydrogen ratio we must observe that one carbon atom has 12 times the mass of a hydrogen atom, so the proportion *by atoms* in the crude is very roughly one to two, carbon to hydrogen. For natural gas, the ratio is more nearly one to four and for coal, not far from one to one.

The molecules of any particular hydrocarbon have the atoms of carbon and hydrogen arranged in a characteristic pattern, and one series which plays a major role in many oil products is the family of *paraffins*. Fig. 6.1(a) shows a few of these: strings of carbon atoms with hydrogen atoms attached, the whole system being held together by forces originating in a rearrangement of the electrons of the constituent atoms. The first four members of the series are gases at room temperatures, natural gas being largely methane. Any such gases contained in crude oil will bubble off when the oil flows from the well, and these were for many years allowed to escape or burned on site. This is of course changing and these valuable products are now collected where economically feasible.

As we progress up the hydrocarbon series, the boiling temperature becomes higher and higher, and with pentane we reach a substance which is liquid at normal temperatures. The progression is the basis for separation by distillation: if the crude is heated to, say, 200°C (400°F) everything with a boiling point lower than this will eventually become vapour, which can then be collected and condensed. At this temperature, just under a fifth of the total contents of the crude are distilled and the result, straight-run gasoline, contains

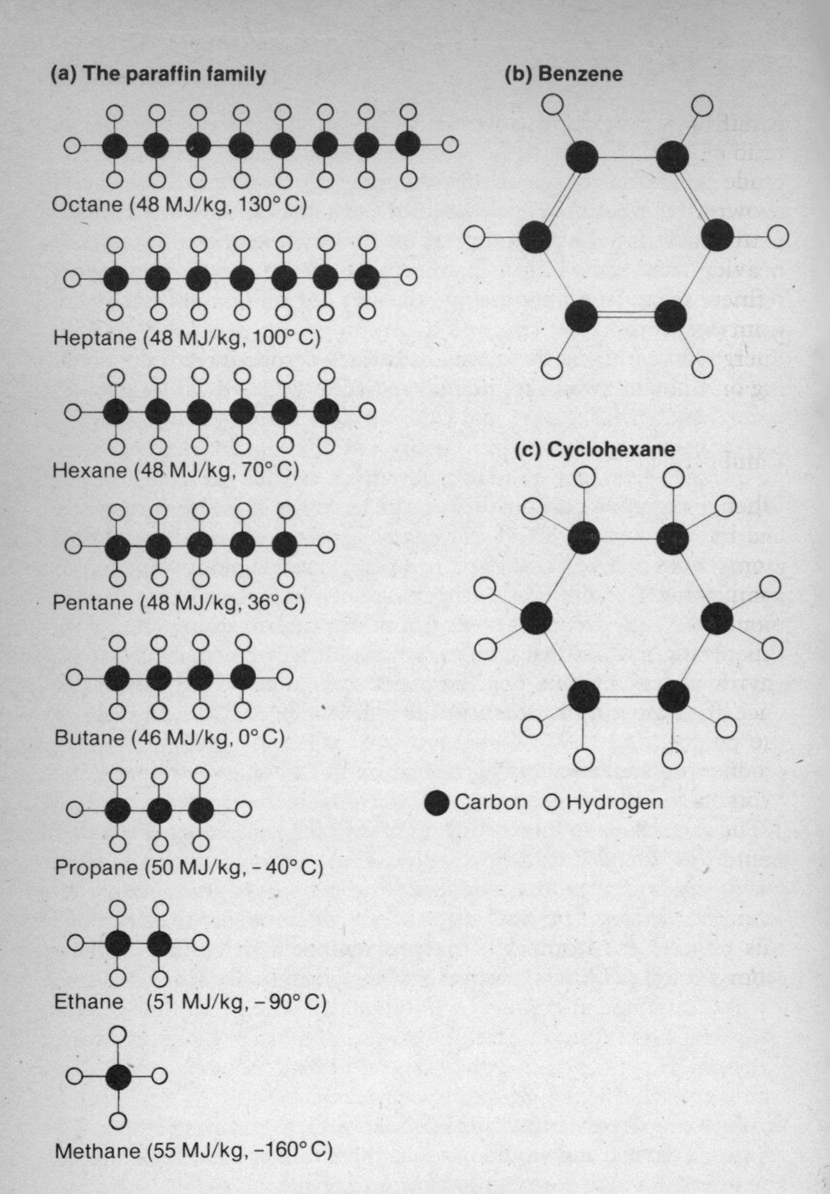

Fig. 6.1 Hydrocarbons

1. The figures in brackets show the heats of combustion and the normal boiling temperatures of the paraffins.
2. Molecules are, of course, three-dimensional entities, so these diagrams must be regarded as schematic versions of the arrangements of the atoms. Some of the paraffins occur also in other patterns, with side-branches.

paraffins from pentane to octane. As we can see from the overall C:H ratio of 7:12, there must be other compounds than paraffins in the crude oil, and a couple of the 200 or so which occur in petrol are shown in Fig. 6.1(b) and (c). The need to increase the proportion of petrol has led to the development of *cracking*, or breaking down the heavier hydrocarbons into lighter molecules. Thus the input to a refinery is crude oil, and its output is a range of products resulting from distillation, cracking and reforming, purification and mixing. Energy losses during these processes are 1–5 per cent or so, depending on the quality of the product required.

Combustion

When a hydrocarbon burns, the bonds holding together the carbon and hydrogen atoms are broken, and each forms new bonds with atoms of oxygen from the air. The products are thus the well-known compound H_2O and some carbon–oxygen combinations. Carbon monoxide (CO) may be present, but full combustion should remove this poison and lead to carbon dioxide (CO_2) which, whatever its environmental effects, is at least not toxic. One result follows at once from this simple account: the higher the C:H ratio, the higher the proportion of CO_2 compared with water in the combustion products. In this respect, natural gas is the 'best' fuel and coal the 'worst' (see Table 7.2, p. 122).

The quantities involved can be calculated by elementary arithmetic. All we need to know is that a carbon atom is 12 times as heavy as a hydrogen atom, and an oxygen atom 16 times. Each 'C' atom combines with two 'O's, and two 'H's combine with one 'O', so this tells us the number of oxygen molecules (each having two atoms) required. Taking pentane as an example, we have

$$\begin{array}{ccccccccc} C_5H_{12} & + & 8\,O_2 & \rightarrow & 5\,CO_2 & + & 6\,H_2O \\ 72 & + & 256 & = & 220 & + & 108 \end{array}$$

So *one tonne* of pentane would need about three and a half (256 ÷ 72) tonnes of oxygen and would produce three tonnes of carbon dioxide and one and a half tonnes of water (or steam).

Environmentalists might like the following idea. A busy six-lane motorway could carry as many as twenty cars per 100 metres. At 50 mph and 20 mpg, each uses two and a half gallons of petrol an hour, making the total hourly consumption fifty gallons, or one-sixth of a tonne. This requires about half a tonne of oxygen an hour from the surrounding atmosphere and produces roughly the same quantity of carbon dioxide. A living plant works in the opposite way, taking in carbon dioxide and water and using them to produce compounds of carbon, hydrogen and oxygen: 'vegetable matter'. Surplus oxygen is given up to the atmosphere, and a large tree can produce perhaps 2 kg an hour. We thus need to plant all motorways with 250 large trees per 100 metres to preserve the oxygen balance. I think that I shall never see . . .

The burning of fossil fuels leads to other combustion products. Any sulphur present will form oxides which are harmful to health as well as being corrosive. The permitted quantity of these is subject to clean-air regulations in some countries, and removal of sulphur or of the combustion products adds to the cost and consumes energy. Then nitrogen, which makes up about four-fifths of the atmosphere, can combine with oxygen in the tumult of a flame or explosion, and the resulting oxides again pollute the environment. Other minor impurities can be poisonous or cause damage and should therefore be removed at some stage. These matters must enter into any debate on future uses of fossil fuels or alternatives, and we shall return to them later.

We still have to see why *energy* becomes available as a result of combustion. The essential starting-point for an explanation is that energy is needed to separate any molecule into its component atoms. (If this were not so, the world about us would be falling to pieces all the time, with wooden chairs turning into gases and a little soot, and table salt into a rather nasty metal and a poisonous vapour.) Looking at the equation above for the combustion of pentane, we see that four different molecules are involved, and the important feature (which is not of course obvious from the equation) is that the energy needed to take apart the pentane and the oxygen is appreciably *less* than that gained in putting together the carbon dioxide and the water. It is this relative ease with which they dissociate which makes the hydrocarbons such good fuels. There is energy to spare in the reaction, so once the process starts – once you light the flame – energy is produced as long as supplies of fuel and oxygen are continued. The energy appears as heat, which as we know means the increased random motions of the molecules. The detailed behaviour of very hot gases is a complicated matter and not entirely understood, but we need not be concerned with this. The important result is that chemical changes can release energy in a form which we can use.

The efficiency with which it is used, even as heat, varies widely.

We have already seen that the boiler in a power-station wastes over 10 per cent of the fuel energy in hot flue gases and steam. Nevertheless, it is a model of efficiency compared with some other processes. Not only hot flue gases, but incomplete combustion, and poor transmission of heat to the place where it's wanted or the substance to be heated, all contribute to the low efficiencies of industrial and domestic furnaces. An open coal fire converts less than 20 per cent of fuel energy into useful heat, and it is claimed that most domestic furnaces (oil or natural gas) are so badly adjusted that nearly half the fuel input is wasted. Then there is the internal combustion engine . . .

A short drive in the country

On a sunny Sunday in June the Smiths decide to visit the seaside. They go by the scenic route over the South Downs and return on the direct flat road – a round trip distance of 100 miles. To simplify our arithmetic we shall assume that they are able to drive at a steady 45 mph all the way and that the car uses a little under three gallons of petrol for the journey. Our problem, then, is the following. The trip has used over 400 MJ of energy but the Smith's have gained none of this. They are back where they started. Where has the energy gone, and did they really need to use so much?

To track the missing joules we'll consider in turn each of the possible energy-consuming processes. First, we know that the energy used to accelerate the car is not returned; the gasoline does not flow back into the tank when the car stops. How many joules are consumed in setting the car in motion at the start of each journey?

The energy which an object has by virtue of its motion – its *kinetic energy* – depends on its speed and how heavy it is. The relationship to the mass is simple: a four-ton vehicle at a given speed has twice the energy of a two-ton one, and an additional pound in the load means the same additional kinetic energy whether you start at two or at four tons. In other words, the kinetic energy is *proportional to* the mass. Not so for the speed. The effect of an additional mile an hour is greater the faster you are moving. To increase your speed from 60 to 61 mph requires twice as much extra energy as an increase from 30 to 31 mph – and this has nothing to do with air resistance. It would be equally true if there were none. If this is so, the energy of the moving vehicle must be proportional not to its speed but to *the square of its speed*. The kinetic energy at 90 mph is not three but nine times that at 30 mph, a fact which has obvious consequences when the energy is dissipated as heat, light and sound as you hit a brick wall.

To compute the kinetic energy of the car in joules we need to know its mass in kilograms and its speed in metres per second. The relationship is then

kinetic energy = one half of the mass times the square of the speed.

Table 6.1 gives some values for a vehicle weighing one tonne, and we see that the kinetic energy at 45 mph is 200,000 J, or one fifth of a megajoule. To accelerate the car for the outward and return journeys would thus require 0.4 MJ, which, as we are trying to account for 400 MJ, does not get us very far.

How about the hill? Suppose it is 200 m (600 ft) high. How many joules are needed to lift the car to the top? The reason that we need any is that the gravitational force of the Earth is pulling vertically down. This force, technically the *weight* of the vehicle, is numerically equal to about ten times the mass in kilograms, and it is also *the number of extra joules needed for one metre of vertical rise.* With 1,000 kg and 200 m, therefore, we need 2 MJ. This is a little better, but still accounts for only about half a per cent of the energy consumed – and we have not allowed for any gain on descending the hill.

Given the results of these two calculations, it seems reasonable to conclude that most of the energy was consumed in keeping the car moving at constant speed. This is of course common experience. Frequent starts and stops do increase petrol consumption, and we do

Table 6.1 Speed and kinetic energy

The kinetic energy of any object, measured in joules, is equal to one half of its mass in kilograms times the square of its speed in metres per second.

Speed		**Kinetic energy (J) for a 1,000-kg car**
mph	**metres/sec**	
20	9	40,000
30	13½	90,000
40	18	160,000
45	**20**	**200,000**
50	22	250,000
60	27	360,000
70	31	490,000
80	36	640,000
90	40	810,000
100	44	990,000

use more fuel in crossing the Alps than in driving the same distance in Holland, but we know that any car steadily consumes petrol when driven at constant speed on a flat road. To see where the energy goes, and how much is used, we need to look at the two main processes which consume the output power from the engine under these conditions: rolling friction and air drag.

The most obvious evidence of *rolling friction* is the temperature of the tyres after a long drive. As a wheel rotates, its tyre is compressed where it makes contact with the road, and this compression needs energy, not all of which is recovered as the tyre expands again. The difference becomes heat, and the amount of energy converted into heat by this and other friction effects associated with wheels and road depends mainly on the weight of the car and the nature and condition of the tyres. A convenient way of looking at rolling friction is to regard it as effectively a force pulling against the motion of the vehicle. We have already seen, in discussing the energy needed to lift the car up the hill, that the force is the additional energy required per metre moved, and we can again make use of this idea here. The rolling friction force is the number of joules needed to move the car forward one metre, and its value is usually 1–2 per cent of the weight of the car. This percentage is the *coefficient of rolling friction*, C_R, and Table 6.2 shows the effect of its value, for the one-tonne car. From the fuel economy point of view, C_R is an important quantity. It depends on a number of factors: radial tyres have lower C_R than cross-ply, high tyre pressure reduces it and – unfortunately – worn tyres have lower C_R than new. Rolling friction can account for half a car's petrol consumption at speeds below about 30 mph, so any reduction in C_R is obviously welcome. (To draw the immediate conclusion and drive on over-inflated worn-down radials is not recommended.) Returning to the Smiths, and assuming that they have average tyres, we find an energy consumption of 24 MJ for the 100-mile journey.

Table 6.2 Rolling friction
Data are for a 1,000-kg car

Coefficient of rolling friction (%)	Energy used per 100 mi (MJ)
1	16
1.5	24
2	32

Air drag results from the fact that, in order to move, the car must push its way through the surrounding air. Not surprisingly, the energy consumed in this process depends very much on the speed of the car, and also increases with the frontal area. Again we have a coefficient: the *coefficient of drag*, which tells us how good (or rather, bad) the car is at getting through the air. Fig. 6.2 shows how the energy consumed depends on the speed for different cars, all with 2 sq m (20 sq ft) frontal area. Present cars have C_D values from about 60 per cent (box on wheels) down to a little under 30 per cent (very streamlined), and if we assume 40 per cent for the Smith's car we find that the energy used is 32 MJ for the 100 miles at 45 mph. Notice that it would have been only half of this at 30 mph, but three times as much at 80.

So what have we got? The total energy in megajoules needed to drive the car is

0.4	+	2	+	24	+	32	= 58.4 MJ
(acceleration)		(hill)		(rolling friction)		(air drag).	

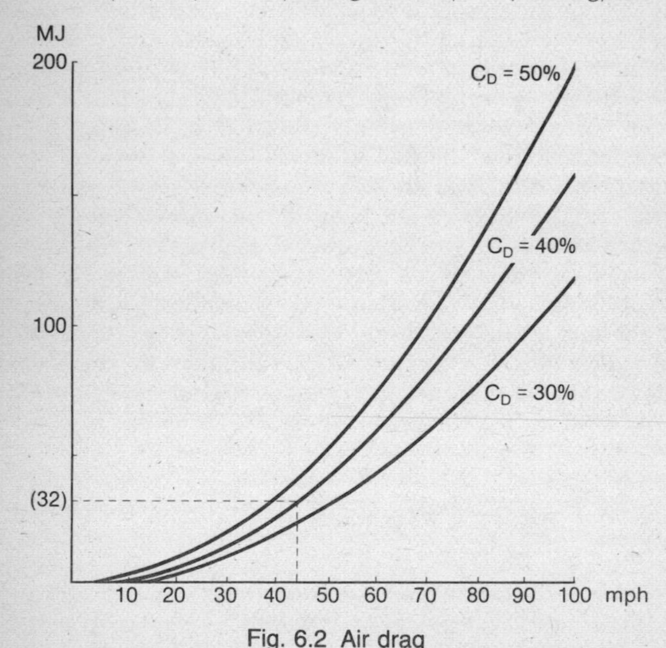

Fig. 6.2 Air drag

The curves show how the energy consumed by air drag depends on the speed of the car and its coefficient of drag, C_D. Data are for a vehicle with frontal area 2 sq m, and for a 100-mile journey at constant speed.

The engine output will need to be a little greater than this because of *transmission losses*, the result of friction in the train of events between engine and wheels. Assuming 80 per cent efficiency, we find that the engine needs to provide 73 MJ in order that 58.4 MJ reaches the wheels. It must also provide energy for all the *accessories*: generator, pumps, fan, lights, etc., and in a modest car this would account for a little under a kilowatt, or say 7 MJ in the driving time of two and a quarter hours. (A large American car with automatic transmission, power steering and brakes and, worst of all, air conditioning, might need twenty times as much.)

We now have a total: *80 MJ of output energy* from the engine was needed for the journey, all of it eventually dissipated as heat. But we have another total: the fuel energy *input* to the engine was *444 MJ* (100 miles at an average 36 mpg, with 160 MJ/gal). The internal combustion engine is a marvellously versatile, flexible, compact and even cheap device, but efficient it is not.

The internal combustion engine

About a sixth of all primary energy goes into internal combustion engines, which makes them collectively the world's second-largest consumer. The leaders, the thermal power-stations, manage at their very best only a rather poor 35 per cent efficiency, but now we find the runners-up gasping along with only about half this, wasting not two-thirds but over four-fifths of their fuel intake. Surely *here* some improvement must be possible? Table 6.3 shows a revised version of the Smiths' day out. Not, unfortunately, a realistic one – you can't go out and buy our 'improved' car. The figures are chosen to illustrate the consequences of a few changes, all of which are technically feasible and many of which are used, for instance, in setting miles-per-gallon records. The effect of even a modest improvement in engine efficiency is obvious, so the chief question we shall ask in this section is why present performance is so poor.

Everyone probably has a rough idea of how a car engine works (Fig 6.3): a little fuel and some air are drawn into the cylinder and compressed, an explosion (high-speed combustion) produces hot gases at high pressure, and these push down the piston, forcing the shaft to rotate. From the energy viewpoint, each cylinder is a small heat engine, converting part of the thermal energy of the combustion products into mechanical energy and rejecting the remainder in hot exhaust gases. So we can start by reducing the problem of engine efficiency to the question: How much of the *heat* energy produced in

Table 6.3 The car journey, and a revised version

Energy-using process	Proposed change	Energy consumption (MJ or MJ/100 mi.)	
		Present car	'Improved' car
Acceleration and climbing	Reduce weight from 1,000 kg to 800 kg.	2.4	1.9
Rolling friction	Reduce weight, as above. Reduce C_R from 1.5% to 1.1% by tyre and suspension changes.	24	14
Air drag	Reduce C_D from 40% to 25% by improved shape.	32	20
	Total energy for driving	58.4	35.9
Transmission	Improve efficiency from 80% to 85%.	$\times \frac{100}{80}$	$\times \frac{100}{85}$
	Engine output for driving	73	43
Accessories	Improve efficiency, eliminate unnecessary fans etc.	7	5
	Total engine output for journey	80	48
Engine	Increase engine efficiency from 18% to 30% average.	$\times \frac{100}{18}$	$\times \frac{100}{30}$
	Fuel energy required	444	160
	Quantity of fuel	$2\frac{3}{4}$ gal	1 gal
	Fuel economy	36 mpg	100 mpg

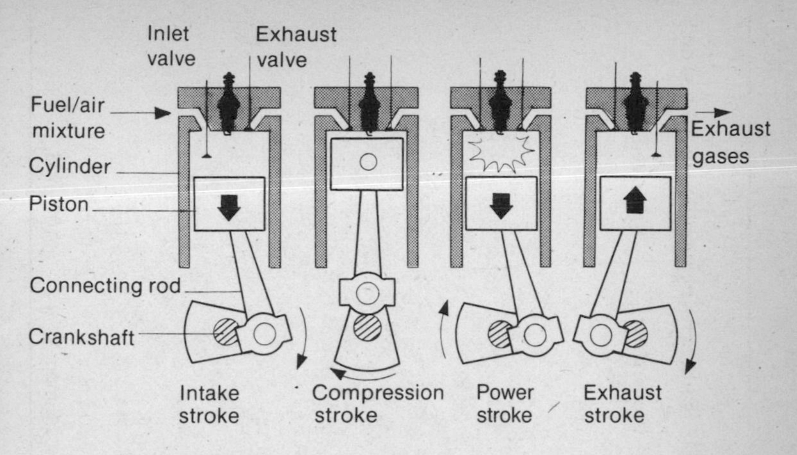

Fig. 6.3 The four-stroke internal combustion engine

each combustion is delivered as *mechanical* energy on each piston stroke? One possible 'ideal' answer might be found by using the heat engine rule (Chapter 4), if the relevant temperatures could be identified; but this would give a very optimistic result, because the system is not really at all like the ideal heat engine. Explosions are complex events, difficult to analyse in detail using simple science (or advanced science either, for that matter). However, *if* it is assumed that all the heat becomes available instantaneously, before the piston starts to move, *and* that there are no heat losses through the cylinder walls, and one or two other unlikely simplifications, then the efficiency can be predicted. We shall not carry out the analysis here, but show the result in Fig. 6.4, and a very important result it is. The prediction is that the efficiency, for this highly simplified case, depends almost entirely on the *compression ratio*. The word 'almost', related to the spread shown on the curve, reflects the fact that the efficiency does depend on the nature of the gases in the cylinder and that the theory has to make some approximations in treating these. However, even with very sophisticated computer simulations, the answer remains essentially the same: greater compression ratio means greater efficiency.

Leaving for the moment the question whether this is the solution to our problems, we note that it certainly can't be the complete explanation of present low efficiencies. The numbers shown in Fig. 6.4 are efficiencies of 40–50 per cent, and even if heat losses or other departures from ideal conditions bring these down by 5 or 10 per

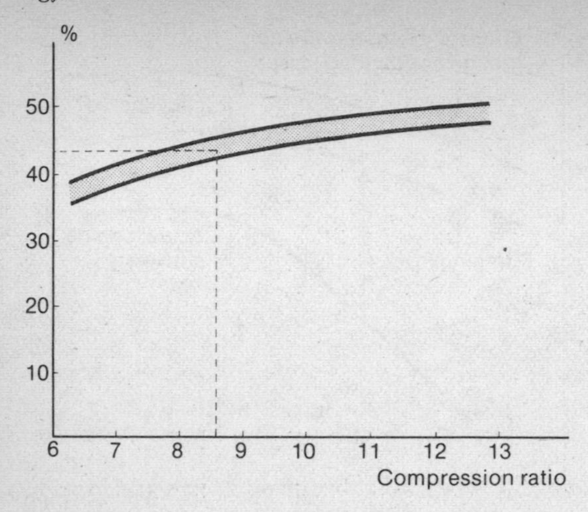

Fig. 6.4 Internal combustion engine: 'ideal' efficiency
The compression ratio is the maximum cylinder volume (at the end of the intake stroke) divided by the minimum volume (at the end of the compression stroke).

cent, they are still much higher than the 18 per cent of the Smiths' car. There must be other losses, and they are not difficult to identify. There is of course friction: the inevitable generation of heat when one surface rubs on another. This is thought to reduce the efficiency, in a well-maintained engine, by about 5 per cent at low speeds, and more as the engine works faster. So, taking as an example the figures for a compression ratio of 8.5, we have 44 per cent 'ideal' efficiency, reduced to an 'actual' value of perhaps 38 per cent before friction losses are taken into account, and to 30 per cent if we include them in the reckoning.

The explanation of the remaining loss is very important, because it shows how the requirements we put on a car ensure that normal driving efficiency will be much less than it need be. The lost energy results from the force which the piston must exert in order to draw in and later push out the vapours and gases; and the important point is the following: this force, and the consequent energy loss, rise steeply *as the throttle is closed*. The reason we have a throttle is of course to allow us to adjust the amount of fuel burned in each cycle, giving more – or less – output energy according to driving requirements, and the unfortunate consequence of the throttling loss is that if you drive with a light foot and keep the speed down to economize you are driving with a closed throttle and a correspon-

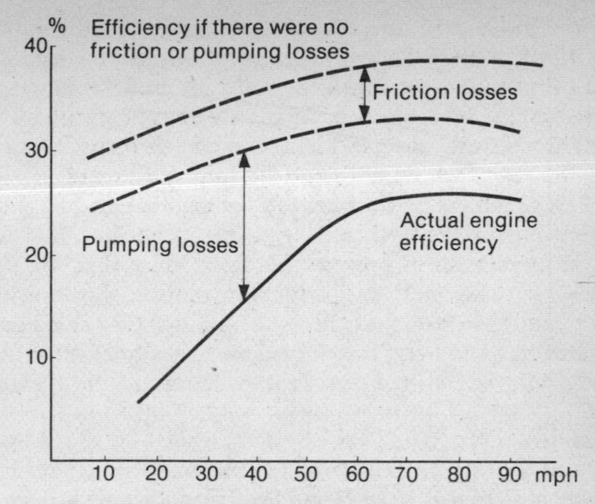

Fig. 6.5 Engine efficiency and losses

1. The data are for level-road driving in top gear.
2. The upper curve already shows a reduction from the 'ideal' efficiency (Fig. 6.4) resulting mainly from heat losses from the cylinder (at low speeds) and incomplete combustion processes (at high speeds).

dingly low engine efficiency. Fig. 6.5 shows the situation. Of course, the over-riding effect of air drag ensures that the *fuel economy* will nevertheless fall off at high speed, even with wide-open throttle, so the combined result is the usual mph curve (Fig. 6.6).

What can be done about all this? Three obvious improvements would be reduced weight, improved design of tyres and suspension to minimize rolling friction, and a better shape to reduce drag. (A

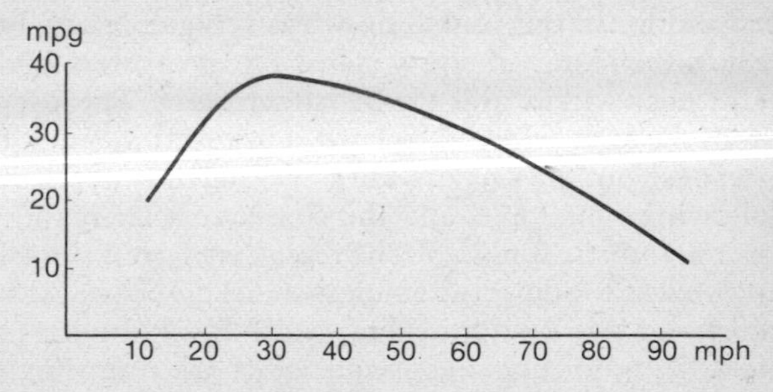

Fig. 6.6 Fuel economy and speed

Data for level-road driving in top gear.

symptom of the lack of interest in fuel economy until very recent times is the fact that General Motors in Detroit commissioned America's first purpose-built, large-scale automobile wind-tunnel *in 1980.*) As far as the engine is concerned, we have identified at least two areas of interest: compression ratio and throttling. Increased compression ratios would undoubtedly improve engine efficiency, but there are problems. High ratios require high-octane fuel, and it is argued that customers will not buy vehicles needing fuel which costs even more than at present. Nor is it clear that the energy balance is always favourable. The fuels are more energy-consuming to produce; and the extremes of pressure in a high CR car are greater, for the same output energy, so the engine needs to be stronger and hence heavier, which reduces fuel economy. Not a clear-cut situation.

The problem of throttling losses is somewhat different. We would be using our car engines much more efficiently in general if they were designed so that at 50 mph on a level road we were driving with wide-open throttle (accelerator pedal fully down). But we do like to be able to accelerate from time to time, and to be able to climb hills without stalling: we like to have some reserve power. It has been argued that at present we enjoy rather too much of this, and from Fig. 6.6 we see that the engine illustrated, not atypically, is most efficient at about 70 mph – for level-road driving. It would probably be working at its maximum *possible* efficiency when driving the car at about 40 mph up a one-in-six incline – with the throttle almost wide open, of course. (Note that this does not imply the maximum miles per gallon, just the most efficient use by the *engine* of the fuel which it consumes.) If we were willing to accept a little less acceleration at speeds which are in any case above the legal limit in most countries, perhaps the Smiths' 18 per cent could be improved to 21 per cent. Not the startling results shown in Table 6.3, but enough to save about forty gallons of petrol a year for the average European motorist, or nearly twice this for the American.

7 Coal

An ignoble fuel

Coal is indeed a particularly unattractive energy source. A large coal-fired power-station can produce enough ash in a year to cover an acre of ground to the height of a six-storey building, while its flue gases may carry several hundred tonnes a day of sulphur dioxide and nitrogen oxides into the atmosphere. The ash, if not carefully isolated, will pollute the nearby ground-water with a variety of unpleasant substances such as sulphuric acid and arsenic, and the flue gases are claimed to contaminate lakes and harm life hundreds of miles downwind. Compared with oil or natural gas, coal produces 50–100 per cent more carbon dioxide for the same useful heat. It is much less convenient to transport, store and use. Its extraction leads to land subsidence and spoil heaps or to the environmental horrors of strip mining, and is responsible, in an average year in Britain alone, for the deaths of fifty miners and serious injury to about 500 others. There is obviously no difficulty in explaining why the share of world energy coming from coal has halved in a few decades.

Yet when we look at the forecasts of the experts, at predictions of world production or of consumption by major industrial countries, we find that coal is to rise phoenix-like from the ashes of its former self. Reversing all trends, it is to provide an ever-*increasing* proportion of world energy. According to current projections, for instance, United States coal production is to grow in the next twenty years by *300 per cent*, and the picture is similar in a number of other countries. Why the sudden enthusiasm for this unappealing and *démodé* fuel?

We don't have to look far for an answer. Almost all experts are agreed that reserves of oil and natural gas are insufficient to meet demand for many more decades, which leaves coal with one supreme advantage: *there is a very great deal of it*. It is thus the one primary resource for which we have fully-developed technology together with plentiful reserves. Coal has been in use for thousands of years and we know how to mine it and how to burn it – in very large quantities if needed. No problems of untested new systems.

No nasty technological surprises. And no nasty international political surprises either, at least for those countries (including Britain, the USA and India) with their own extensive reserves.

Coal, nuclear energy or conservation? We may or may not agree – and we'll be looking at other alternatives later – but the overwhelming majority of energy planners in industrial countries see these as the only realistic large-scale alternatives for the next fifty years. If we also accept the view that the rise in demand which goes with economic growth will continue, despite occasional fluctuations, to exceed any saving due to conservation, we are left with a simple problem: Do we build coal-fired power-stations or nuclear power-stations? The need for serious, informed discussion of this issue determines the main questions which we must now try to tackle. Can mining be made safer and less harmful to the environment, and can we burn huge quantities of coal without asphyxiating ourselves? Coal is a very complex material, by no means fully understood; but we can try, by looking a little at its nature and composition, to understand how these problems of extraction and combustion arise. Then we can ask how far present technology is able to deal with them, and whether new methods are likely to bring improvements in the near future.

Predictions for the year 2000 don't limit the energy contribution to electric power and direct heat. Almost all plans include synthetic substitutes for oil and gas produced by the conversion of coal: the *synfuels*. In the USA and Western Europe alone, nearly half a billion dollars a year is currently being invested in some seventy different schemes for converting the plentiful *solid* fossil fuel on a large scale into *fluid* hydrocarbons – in the hope of thus postponing at least for a couple of centuries the problems of the coming decades. A detailed analysis of the many and varied current approaches would require a treatise to itself. All we can do here, as with other major technologies, is to try to see the underlying principles so that we can understand what is possible and where the difficulties lie, and then look at just a couple of the many approaches in a little more detail.

Mining

Consider the ideal coal-mine. Fully automatic cutters, remotely controlled, shear off the coal at the underground face and load it continuously on to conveyors which bring it to the surface, where it is fed directly to a preparation plant and then to a minehead power-station or conversion plant. The output is clean electric power or fluid fuel, and the entire operation – fully enclosed and controlled from the surface – is both safe and environmentally harmless. The answer, surely, to many of our energy problems. But how near are we to this ideal, and are we likely to move closer to it in the next few years? Later in the chapter, we'll deal with the output end of the system: the power-station or conversion plant; but here we want to know about the extraction process. Or rather, we want to know what determines the main things that matter in comparing coal with other energy sources: cost, safety and pollution. We'll start with the coal itself.

Coal seams

Coal is the result of geological forces and chemical changes acting on plant material for hundreds of millions of years. The seams come from the accretion of layers of dead trees and other plants, initially protected from atmospheric decay by water, and ultimately compressed to a tenth or less of their original thickness. They range from a few inches to hundreds of feet thick, and can lie thousands of feet under the surface or so near that they penetrate it. Usually, but not always, coal occurs in *beds*: extended systems of seams. In Britain, for instance, the new Selby mine will reach for some ten miles underground, covering an area the size of the Isle of Wight at depths of 1,000 ft and more. On a somewhat different scale, the Wyodak bed in Wyoming and adjacent states stretches for well over 1,000 miles in seams up to 150-ft thick very near the surface: some fifteen billion tonnes of strip-minable coal. The *ages* of coals in different places vary considerably too, from under 100 to over 500 million years – though the different compositions (the ranks discussed further in the next section) are thought to be more the result of different geological conditions than of sheer age variations.

Unfortunately for the mining engineer, coal seams only rarely stay as nice, uniform, parallel-sided slabs. The original layers of plant material may have been like this, but we have only to reflect that, at the time when the coal forests flourished, the land which is now Britain was probably in the tropics thousands of miles from its present position, to appreciate that certain changes will have occurred. In the restless shifting of continents the flat seam is only too likely to have been tilted, buckled, sheared and otherwise distorted. So the fully automatic cutter needs to deal with a range of thicknesses and angles as well as a variety of compositions. If it is to deliver the coal, all the coal and nothing but the coal, it will need a sensitivity and versatility close to those of the human miner it displaces.

Technology and productivity

We have however come some way from the miner with nothing but pick and shovel, though it has taken a long time: just about 200 years since the first steam-engine was used to wind coal to the surface. As recently as a generation ago, most coal was still hand-loaded on to conveyors in Britain's mines, and although a single miner might load as much as ten tonnes in a shift, the long process of cutting and blasting, and placing support walls and props by hand meant that productivity for underground workers was no more than two tonnes per man-shift.

In discussions of mining costs, output per man-shift is a vexed question. Which 'man' for a start? Recent figures range from a little over two tonnes through a little over twenty tonnes to considerably over 200 te. The first two are the British industry average for all workers and the average for face workers in a modern mine, while the third is for a large surface mine in the western USA.

In the newest underground mines the coal is cut and loaded continuously into conveyors, with the whole system, including hydraulically-operated roof supports, advancing automatically as the face is cut away. The cutters themselves are complex machines now, and become more versatile each year; but they are still a long way from the automaton described at the start of this section. Probes which test the extent of the seam and guide the cutter are being developed, but at present men on the spot must still control the operation – and also maintain all the machinery, another aspect which raises problems in the hostile environment of a mine.

For our present concerns, mechanization matters in two ways: it affects costs and it affects safety statistics. In the past thirty years the contribution of wages to the cost per tonne has fallen from just over a half to about a quarter in the USA and from three-quarters to just over a half in Britain. Predictions for Britain's new Selby mine give an indicator for the future: 10 million tonnes a year with 4,000 workers in all, which suggests a wages bill contributing only 10 per cent of the cost of the coal.

Of course the capital costs are then high. Hundreds of millions of pounds or dollars for a new mine, and millions just for the cutting equipment for one face. Surface mines are much cheaper: about a third of both capital and labour costs for the same output; though more stringent land reclamation regulations may increase this figure in the future.

Safety

The relationship between mining technology and accident rates is not simple. Improved methods mean more coal per man-shift, so the 'deaths per thousand tonnes' can fall even though the chance of a fatal accident remains the same for an individual miner. Then, some improvements – better ventilation, improved monitoring and control of dangerous gases and dust accumulations, safer underground transport and hoisting equipment – are relatively independent of the actual cutting and loading technology. In Britain, the number of deaths per million man-shifts was reduced from about twenty in the 1930s to between five and ten in the post-war period and is now about one per million: roughly one chance in a hundred of being killed in a working life. The figures for the USA are slightly worse, despite rapid improvement in the past decade. With present productivity and fatality rates, each large coal-fired power-station in either country accounts for the death of one miner every eighteen months, while the proposed growth in US coal consumption would mean an additional 200 deaths a year by the year 2000.

Most of us would surely regard these figures as unacceptable as part of a long-term plan for energy, and they must weigh heavily in an assessment of the future for coal. However, the number of deaths would fall to a fifth of the present figure if the whole industry were to achieve the safety level of the best mines. Further, the predicted four-fold increase in productivity with newer methods would mean a corresponding reduction in fatalities per tonne of coal mined. There thus seems to be no *insuperable* reason why the rate of deaths and serious accidents should not be brought down to the 'normal' industrial level.

The environment

Air quality is affected in so many ways by so many different energy processes that we'll devote a separate section to it later. Here, we shall look briefly at three problems which are specifically related to coal-mining: subsidence, surface disturbance and wastes. The new extraction techniques unfortunately do nothing to solve these. If anything, they exacerbate them. Ground subsidence is of course a result of underground mining, and it is customary with the modern 'longwall' method that the roof is allowed to fall in behind the working. It is claimed that detailed knowledge of soil mechanics makes this a controlled process, rather than the unpredictable cavings-in above old mines. And special cases get special treatment: the

900-year-old Selby Abbey will be left sitting on an extended pillar of undisturbed coal. Only time will tell whether the calculations are right.

With open-cast or strip mining, surface disturbance is of course the main concern. As much as a square mile at a time of top-soil and sub-surface material, down to perhaps a few hundred feet, is removed in order to reach the coal. (Those magnificent drag-line machines towering 300 ft high are removing overburden, not the coal, which is dug out by much more modest excavators.) In most of the world's mining regions it is undoubtedly *possible* to restore the land. The question is whether the mining bodies – and the consumers – are willing or are forced to spend the money. Estimates of cost vary but are of the order of a few thousand pounds or dollars an acre, and the effect on the cost of a tonne of coal then obviously depends on the thickness of the seam. It could be as little as a few cents for some of the thick seams of the American West or many times this elsewhere. The total cost of reclamation, including water treatment, is claimed to add about a fifth to the cost of the relatively cheap surface-mined American coal. However, there is a more serious problem in the arid South-West, in Arizona and New Mexico and southern Utah and Colorado, where the fragile semi-desert vegetation may take centuries to re-establish, if indeed it ever does. Water shortage adds to the problem, and it is by no means clear whether there is an environmentally acceptable method of extracting the enormous coal reserves of this region.

For people who live in Wales, the Midlands or the North of England, or Scotland, or the Appalachian mountains, coal-mines mean waste tips. And for anyone in Britain over the age of thirty, waste tips mean Abervan, where in 1966 an avalanche of sludge engulfed not only houses but a primary school, killing 116 children. Catastrophe brought in its wake investigation and regulation, designed to ensure that there should be no repetition of the tragedy; and subsequent pressure on environmental grounds has led to more severe constraints on the treatment of wastes from new mines. Unfortunately there is no foreseeable way of extracting coal without wastes, and increasing mechanization has increased the quantity per tonne of useful coal. The present universal method of separation is by washing, which leads to the undesirable sludge of coal dust, water and mineral matter.

For the future there seem to be three options: to ensure that tips are physically and chemically safe and then to establish vegetation so that they become an acceptable part of the landscape, to return the material underground, or to burn the wastes, giving a clean ash residue which might be used by the construction industry. There remain many unknowns – including costs – for all three methods, as well as some known difficulties. (At the simplest, returning wastes underground is hardly possible if the roof has been allowed to fall in.) The possibility of burning the wastes is linked with the development of fluidized-bed furnaces, which we discuss later, and a pilot plant to study this is being operated in Britain, but the capital costs of dealing with the output from even the present mines would certainly be very high.

What do we conclude then, from this very limited survey of coal-mining? First, that although there are many problems, there are probably few *technically insuperable* problems. However much coal we use, it is not *necessary* that hundreds of miners die each year, that thousands of miles of rivers are polluted, or the landscape marked for ever by the scars of surface mines or mountains of poisonous wastes. It is not necessary; but to avoid all these ills will without doubt be expensive, and our second conclusion must be that we don't know how expensive. The planners and forecasters, with their unhappy task of producing figures for the year 2000, have made their guestimates, and we shall return to these later. For the remainder of this chapter, however, our question becomes how, and at what cost, the coal which the miners extract can be used.

A nice coal fire

The British taxpayer might well feel aggrieved to learn that his government spends a few million of his pounds each year finding out how to burn coal. After all, his ancestors have been burning it for about 1,000 years, and it's a fairly simple process. You pile coal on a grate, raise its temperature until it ignites and then keep it going by adding more lumps of coal. True, but perhaps we should look at a few figures.

If you do burn lumps of coal in an open fireplace, and if your chimney is on an outside wall, the useful heat you obtain is probably about one-tenth of the heat content of the coal, and your consequent heating costs only a little less than if you used electricity, the most expensive of present domestic energy sources. With a modern closed stove, and the chimney on an inside wall, you can do better, raising the efficiency to over 60 per cent under ideal conditions – though your annual average is unlikely to exceed half this value. The best industrial or power-station furnace, on the other hand, can convert almost 90 per cent of the energy of coal into useful heat.

To see how such wide variations arise, we'll start with a list of conditions for efficient combustion:

- Combustion should be *complete*. Neither the flue gases nor the ash should contain unburnt fuel.
- The air flow should be the minimum needed for complete combustion. Surplus air will carry more heat out of the chimney (and also increase production of nitrogen oxides).
- As much heat as possible should be extracted from the flue gases and the ash.
- The heat should go where it is wanted: to the water in a boiler or to circulating air or water in a space-heating system.
- Flexibility is important for efficiency, the ideal being instant turn-down from full output to a low level or vice versa, with no rise in conversion losses.

These criteria are all to do with efficiency, but we should add two other requirements. First, the flue gases (and the ash) should be as *clean* as possible, free of harmful pollutants; and then, at least in a system of any size, *continuous operation* is desirable.

Unfortunately, the domestic open fire falls short in almost all the above respects, and perhaps the time has come when we should recognize it for what it is: a luxury. Like the Sunday drive in the country, it is something on which we may be willing to spend energy because it gives us pleasure, but for the serious purpose of heating houses it is no longer acceptable.

Returning to the main question, how to burn coal efficiently and cleanly in very large quantities, we must relate the above list of conditions to present knowledge of coal and its combustion. We start by recalling the general introduction to hydrocarbons in Chapter 6. We identified natural gas as largely methane (CH_4) and crude oil as a mixture with a C:H atomic ratio of about one to two. The molecular structure of coal is much less well-understood. Rings of six carbon atoms certainly play a role, and groups of these are linked to form extended arrangements which also include hydrogen, oxygen, nitrogen and sulphur atoms, but the detail is complex and varies from sample to sample, so it doesn't really make sense even to talk of a 'coal molecule'. The basic atomic proportions can however be found relatively easily, and Table 7.1 shows the analysis for a particular sample. We see that the C:H atomic ratio is about five to

Table 7.1 Ultimate analysis of a coal sample

Element	C (%)	H (%)	O (%)	N (%)	S (%)
Percentage by weight	83	$5\frac{1}{2}$	9	1	1
Percentage by atoms	53	42	4	$\frac{1}{2}$	$\frac{1}{5}$

four, or a little over one to one. This is fairly typical for coal, although in anthracite, most nearly pure carbon, it can be three to one or even higher.

On pp. 101–2 we saw how in combustion a hydrocarbon reacts with oxygen from the air, producing carbon dioxide and water vapour and liberating energy as heat. If we assume that the burning of coal is such a process we can calculate without further knowledge of the structure the quantities of CO_2 and H_2O produced. Table 7.2 shows the results, and a comparison with natural gas and fuel oil. The 'ultimate' analysis into atomic constituents does therefore allow us to predict the main final products, but unfortunately it tells us nothing at all about the way coal actually burns, and it is this which the furnace designer needs to know. So suppose we start again with a piece of coal and observe carefully what happens as it is heated. We find the following sequence of events:

1 The coal gives off water vapour. All coal (even if it has not been left out in the rain) has some *moisture content*.

2 As the temperature rises a whole range of gases is evolved. These come from thermal dissociation of the coal and include hydrogen, carbon monoxide, methane and other heavier hydrocarbons, and they are important because they are *fuels*. As much as half the heat produced when coal burns may be generated in the combustion of these gases, the so-called *volatile matter*, VM. (Anyone who has watched a coal fire will have seen the little spurt of intense flame which follows the sudden release of a jet of gases.)

3 The useful part of what is left after the VM has gone is the *fixed carbon*, FC. When heated in air or oxygen this will burn to produce carbon dioxide:

   $C + O_2 \rightarrow CO_2$.

4 With all the fuel – gaseous and solid – burnt we are left with the residue, the *ash* content, which comes from the minerals in the coal.

Table 7.2 Carbon dioxide and steam produced in combustion of fossil fuels

The figures ignore any small contribution from constituents other than carbon and hydrogen, and assume that the only products are carbon dioxide and steam.

	Energy per kg of fuel burnt (MJ)	Quantities for an output of 1,000 MJ (10 therms), all in kg				
		Total fuel needed	**Carbon content**	**Hydrogen content**	**Carbon dioxide produced**	**Steam produced**
Coal	29	34	32	2.2	120	20
Oil	42	24	20	3.5	75	30
NG	55	18	14	4.5	50	40

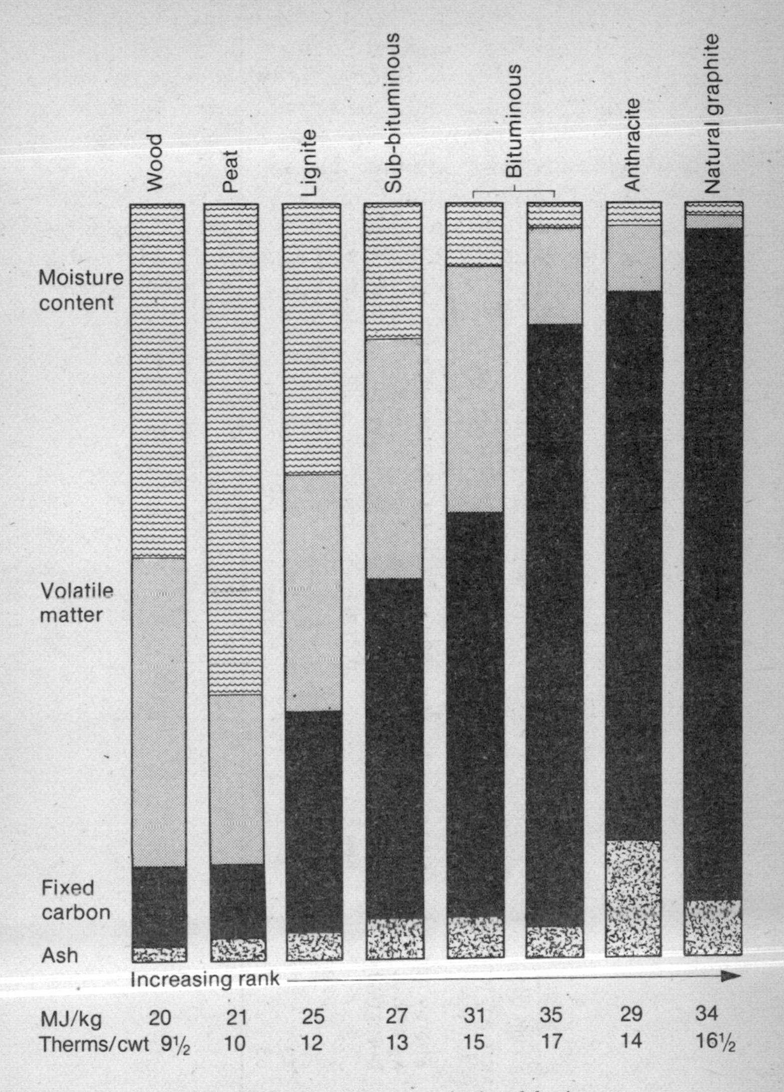

MJ/kg	20	21	25	27	31	35	29	34
Therms/cwt	9½	10	12	13	15	17	14	16½

Fig. 7.1 Proximate analysis of fuels

Fig. 7.1 is a summary of results obtained in this *proximate analysis*, and is a useful way of characterizing different coals. We note that the FC content increases from left to right, with increasing *rank* – a rather loosely-defined term which indicates how far the coal is from the original plant material. Coals from different sources vary widely, so the data (and those for heat content) should be taken as general indicators rather than exact values.

What does the furnace designer learn, then, from Fig. 7.1? Certainly that, unless a power-station can rely on one source of coal for the whole of its thirty-year life, it had better be versatile. If it can burn anything from lignite to anthracite it is in a safer supply position

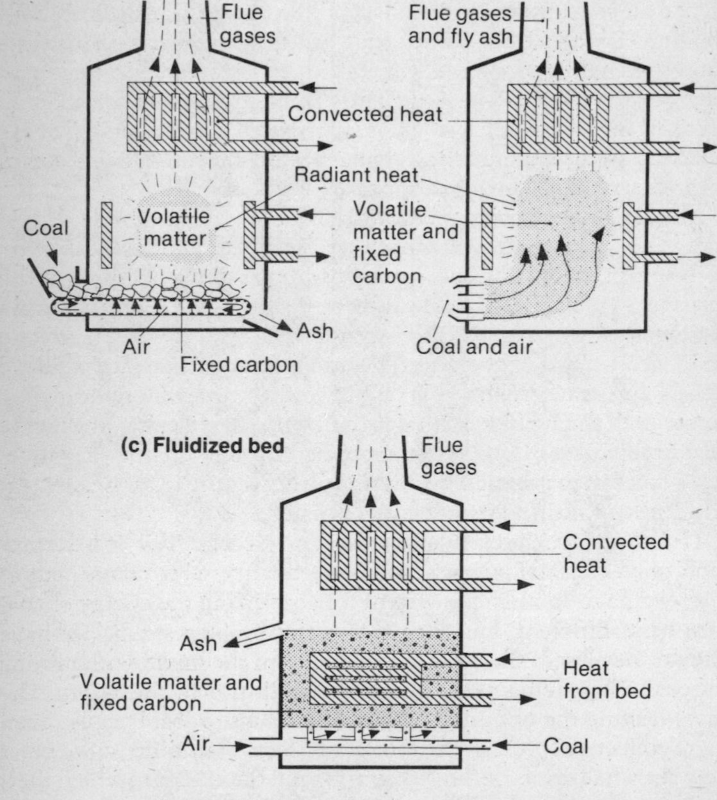

Fig. 7.2 Three types of coal-fired boiler

than if it needs precisely 40 per cent VM. Then there is the question of heat transfer. How do you deal with a fuel which produces part of its heat in a mass of solid material and part in the gases which are rapidly disappearing up the chimney? And how is the air flow adjusted so that both the solid and the gases receive enough oxygen for full combustion without the heat losses associated with quantities of excess air? To make matter worse, there are factors not even mentioned in the proximate analysis. Some but not all bituminous coals, for instance, soften and become sticky when heated. At very high temperatures, the ash residue itself becomes molten. Nitrogen from the air combines with oxygen, as in any furnace. Bearing all these factors in mind, let us look at the ways we burn coal now, and then at one possiblity for the future.

Fig. 7.2 shows in schematic form three types of furnace, all designed for continuous operation. (They do not include the man with the shovel, who left the scene a long time ago together with his large lumps of coal.) In *fixed bed* systems, there is a moving grate which carries the coal. Air flows upwards through this, while the coal, crushed into half-inch pieces, is distributed continuously. These furnaces are used mainly in middle-sized boilers, producing steam for industry, including small local power-stations.

Almost all large power-stations today use the *entrained bed* system. The coal is pulverized into particles about a thousandth of an inch across, a dust fine enough to float in the rising hot gases of the furnace. Not only is all the heat then produced in roughly the same region of the furnace, but the tiny particles burn more quickly and completely than larger pieces. The furnace runs with relatively little excess air, heat transfer is good (up to fifty times more heat flow across each square foot of surface than with a fixed grate), the rate of heat production can be varied more rapidly, and finally, the pulverized fuel can in principle be anything from lignite to anthracite. An efficiency of nearly 90 per cent is possible.

This then is the present situation. The 'Second Law' losses common to all thermal power-stations are unaffected of course, but at least we have an efficient system for converting the energy of coal into heat. Efficient, but with two major problems. One we have already discussed: the costs associated with the *input*, with mining the coal. There remains the question of pollution by the *output*. The gases leaving the chimneys of power-stations probably cause even more concern at present than the side effects of mining, so we must now see what these are, how they arise and the extent to which their emission can be controlled.

Flue gases

The exact composition of the flue gases will vary with the type of coal and the furnace used, but the following inventory might be the hourly output from a large power-station:

- 2,500 te of *nitrogen*, representing four-fifths of the air drawn in for combustion. This has passed unchanged through the whole system, except that heating it accounts for about half the energy loss. Four-fifths of the air we breathe is in any case nitrogen, so it presumably remains harmless.
- 600 te of *carbon dioxide* produced in combustion, a minor constituent of normal air and not thought to be toxic in the concentrations generated by power-stations. (For its possible effects on climate, see Chapter 12.)
- 200 te of *steam*, responsible for the other half of the energy loss.
- 10–20 te of *fly ash*. If you start with pulverized coal so fine that it floats, then you will produce ash which also floats – straight up the chimney and into the atmosphere, unless it is collected.
- A tonne or so of *oxides of nitrogen*. These compounds (N_2O, NO, NO_2 etc., known generically as NO_x) are produced mainly by the interaction of nitrogen from the coal with any excess oxygen, though they will be formed in any furnace which uses air. The amount increases with increasing furnace temperature and with increasing surplus air.
- Anything between one and twenty tonnes of *sulphur dioxide* (SO_2). The presence of sulphur in the furnace is not a matter of design, but an 'accident' of Nature. Sulphur is present in fossil fuels and, at least in the case of coal, cannot be removed entirely because some of it lies within the molecular structure. Sulphur content varies from less than 1 per cent to over 5 per cent by weight, and the power-station of our example would produce about four tonnes of SO_2 an hour from a coal with 1 per cent sulphur.

No other constituent should be present in more than a few parts per million if furnaces are properly maintained. If not, incomplete combustion can lead to soot and other forms of unburnt carbon, to carbon *monoxide* (CO, a toxic gas) and to hydrocarbons which may be carcenogenic. The first of these items appears in the 'particulates' discussed below, but the others are not specific problems of coal. At present they are much more likely to result from internal combustion engines than from power-stations, and we shall therefore return to them in the general treatment of environmental effects.

There are two ways of dealing with flue gas pollutants: remove them, or don't produce them in the first place. The second is considered technologically the more elegant solution. It is also the only present means of dealing with the NO_x problem, which therefore reduces to the question: Can furnaces be run at present efficiencies but at lower temperatures? This is one aim of the fluidized beds, which we discuss later.

Apart from partial removal of sulphur by washing the coal, all present methods for dealing with fly ash and SO_2 involve removing them from the flue gases. There are many techniques in principle capable of doing this, but before we are too critical of 'them' for not using them, let us be sure that we understand the problem. 3,000 te an hour of hot gases leave the boiler. They are hotter than the hottest of domestic ovens, with a particulate level some thousand times that of the worst London smog, an SO_2 concentration a thousand times greater than downtown Los Angeles on a bad day, and enough moisture to cause it to start raining in the gas stream if the temperature falls below that of a moderate oven. Any cleaning system must be able to handle this entire hot, dirty, corrosive mass on a continuous basis. It should remove most of the pollutants (about 90 per cent of SO_2 and over 99 per cent of particulates are current aims), use as little energy as possible in the process, produce environmentally acceptable residues – and be cheap. There is of course no such system.

Mechanical removal of particulates

Current methods include (Fig. 7.3) allowing the particles to fall under gravity out of the gas stream, swirling the gases to throw the particles sideways (as in a spin-drier) and pushing the gases through

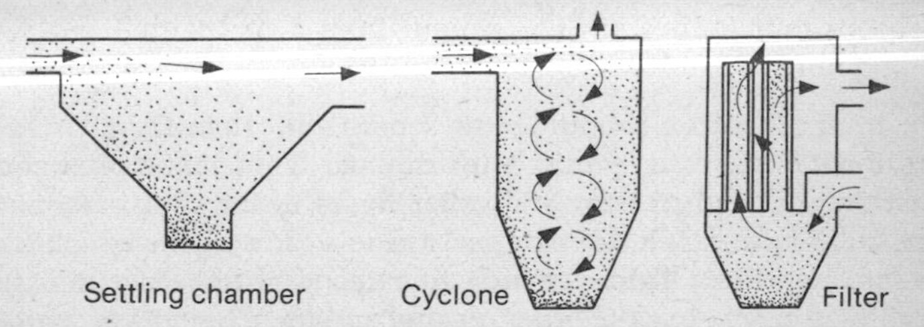

Fig. 7.3 Methods for mechanical removal of particulates

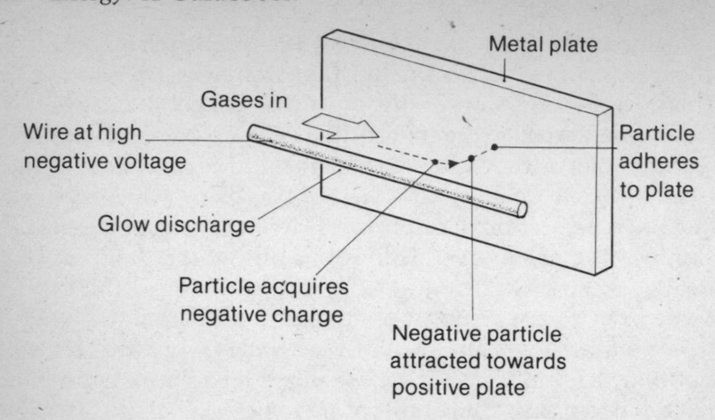

Fig. 7.4. Principle of the electrostatic precipitator

a filter (like the dust bag in a vacuum cleaner). The first two, while relatively cheap and using little energy, are inefficient for particles less than a few thousandths of an inch across. This fine ash, thought to have serious health effects, is removed very effectively (99 per cent) by filters, but there is the problem of finding suitable materials to withstand the heat and corrosive gases; and appreciable energy is consumed in pushing the flow through. All these mechanical methods do have the advantage of producing a *dry* residue.

Electrostatic precipitators

Particles given a negative electric charge by passing near a high-voltage wire (Fig. 7.4) are pulled sideways out of the gas stream towards a plate held at a more positive voltage. Effective even for small particles, precipitators produce a dry residue and consume much less energy than filters, but they are not efficient for all coals, and the capital cost is still high. Modern British power-stations are equipped with electrostatic precipitators, and although there were teething troubles in their first decade or so, it is now claimed that they do remove (as planned) over 99 per cent of particulates.

Wet scrubbers

A fine spray of water mixed with calcium carbonate (limestone, $CaCO_3$) or other suitable material is blown through the gas stream (Fig. 7.5). The SO_2 reacts with this to produce the sparingly soluble calcium sulphate ($CaSO_4$) which together with the ash brought down by the spray is collected as sludge. Wet scrubbers have the obvious advantage of dealing simultaneously with two major pollu-

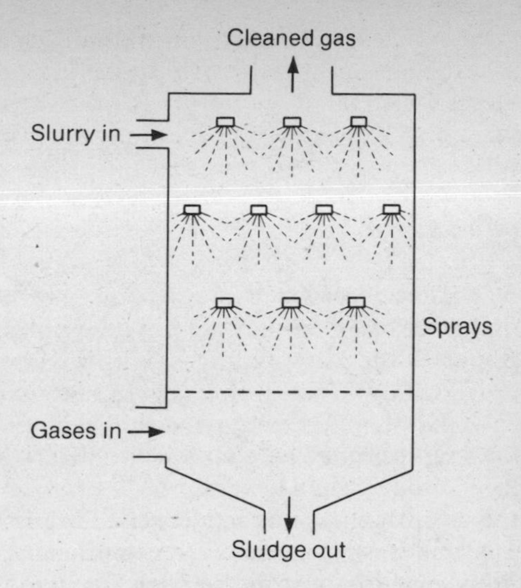

Fig. 7.5 Wet scrubber

tants, but they consume about a tonne of water for each tonne of coal burned, suffer from serious corrosion and produce an undesirable *liquid* effluent requiring, over the life of a large power-station, a storage 'pond' covering some 500 acres. Their energy consumption reduces efficiency by 5 per cent or so and it is considered that their general introduction would increase the price of electric power by perhaps a fifth.

Other methods

Many 'dry' systems for SO_2 have been studied. The idea is to introduce a substance which will react with the SO_2 to produce a solid which can then be removed and, if possible, recycled. No present method achieves on a large scale the required combination of effectiveness, energy efficiency, reliability and low cost.

By now the reader may feel that the expenditure of so many words on the unexciting topic of dirty gases, only to conclude that cleaning them is difficult, indicates a certain lack of balance. If it does, it is in the rather specific sense that we allow present issues to weigh more heavily than less immediate questions. The coal versus nuclear battle is being fought *now*, on the two battlefields of cost and pollution, and it is arguable that our decisions about these alternatives could have more serious and long-term consequences than any other action we may take on energy in the coming years. If so, the unglamorous scrubbers and precipitators appear in a different light, and the problems which they pose are surely worth at least as much of our attention as the latest hypothetical hi-tech concept for the next century.

Burning clean

To introduce fluidized beds as the new white-hot hope for coal would be wrong, because first they are not new and secondly they are not white-hot. While present furnaces burn well into the white-hot range (at over 2,000°C) one of the most important points about fluidized beds is that they operate at red or dull yellow temperatures (750–950°C), with the immediate consequence that NO_x production is much reduced. Add to this a lower fly ash content, the same heat output for only a tenth of the furnace size and the possibility of dry removal of SO_2, and it is not surprising that millions of pounds and dollars are now being invested in their development.

The principle has been known for a long time, and many fluidized beds are already in use, though none yet for large-scale burning of coal. Fig. 7.6 shows the idea. Take a layer of sand or gravel up to a foot or more deep and blow air through it from many small holes in the base. At a certain air speed, the whole mass expands and starts to behave just like a liquid. Objects will float on it or sink into it, it will flow and – very important for combustion and heat transfer – the solid particles of the bed bounce around as though they were indeed particles of a boiling liquid. In a furnace the main bed material consists of inert material (ash itself is suitable), and the coal, also in small pieces, is fed in continuously (Fig. 7.2(c)). Pre-heating starts the combustion, which then continues as long as coal is added and

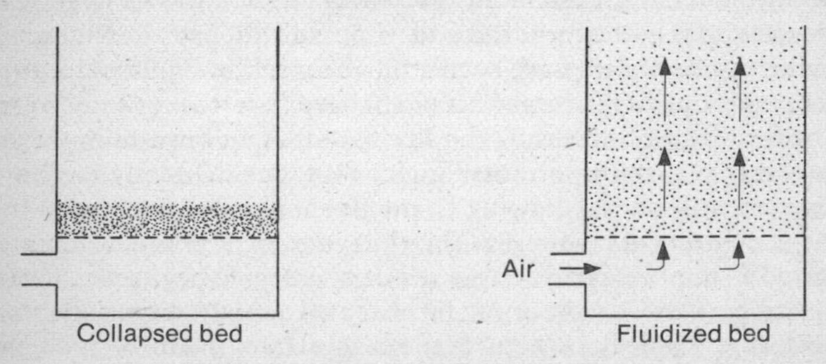

Fig. 7.6 Principle of the fluidized bed

the air flow maintained. The turbulence ensures good air supply to each particle so that ideally both the volatile matter and the fixed carbon burn within the bed. Lower temperatures are possible because much better thermal contact is made between the hot particles and boiler tubes buried in the bed than can be achieved by radiation from the incandescent coals of a fixed bed or by transfer from the more diffuse gases of an entrained bed.

The two big advantages foreseen for fluidized beds are a reduction in flue gas pollution and an increase in thermal efficiency: more power with less dirt. NO_x is less because the temperature is lower, fly ash is reduced because the system does not use pulverized coal but larger particles, and, if limestone is added to the bed, the SO_2 can react to produce calcium sulphate which is removed with the solid ash.

Moreover, if the bed can be operated at high pressure (a few hundred or so pounds per square inch) the corresponding high-pressure output gases could go to the gas turbine of a combined-cycle system (Fig. 7.7. See also p. 93). This could mean a 5 per cent or so increase in thermal efficiency. With flue gas desulphurization reducing the overall efficiency of conventional power-stations to under 30 per cent and still leaving the problem of disposal of wet sludge, the pressurized fluidized bed, offering perhaps 35 per cent overall after removal of residual pollutants, must surely look interesting. You don't lightly ignore a fifth more electricity for the same fuel consumption.

How is it then, that all large-scale coal-fired power-stations being built or even planned are of the conventional type? The answer brings us to an important fact of technological life which will recur with increasing frequency as we look more and more at the future rather than the present: to translate the bright idea of the research lab into a reliable system for widespread use on a very large scale requires massive expenditure of time and money. Low-pressure fluidized-bed boilers have been studied since the 1950s and a number of pilot plants (a megawatt or so) have operated for some years. The next stages, increasing the size to tens of megawatts and trying pressurized systems, are now under way, but the really large systems are still on the drawing-board. (It should perhaps be said that the supporters of coal regard the thirty- or forty-year development period as unnecessarily long, a result of culpable negligence during the time when we were drunk on cheap oil or high on the delights of nuclear technology.) Whatever the causes, there are many problems still to be solved. Adjusting the heat output is not easy, if the bed is

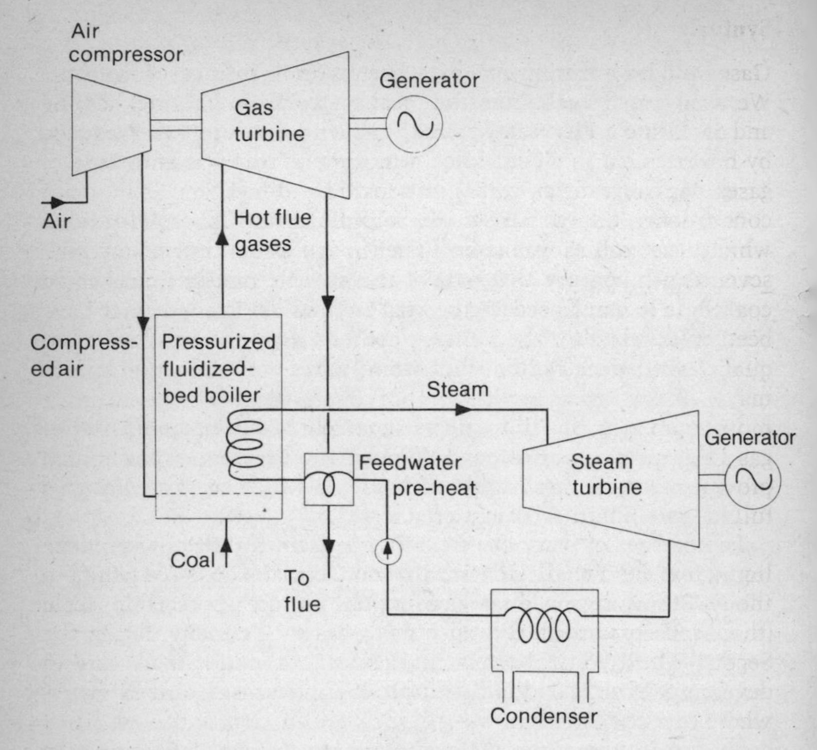

Fig. 7.7 A combined-cycle system using a pressurized, fluidized-bed boiler

The system has two generators, one driven by a conventional steam turbine and the other by the gas turbine, which also drives the air compressor for the fluidized bed.

not to collapse. And there are many known and probably some unknown difficulties with the pressurized systems. (The blades of the first turbine to run on gases from coal looked after a few hours as though they'd been fitted with black fur coats.) No one seems to think that there are insuperable problems, and with some dozen atmospheric-pressure furnaces and about half that number of pressurized systems currently under test, progress should be more rapid. But the fact remains that in assessing future prospects for clean power from coal, particularly in the important matter of cost, there are once again uncomfortably many unknowns.

Synfuels

Gases and liquids from coal have been with us for a *very* long time. We used 'town's gas' from the local gasworks in lighting, heating and cooking for 150 years before the advent of natural gas. Produced by heating coal in a controlled atmosphere, this was a mixture of gases, with hydrogen, carbon monoxide and methane as its main constituents, a large part of the coal remaining as a solid residue which was sold as 'gas coke'. Then it has been known since the seventeenth century that part of the volatile matter from heated coal could be condensed to give 'coal tar' liquids. These seem to have been valued less for their energy content than for their medicinal qualities, with claims that they would cure everything from housemaids' knee to a cold in the head. (Of course, these days we are much more sophisticated. All we now ask of coal is that it should fill our gas pipelines and petrol tanks, supply us with electric power, and provide raw materials for plastic bags or – which seems to bring us full circle – pharmaceutical products!)

The present century has seen intermittent attempts to produce liquid fuel from coal. Both Britain and Germany operated plants in the 1930s and the first production on an appreciable scale (thousands of tonnes of coal a day) was in Germany during the Second World War. Cheap oil in the post-war period made further development uneconomic, and production ceased almost everywhere. South Africa, however, with cheap coal and a sensitive dependence on imported oil, has brought its SASOL plant during the past 25 years to a scale where, consuming as much coal as several large power-stations, it will soon produce 40,000 barrels a day of petrol and fuel oil. This is the world's only large-scale liquid synfuels operation. All others are somewhere between the laboratory bench and the pilot plant (between a few tonnes and a few hundred tonnes a day) with a very few larger demonstration plants in the planning stage. There is a long way to go before they make any dent in world oil needs, equivalent at present conversion efficiencies to an input of perhaps 50 Mte of coal a day.

Four different aims can be identified in current conversion projects:

- Production of a gas for burning in *power-stations*, as a possibly cleaner and more efficient alternative to the direct combustion of coal.

- Production of the rather oddly named *substitute natural gas* (SNG) for use in existing gas distribution networks and equipment.
- Production of *synthetic petrol* for use in internal combustion engines. Two routes are being followed here: synthesis of liquid fuel from the products of a gasification process, and direct liquefaction leading to a *syncrude* which could be distilled like crude oil.
- Production of *chemical feedstocks* to replace those at present derived from oil.

The fourth endeavour, a relatively minor but growing field, is not strictly a 'fuel' use of coal and we shall not discuss it further.

The first item on the list may well prove very important indeed. The idea is to produce gas from coal and then burn this gas in the boiler of a combined-cycle turbine system like that shown in Fig. 7.7. It may seem a rather roundabout way of using the coal, but it has two potential advantages. First, it may be cheaper to produce gas and clean it before burning it than to remove pollutants from the combustion products of coal, and secondly, the efficiency of a gas turbine run with flue gases from a fluidized bed is limited by the low bed temperature, whereas cleaned gases could be burned much hotter. Optimists foresee temperatures above 1,500°C at the turbine intake and efficiencies as high as 50 per cent, an enormous increase over the present. The sole existing coal gasification combined-cycle power-station is at Lünen in Germany. Built in 1970–1 and using the well-tested Lurgi gasifiers (see later) and the 'low' temperature of 820°C, it nevertheless suffered from teething troubles for a couple of years before operating more or less as intended. A plant efficiency of 38 per cent is now claimed, and a 400-MW system is planned.

Several dozen systems are currently under development to produce electric power from coal with gasification as an intermediate step. They produce (or in most cases, will produce when built) gases at temperatures from 500°C to over 1,500°C and pressures from normal atmospheric to seventy or eighty times this. They also produce gases of a very wide range of heating values and compositions. This is an important point: unlike SNG, gas for power generation need not be identical with present natural gas, because furnaces can be purpose-designed for the intended product.

Means and ends

At this point, before moving on to other synfuels, we might usefully try to establish a framework to help us find our way around a varied and rather complex area of technology. We'll start by distinguishing three questions:

- What is the end product?
- What is the series of reactions leading from coal to this end product?
- What physical systems are used to bring about the reactions?

Coal gasification and liquefaction are chemical processes, so let us try to see how the particular expertise of the industrial chemist enters. Assume that the aim is to produce methane. Any sensible technologist will choose the cheapest raw materials, all other things being equal. But of course all other things are hardly ever equal.

Consider the following splendid scheme. We want methane, CH_4, so why not start with two *very* cheap raw materials, carbon dioxide and water?

$$CO_2 + 2H_2O \rightarrow CH_4 + 2O_2.$$

There is just one flaw in this, which becomes evident if we reverse the arrow. Our proposed process is simply the combustion of methane run backwards, and it follows that we would have to supply *at least as much energy* to produce the methane as would be obtained in burning it. If this were not so energy would not be conserved; we would have a type of perpetual motion machine, and if *they* were possible there'd be no energy problem in the first place!

Back to the drawing-board then, with a change of raw materials. How about coal and water? The reaction with the carbon might be

$$C + H_2O \leftrightarrows CO + H_2.$$

No methane, but as we'll see, the CO and H_2 will be useful. The process does in fact need some energy to run, but only as much as could be obtained by burning about a third as much again coal. It is the double arrow which raises the question here. Like any chemical process this one can go either way. As soon as some CO and H_2 are produced they will start reacting to give carbon and steam, and there will always be a competition between opposing processes. How can we ensure that the end result is what we want: more product than raw material? The answer is that we can't, without further data; but we do know one simple and important rule: if the process is one which *uses* heat, then it is favoured by higher temperatures. (One which *produces* heat is, conversely, favoured by lower temperatures, and the rule is a direct consequence of the Second Law of Thermodynamics, in what might be called its 'Nature is cussed' version.) The *carbon-steam reaction* above leads to ten times as much product at 800°C as at 700°C, for instance, an important point in practice.

Suppose the H_2 is now used, with more carbon, to form methane:

$$C + 2H_2 \rightleftharpoons CH_4.$$

This reaction actually produces heat, so it looks an excellent prospect. But we must be careful. The rule quoted above refers to the *final* proportions, after everything has settled into equilibrium, but what it does not say is how long, starting from C and H_2, it takes to get there. Data on *rates* of processes are obviously also needed, and again there is a simple rule: *hotter means faster*. Cold carbon and hydrogen may indeed change into methane, but the reaction is of little use if it takes geological ages to happen. A compromise between final proportions and speed must evidently be sought.

Any commercially useful process must satisfy these three criteria then:

- It must be a *net producer* of energy.
- All its reactions must go in the *right directions* at accessible temperatures.
- They must do so at *reasonable speeds*.

Of course all unwanted reactions, of which there are many, must be discouraged.

Processes and products

Many present systems do use the carbon-steam reaction for initial gasification. In a common arrangement (see, for instance, Fig. 7.9(a)) coal and steam are fed in together with air or oxygen, so that three reactions take place together: volatile matter is given off by the heated coal, heat is suppled by combustion of some of the coal, and the carbon-steam reaction also proceeds. The product will be some mixture containing H_2, CH_4, CO, various hydrocarbons and, depending on circumstances, different amounts of CO_2, nitrogen and other compounds. Fig 7.8 summarizes just a few of the processes and products.

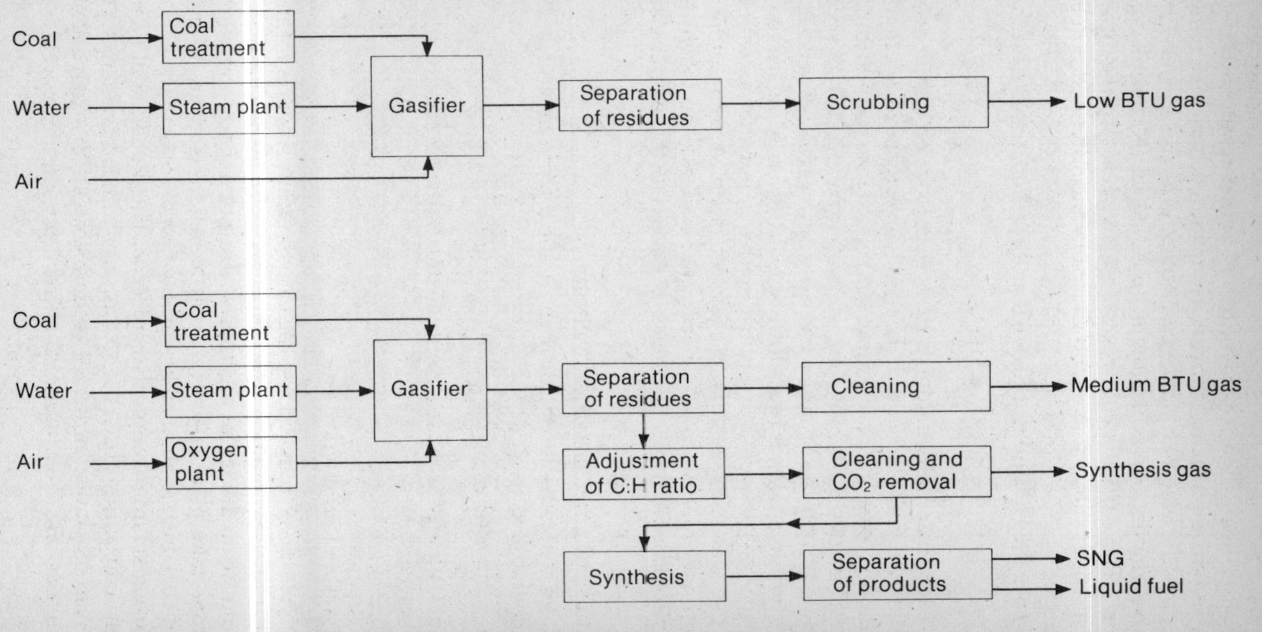

Fig. 7.8 Coal gasification processes and products

Table 7.3 Gas compositions and heating values

Figures in brackets give the temperature and pressure of gas leaving the gasifier. All heating values are for a cubic metre or cubic foot at normal atmospheric pressure (1 bar) and room temperature.

	Dry gas composition by volume						Heating value	
	H_2 (%)	CO (%)	CH_4 (%)	CO_2 (%)	N_2 (%)	Other (%)	MJ/m³	BTU/SCF
Low BTU gas	15	25	3	5	52	—	6	160
Medium BTU gas								
Lurgi (500°C, 20 bar)	40	18	9	30	1	1		
Koppers-Totzek (1,500°C, 1 bar)	56	36	—	6	1	1	11	300
Winkler (900°C, 1 bar)	40	35	3	20	1	1		
Natural gas	—	—	93	1	1	5	37	1,000
Hydrogen	100	—	—	—	—	—	12	330

Plant

The design of gasifiers has much in common with the design of furnaces, because the aims have much in common: to achieve the best possible conditions for controlled reactions between coal and some gas or gases. So it is not surprising that there are three main types of gasifier: fixed bed, entrained bed and fluidized bed. The best-known of these are the Lurgi, Koppers-Totzek and Winkler gasifiers, all of which have been used for many years. Table 7.3 shows their operating conditions and product compositions, and it is obvious that none can provide the combination of temperature and pressure needed for the highly efficient turbines discussed previously. However, they do have one great advantage over more sophisticated systems: *they exist*. It has been argued, for instance, that if the USA were faced with a serious threat to its imported oil supplies, the only contribution which synthetic oil could possibly make in less than a couple of decades would be from a crash programme of SASOL plant construction. We shall finish, therefore, with brief accounts of the two systems shown in Fig. 7.9. In both, the aim is the production of *liquid* fuel from coal, but their approaches differ greatly. The SASOL method, well-tested but not very efficient, starts by breaking down the coal to give CO and H_2 and then proceeds to build up the required hydrocarbons from these. The EDS process, potentially more efficient but as yet untried on a large scale, 'dissolves' the coal to produce synthetic crude oil which is then distilled.

SASOL

Synthesis gas, the mixture of CO and H_2, is made in a Lurgi gasifier. This is cleaned and then reacted with more steam:

$$CO + H_2O \rightleftharpoons H_2 + CO_2.$$

This is the very important *water-gas shift reaction*, whose effect is to increase the proportion of hydrogen to carbon monoxide; a necessary step in any process going from coal, with its H:C ratio of about one, to a hydrocarbon with a ratio of two or more. Once the correct ratio is established the gases pass to the synthesizer where they combine to produce a range of hydrocarbons from methane to heavy oils. These are then separated out to give SNG, petrol, fuel oil etc.

Judged by its petrol production, the SASOL process is very inefficient, giving not much over one barrel per tonne of coal: an energy conversion efficiency of under 30 per cent. If credit is given

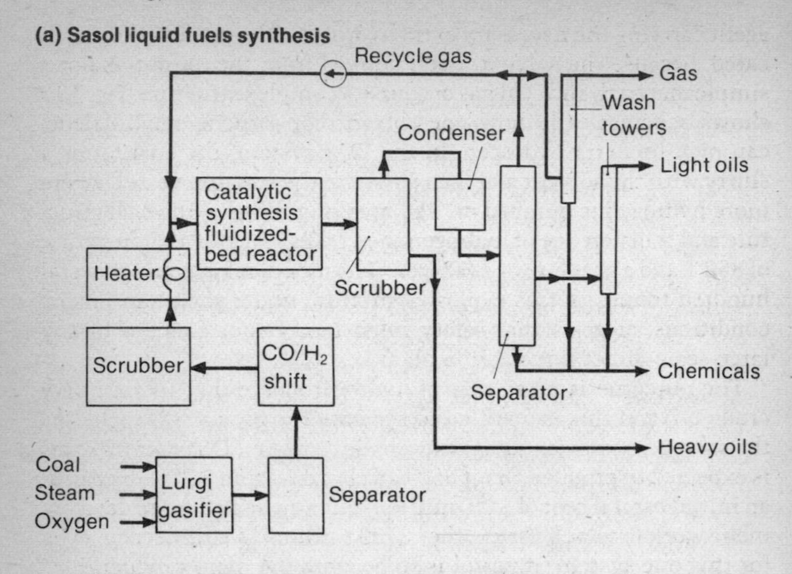

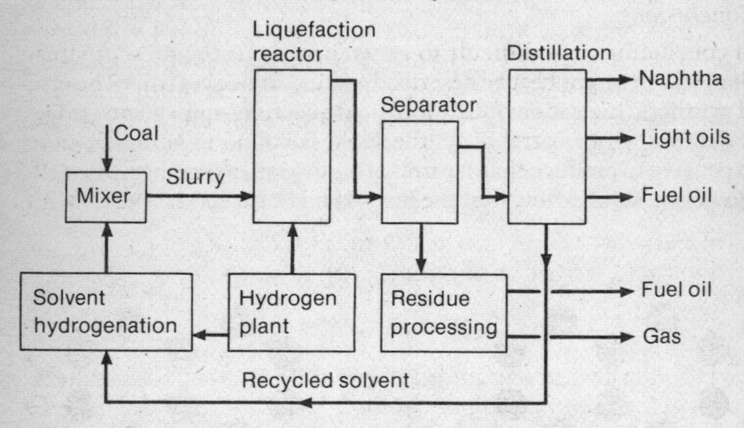

Fig. 7.9 Two coal liquefaction processes

for other products the figure rises to perhaps twice this, but the economics of the system then depend on the demand for these.

EDS

As the name suggests, the main step in the Exxon Donor Solvent process is to dissolve the coal in a suitable solvent which is also the agent carrying the necessary extra hydrogen. The process is complicated, because the solvent, itself derived from the output, is not a simple material, and coal has of course a complex structure. Fig. 7.10 shows schematically how one solvent constituent, naphthalene, can pick up extra hydrogen. In the EDS system, the coal forms a slurry with the solvent and then passes to the reaction vessel, where more hydrogen is pumped in. The breaking-down of the coal structure and transference of hydrogen then takes place, at a temperature of 850°F and a pressure of 2,000 psi. (The idea of maintaining several hundred tonnes of this explosive mixture under such hair-raising conditions suggests that safety must be a major concern in any large-scale direct liquefaction plant.)

The outcome is a mixture of hydrocarbons rather like a heavy crude oil, and this is distilled to separate the products – including the solvent, now without its extra hydrogen. An EDS plant in Texas is expected to produce about 650 barrels a day from 250 te of coal, at an initial capital cost of $120 million; but a great deal more development work (and at least another $200 million) is still needed, even for this one system, if petrol is to become the main product.

Conclusions

In concluding, it is difficult to generalize about the present situation, which might best be described as fluid, if not volatile. The cost of synthetic fuels is obviously a critical question, and a glance at Fig. 7.9 shows why a barrel of synthetic oil is bound to be much more expensive to produce than a barrel of crude from a good healthy well. However, we all know that the *price* of oil or gas has less to do with

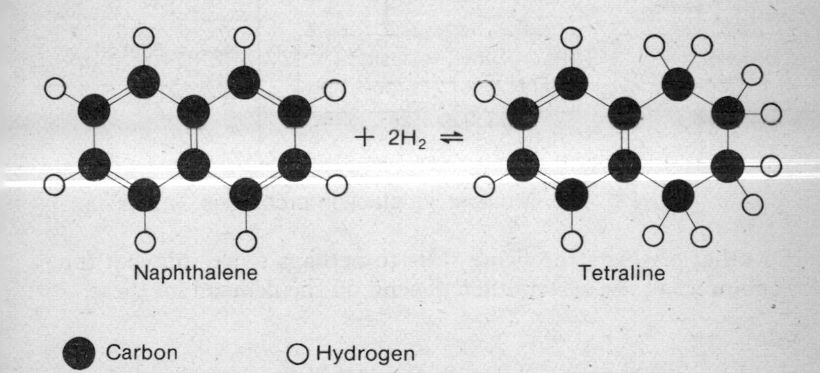

Fig. 7.10 A possible hydrogen-donor solvent

the cost of extraction than with the supply situation or with international politics. Cost estimates for synthetic gasoline have increased at about the same rate as the price of oil during the past decade, partly because coal prices have risen but also through that seemingly inevitable escalation which overtakes a project for cheap anything as we learn more about the difficulties. So the $12 a barrel which, according to a 1974 spokesman, 'would make synfuels attractive if oil prices continue to rise' has become $40 and more, and the gap remains. Nevertheless, as we saw at the start of the chapter, a great deal of money is being invested in development. Most of this is in the USA, but even Britain, with its present oil self-sufficiency and its rather expensive coal, has a few projects under way.

There are of course other competitors for the future oil market: new methods for extracting oil from the ground, and unorthodox sources such as shale oil or tar sands; just as there are other fuels for electric power generation. The important question, to which we must return, is how the potentialities and drawbacks of the technologies we've been discussing here stand in comparison with these alternatives. The answers to this question should then determine the future for coal, at least for the next half century. (The cynic, noting that rational discussion has never yet decided energy policy, may disagree.)

8 Nuclear Power

Introduction

The past forty years have seen an interesting extension in the way we use two words at the very centre of the energy vocabulary. There doesn't seem to have been a formal announcement, but somehow during the time since the first man-made nuclear reactor was built, the terms *fuel* and *burning* have come to refer to materials and processes very different from the hydrocarbons and chemical reactions we've been discussing so far. The 'fuel' in a nuclear reactor is totally different from a fossil fuel, and its 'burning' doesn't involve a reaction with oxygen, nor indeed any chemical process in the usual sense. When a uranium nucleus splits into fragments the change is much more profound than a mere rearrangement of atoms in molecules. It is in a sense the realization of the dream of the alchemists down the ages: the transmutation of one chemical element into another. An element is characterized by the number of electrons making up a complete, electrically neutral atom (Chapter 5), and the total negative charge of these must exactly balance the positive charge of the nucleus. So if the nuclear charge is altered, a different number of electrons is needed and you have a different atom, a different chemical element.

To understand how changes in nuclei can occur and why they release energy we must penetrate more deeply into the atom and ask about the structure of the nucleus itself. This will be the first task of the present chapter, and the remainder is then a gradual progression from the general to the particular; from the nucleus to the fission process, then to the reactor using this process and ultimately to specific types of reactor. Safety must feature largely in any account of nuclear power, and a section on radioactivity establishes the background for the discussion of its effects in a later chapter. We finish with a brief look at the 'other' nuclear power process: fusion.

Nuclear power – like education, or modern architecture – is one of those subjects on which *everybody* has views. The experts are just as fervent and polarized in their opinions as the laymen, and to discuss reactors at all is to invite a hail of response from the committed of both extremes. To embark on our usual journey, from basic ideas to current technology in a few short stages, must be rash indeed, and the missiles are surely waiting: 'over-simplification', 'omission', 'bias'. Each will no doubt find its target, and the only defence must be that this chapter is no more and no less vulnerable to such attacks than the rest of the book. At least there is no shortage of publications where the reader may find detailed accounts of the excellent safety record of nuclear reactors – or conversely, of the many accidents known to have occurred. Books offering both are unfortunately rare, so perhaps a minimum measure of the objectivity of this account will be the extent to which it attracts equal fire from the cannons to the left and to the right.

Nuclei

Nuclei have positive electric charges equal to *exact* multiples of the charge of an electron, and they account for most of the mass of the atoms. Consider the implications of the first of these statements. The child whose building-blocks are two-inch cubes will make towers whose heights are multiples of two inches, and a parallel argument suggests that, if nuclei have charges which are multiples of a particular charge, perhaps all nuclei have a common 'building block', an entity with just this basic charge.

A nice tidy idea, which unfortunately becomes less tidy when we look at the masses of the nuclei. Table 8.1 shows a few examples. The *atomic number* is the number of basic charge units. (The masses are given in 'atomic mass units' to bring out the relationships more clearly than if we used kilograms.) The data reveal four important features:

- Cases exist of several nuclei with the same charge but different masses: a particular element can have several *isotopes*.
- Unlike the charges, the masses are not *exact* multiples of the smallest mass.
- They are, however, *very nearly* multiples. (The nearest whole numbers, called the *mass numbers*, are shown in the third column.)

The first feature at once eliminates the possibility of a *single* building block; but if we ignore for the moment that slight but very important discrepancy, the third observation suggests an alternative. This is that there are *two* basic particles, with the same mass

Table 8.1 Nuclear masses

The nuclear masses are given in *atomic mass units (u)*. 1 u = 1.66 × 10^{-27} kg, and the masses of a single proton and a single neutron are respectively 1.0073 u and 1.0086 u.
For the elements up to bismuth, the masses of the most common isotopes only are shown. (See the text for further discussion.)

Element	Atomic number	Mass of nucleus	Mass number
Hydrogen	1	1.007	1
		2.014	2
Helium	2	4.002	4
Carbon	6	12.00	12
		13.00	13
Iron	26	55.92	56
Copper	29	62.91	63
		64.91	65
Krypton	36	83.89	84
		85.89	86
Strontium	38	85.89	86
		87.88	88
Iodine	53	126.87	127
Barium	56	137.87	138
Gold	79	196.92	197
Lead	82	207.93	208
Bismuth	83	208.93	209
Uranium	92	232.99	233
		234.99	235
		238.00	238
Plutonium	94	239.00	239

but only one of them having electric charge. The charged particle we call the *proton* and the neutral one the *neutron*, whilst collectively they are referred to as *nucleons*. The atomic number of a nucleus indicates then its number of protons and the mass number the total number of nucleons. The nucleus of the uranium isotope U-235, for instance, has 92 protons and 143 neutrons while U-238 has three more neutrons. With this picture, we can now leap directly to the idea of nuclear fission as the splitting apart of the original cluster of nucleons to form new clusters, new nuclei.

All very well, but is there any reason to expect it to happen? Most of the nuclei about us are after all rather stable. Elements are not continually changing into other elements. Iron Age tools may have rusted – a chemical process – but we firmly believe that they were indeed originally made of iron, that today's iron nuclei are the *same* nuclei which existed thousands of years, or even thousands of millions of years, ago. Nevertheless there *is* evidence of unstable nuclei. There is the phenomenon of *natural radioactivity*, the spontaneous emission of particles by certain nuclei. This is not yet fission (a term normally reserved for splitting into roughly equal pieces), but the important point is that *all* the heavier nuclei are radioactive. Although the heaviest found in nature is U-238, the heaviest truly stable nucleus is bismuth-209, with 83 protons and 126 neutrons.

The reason that we can in fact go mining for uranium and find some is that the isotope U-238 decays very, very slowly indeed. If you were to start now with a piece of U-238, you would have to wait nearly 5,000 million years for half its nuclei to have emitted particles and changed into thorium-234 nuclei. (With the less common U-235, the time would be 'only' 700 million years.) These facts really explain why uranium is at the centre of our interest. On the one hand it exists, at least a few million available tonnes of it; it hasn't decayed away. But on the other hand it is certainly unstable, which means that the disintegration must be releasing energy. If it didn't, how could it happen without outside assistance?

Fission

Natural radioactivity is important because it shows that some nuclei are unstable, but it is fission which offers the prospect of large-scale power. To see why, we need only look at two facts:

- Fission can be *induced* by a neutron.
- A fission event *produces* surplus neutrons.

Together, these suggest the possibility of a *chain reaction* (Fig. 8.1). Neutrons from the first fission induce further fissions, and the neutrons from these induce yet more, and so on. Then if each fission event releases energy we have a process which will generate ever more power until the supply of fissile material runs out. Of course there are questions. Will the chain reaction actually happen in practice? If it happens, how much energy is produced? Then – the crucial question – can we control the process? Can we govern the rate of multiplication? After all, we control combustion by limiting the oxygen supply: no more air and the fire goes out. But a nuclear chain reaction doesn't need an outside supply of anything. How can we ever switch it off?

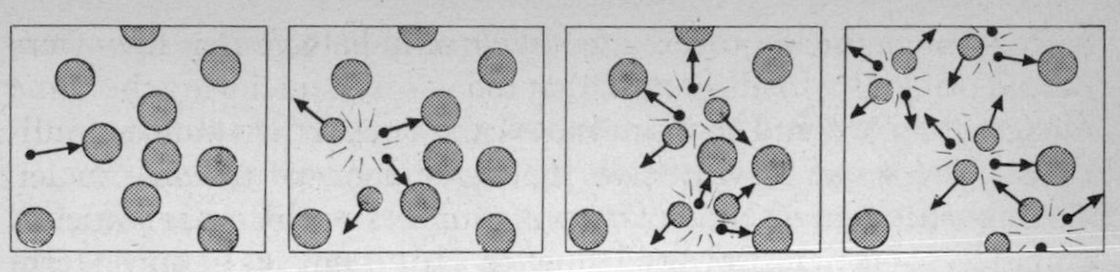

Neutron
Fissile nucleus
Product nucleus

Fig. 8.1 A chain reaction

Nuclear processes can be represented by equations rather like those used for chemical reactions. We can write the equation for the fission of the nucleus U-235 in the form

$$^{235}_{92}U + ^{1}_{0}n \rightarrow X + Y + \text{energy} + \text{neutrons}.$$

Notice that, at least on the left-hand side, we keep track of the mass numbers and atomic numbers by superscripts and subscripts. The right-hand side as it stands, however, is a bit vague and imprecise. The trouble is that 'fission of U-235' is not a single well-defined reaction at all, but a process leading to a range of possible results. In order to understand enough about these to follow the events in a nuclear reactor, let us take each item of the above equation in turn.

The uranium

The isotope U-235 is the only naturally-occurring fissile material, 'fissile' meaning that the nucleus can be induced to split by an encounter with a not-particularly-energetic neutron. About 0.7 per cent of natural uranium is U-235, roughly one atom in 140. If a reactor needs more concentrated fuel than this, the uranium must be *enriched* to increase the U-235 content.

Two other fissile nuclei are known, plutonium-239 ($^{239}_{94}Pu$) and uranium-233; but these do not occur naturally and must be *bred*, a process we'll return to later.

The initiating neutron

Spontaneous fission does occur but is a rare event, whereas if a suitable neutron encounters a U-235 nucleus the chances of induced fission are high. To discuss exactly what is meant by an 'encounter' or by a 'high' chance would take us into rather too much detail, but the word 'suitable' must certainly not pass without comment. A fact which is crucial in the design of reactors is that a slowly-moving neutron is *at least 100 times* more likely to induce fission than a fast one.

What do 'slow' and 'fast' mean? A better term than slow is *thermal*, in the sense in which we discussed thermal energies in an earlier chapter. A *thermal neutron* is simply one which has bounced around in some material until it has the same average energy as the surrounding atoms or molecules. If the temperature inside a reactor is a few hundred degrees, the average speed of a thermal neutron will be a few kilometres a second. In contrast, the neutrons *produced* in fission move 10,000 times faster, with 100 million times the kinetic energy of a thermal neutron. To be as effective as possible they must be slowed down, and we return to this shortly.

The fission products

X and Y in our equation (known to the flippant as fish 'n chips!) are new nuclei resulting from the splitting of U-235. There are usually two of these and they account for all the protons and most of the neutrons of the original nucleus. On the small scale of the nucleus, fission is a violent process and we cannot predict its outcome in detail, so X and Y must stand for a variety of possible new nuclei. We can however make a few remarks about the average result from many fission events.

X and Y tend to be unequal, one with an atomic number in the high thirties and one in the low fifties, with the mass split in a ratio of about 95:140. (A glance at the table of elements on pp. 82f. shows why we find the gases krypton, iodine and xenon and the solids strontium, cesium and barium in the inventory of reactor products.) It would be surprising if the initial explosive fission led to fragments with just the right proton-neutron combinations for stable nuclei, and indeed it doesn't; so the products are almost always radioactive, emitting particles and changing through a series of forms until they reach a stable state.

Energy

Energy is released, appearing in the first place mainly as kinetic energy of the fission products and the excess neutrons. In terms of the large-scale world, the energy released in a fission event is extremely small: about one-thirtieth of a billionth of a joule; but a few comparisons might show why words like 'enormous' tend to creep in. The fission of one nucleus releases about 20 million times as much energy as the combustion of one natural gas molecule. Complete fission of a pound of pure U-235 would release the energy equivalent of about 1,000 te of oil, or 1,500 te of coal. A 600-MW power-station could run for a year on the U-235 in 100 te of natural uranium. The energy equivalent of a ten-gallon tank of petrol is a sphere of U-235 one-twentieth of an inch in diameter.

What is the source of all this energy? One explanation is much like the answer to the similar question about the energy released in combustion. Energy would be needed completely to take apart a nucleus into its constituent nucleons; and conversely the same energy, the *binding energy*, would be released if the nucleus were to be assembled from its constituents. It follows that if the binding energy of the original uranium nucleus is *smaller* than the combined binding energies of the product nuclei, there will be energy to spare when fission occurs. (The actual process does not of course involve complete dissociation and reassembly; it is only the net difference that matters.)

So what *is* the source of the energy? The idea of binding energy is useful, but it doesn't really *explain* anything. (Rather as though we ask why things fall to the ground and are told, 'It's because the ground is lower down.') We know that the origin of the energy released in a *chemical* process lies in the electrical forces holding molecules together: chemical energy is electrical energy. Electrical forces play a role in the nucleus too, but they certainly don't hold it together. Observing that particles with the same type of electric charge *repel* each other, we can calculate that there is a great deal of electrical energy waiting to be released if the 92 protons of a uranium nucleus were allowed to spring apart. Opposing this tendency, providing as it were the elastic band which holds the spring compressed, is the *nuclear force* of attraction between nucleons. This is a third force, different from the other fundamental forces of gravity and electricity, and it is this which makes a stable nucleus possible. When a thermal neutron interacts with a U-235 nucleus, a delicate balance is tipped, the nucleus splits and the fragments fly apart, with kinetic energy which comes largely from the stored electrical energy of the original nucleus.

How we measure the binding energies of nuclei is in itself an interesting piece of physics. Because fission is not a single process, detailed energy changes cannot be measured directly. They can however be calculated. Einstein provided the method when – over thirty years before the discovery of fission – he developed the Special Theory of Relativity, an essential consequence of which is that *mass and energy are equivalent quantities*. The mass of a system is not a fixed, unchanging quantity, but depends on its total energy. The measured mass of the U-235 nucleus is nearly one per cent less than the total mass of 92 protons and 143 neutrons, and this difference is the mass equivalent of the binding energy.

The conversion factor relating mass and energy appears in the well-known equation

$$E = m c^2$$

in which the quantity 'c' is the speed of light in empty space, a universal constant which plays an important role in the theory. Its value is about 300 million metres a second, so the conversion is

$$\text{energy in joules} = 9 \times 10^{16} \times \text{mass in kilograms.}$$

Looking again at Table 8.1, we can predict that if a U-235 nucleus splits into two roughly equal parts, together with a few spare neutrons, the total mass will decrease by about one-fifth of an atomic mass unit. It is the conversion of this into energy units which gives the figure quoted at the start of this section, for the joules released per fission. (We should note that there is nothing special here about *nuclear* reactions. Mass changes will accompany *any* reactions, but for chemical processes they will be extremely small: combustion of one tonne of coal, for instance, leads to a mass decrease of less than one-millionth of a pound!)

The product neutrons

There is a simple reason why extra neutrons are released in fission: the ratio of neutrons to protons is greater for the heavy nuclei than for the stable light ones. In practice, as we've seen, the product nuclei are not stable; they are radioactive precisely because they start life with too many neutrons. Nevertheless, there are usually about two free neutrons released in each fission event.

These survive as free particles for a ten-thousandth of a second or less, during which time one of them on average must initiate a further fission if a constant rate of energy production is to be maintained. This is however only one of many possible fates awaiting a free neutron. It could for instance escape altogether, and be lost in the surrounding material. It could, as we'll see, interact with the 'wrong' isotope, U-238, or with other nuclei present in the fuel. It could even interact with U-235 without inducing fission. All these processes compete for the available neutrons.

One way of increasing the chances of fission is to slow down the neutrons, and this is the method adopted in most present reactors – the so-called *thermal reactors*. If this is not done, the reactor must be designed to maintain a constant fission rate using the less efficient high energy or *fast* neutrons direct from the fission process. It is then called a *fast reactor*. Whatever the design details, it is evident that control of the output of a reactor can be achieved if a way is found of controlling the numbers of free neutrons and hence the rate at which fission occurs and energy is produced. When the desired condition is reached, with exactly one neutron from each fission event inducing a new fission, the reactor is said to be *critical*.

A significant side-effect

Absorption of a neutron by U-238, dismissed above as a loss process, is in fact the start of the extremely important series of events shown in Table 8.2. We see that the resulting U-239 is radioactive, emitting an electron (negligible mass but one unit of negative charge) and turning into neptunium-239, which is similarly radioactive and turns into plutonium-239. The significance of this is that *Pu-239 is fissile*. The interaction of the neutron with non-fissile U-238 has bred a new fissile nucleus. It has increased the fuel content of the reactor. A thermal neutron can now induce fission of the Pu-239, releasing about the same energy as in a U-235 fission.

Materials like U-238, whose interaction with neutrons leads to new fissile nuclei, are called *fertile*, and the second important instance is thorium-232, which breeds fissile U-233. Present commercial reactors all use uranium fuel however, and consequently breed plutonium. The term 'breeder reactor' is reserved for systems which actually produce more fissile material than they consume, and this is not the case for current thermal reactors. Let us take a simplified example to see how the figures work out.

We'll follow the fate of 1,000 uranium nuclei in a reactor during the period of one year between insertion of a fresh fuel element and removal of the spent fuel. Table 8.3 shows the figures. With enriched

Table 8.2 The breeding process

U-238 captures a neutron and becomes U-239	$^{238}_{92}U + ^{1}_{0}n \rightarrow ^{239}_{92}U$
U-239 emits an electron and becomes Np-239	$^{239}_{92}U \rightarrow ^{239}_{93}Np + ^{0}_{-1}e$
Np-239 emits an electron and becomes Pu-239	$^{239}_{93}Np \rightarrow ^{239}_{94}Pu + ^{0}_{-1}e$

Table 8.3 Uranium and plutonium in a thermal reactor

Data: Fresh fuel contains 3.2% U-235 and no plutonium.
Spent fuel contains 0.9% U-235 and 0.6% Pu-239.
80% of interactions between neutrons and U-235 nuclei lead to fission.
For every 10 neutrons absorbed by U-235, 6 are absorbed by U-238 to produce Pu-239.

Analysis of a fuel sample containing initially 1,000 uranium atoms		**Fission events**
U-235 nuclei		
At the start: 3.2% of 1,000	32	
At the finish: 0.9% of 1,000	9	
Consumed	23	
Of which 80% undergo fission:		18
Pu-239 nuclei		
At the start	0	
Bred from U-238: $\frac{6}{10}$ of 23	14	
At the finish: 0.6% of 1,000	6	
Consumed in fission events		8
Total number of fission events		**26**

fuel, about 32 of the original nuclei are U-235 and the rest U-238. At the end of the year, analysis of the spent fuel reveals about nine remaining U-235 nuclei in our sample, and six Pu-239 nuclei. Table 8.3 also incorporates a further piece of information: that for every *ten* neutrons absorbed by U-235, about *six* are absorbed by U-238 to breed plutonium. With these numbers, we see that nearly a third of the total fission events are not U-235 at all but Pu-239 – a third of the energy comes from the plutonium process.

The significance of this little analysis lies not in the precise figures, which will vary from reactor to reactor and which are in any case over-simplified, but in three specific implications. There is the fact that any uranium reactor is a plutonium producer, and if our figures are at all typical, the total fissile content of the spent fuel can be as much as *twice* that of natural uranium. This of course reinforces the argument for reprocessing spent fuel in order to make better use of resources. Then there is the possibility of increasing the efficiency with which uranium is used by designing reactors to produce *and consume* as much plutonium as is practicable.

Finally, we have the true breeder reactor, designed to convert so much U-238 into plutonium that it is a net producer of fissile material, which can then be extracted for use in other reactors. In principle this would increase the fuel potential of uranium more than a hundred-fold, from 0.7 per cent to nearly 100 per cent.

Radioactivity and radioactive substances

Nearly all the special problems of the nuclear industry can be traced to a single essential feature of fission reactors. This unique characteristic, not found in any other power system, is the unavoidable presence of intense radioactivity. Radioactivity brings complex new problems for engineers, adds considerably to costs and is at the centre of public concern about reactor safety and waste disposal. Questions of industrial safety and the disposal of dangerous wastes are not unique to the nuclear industry of course; they are equally important in coal-mining or the production of toxic chemicals. A significant difference, however, is that mines or chemical plants are not designed for the production of thousands of megawatts of power. If there is a justification for what the nuclear industry sees as the disproportionate attention paid to its slightest mishap, it must surely lie in this conjunction of very large power potential and the presence of an invisibly lethal material for which there is no neutralizer or antidote.

It is by no means easy to assess the potential hazards of nuclear reactors or the effectiveness of measures to make them safe. We'll start optimistically with a series of questions we'd like to answer:

- What is radioactivity?
- What causes it?
- Why are radioactive substances produced in reactors?
- What is produced, how much and what happens to it?
- What are the relevant effects of radioactivity?
- How much produces what sort of effect?
- How can radioactivity or its effects be prevented?
- What are the safety measures in present systems?
- Have they been successful?
- Will they be successful in the future?
- How about new systems?

Some of the answers are straightforward enough and we'll deal with them at once. Others involving matters of detail about reactor design and safety systems are treated briefly here and taken up again in other sections of this chapter. Biological effects are discussed later together with other environmental issues, and as we'll see, controversy over the facts then begins to emerge. Finally, there are really difficult questions to do with prediction and the assessment of risk, and we shall try to analyse some of the many conflicting answers to these in later chapters too.

Radioactivity

We've already seen that this is a natural phenomenon; that certain nuclei emit particles and turn into different nuclei. Let us start with the particles. The heaviest object emitted is a complete helium nucleus, known as an alpha (α) particle. As one might expect, alpha decay is particularly common in the heavier nuclei and an example we've already met is U-238 which decays to produce thorium:

$$^{238}_{92}\text{U} \xrightarrow[4.5 \times 10^9 \text{ yr}]{} {}^{234}_{90}\text{Th} + {}^{4}_{2}\text{He}.$$

The time shown under the arrow is called the *half-life* and is the time needed for half the U-238 in any sample to decay. It is a very important quantity and from it we can deduce a number of other facts about the radio-isotope concerned. The essential feature of radioactive decay is that the number of nuclei which will emit a particle in the next second, hour, year or century is proportional to the number present when we start counting. If half the nuclei in some sample decay in the first eight days then half the remainder will decay in the next eight, and so on.

Let's take U-238 as an example – though the reasoning is equally good for any half-life, whether it is as long as the age of the universe or as short as a millionth of a second. It's not difficult to show that if half the nuclei are to decay in the next 4,500 million years, five in every million, million, million must disintegrate in the next second (Table 8.4). Now five per 10^{18} sounds a rather tiny number, but we must remember that there is a *very large* number of atoms in even a small piece of material. Consider one gramme of uranium, a sphere less than a fifth of an inch across. This will consist of about 25×10^{20} atoms, so five per 10^{18} means a total decay rate of about 12,000 in each second. (A complication in practice is that the thorium decays in turn, and so on through a series of isotopes; but none the less the figure gives an idea of the activity.) A point to note is that this is a *weak* source. Each gramme of spent fuel direct from a reactor produces several million, million particles a second.

We can use the U-238 example again to bring out a crucial characteristic of radioactivity. Suppose we take a tiny fragment of uranium

Table 8.4 Radioactive decay rate and half-life

The basic rule

The number of nuclei which decay in any one year is always a certain fraction of those present at the start of that year.

The number remaining at the end of the first year is *one minus one Nth* of the original number:

$$(1 - \tfrac{1}{N}) \times \text{original.}$$

At the end of the second year it is one minus one Nth of the '*one year*' number:

$$(1 - \tfrac{1}{N}) \times (1 - \tfrac{1}{N}) \times \text{original}$$

and after a third year

$$(1 - \tfrac{1}{N}) \times (1 - \tfrac{1}{N}) \times (1 - \tfrac{1}{N}) \text{ times,}$$

or more concisely

$$(1 - \tfrac{1}{N})^3 \times \text{original.}$$

The number continues to fall in this way, until after 4,500 million years there remain

$$(1 - \tfrac{1}{N})^{4.5 \times 10^9} \text{ of the original number.}$$

However, if 4.5×10^9 years is the *half-life* (for U-238), then the number remaining must be just one half of the original number:

$$(1 - \tfrac{1}{N})^{4.5 \times 10^9} = \tfrac{1}{2}.$$

We want to know what N must be, for this to be true. What fraction must decay each year if one half remains after 4,500 million years?

Provided the fraction one Nth is *very small* there is a simple rule for calculating N. It is

$$N = 1.44 \text{ times the half-life.}$$

For U-238, N is thus about 6,500 million: one nucleus in 6,500 million decays in a year, or five per million, million, million in each second.

containing 'only' 1,000 million atoms, and watch it for 6,000 years. We can confidently predict that about 1,000 nuclei will decay in this time, but what we cannot predict at all is *which* nuclei. Right up to the moment it happens there is absolutely no way to tell that a nucleus is about to decay. And this is not just a matter of poor technology. It's not even that the physicists haven't found how the hidden clock ticks. On the contrary, they firmly believe that there *is* no hidden clock. Unlike any other process we've studied, radioactive decay is *an effect without a cause*, and the practical implication is that we ourselves can neither cause nor prevent the decay of a particular nucleus. We could of course smash it with a sufficiently fast sub-nuclear projectile, but this is not a practical way of dealing with large quantities of material. Otherwise, once we've produced radioactive material there's no way to stop the radioactivity. We can only wait for enough half-lives for it to die out naturally.

Not all radioactive nuclei emit alpha particles. Some produce beta (β) particles, and these are a little surprising because they are simply electrons. A negative particle coming from a positive nucleus? It may seem odd but it's not impossible provided that the positive charge *increases* by one unit to conserve the overall net charge. The electron carries away very little mass so the mass number is unchanged, and effectively we have the conversion of one neutron into one proton. Two beta emitters of some importance are strontium-90 and tritium:

$$^{90}_{38}\text{Sr} \xrightarrow[28\text{ yr}]{} {}^{90}_{39}\text{Y} + {}^{\ 0}_{-1}\text{e}$$

$$^{3}_{1}\text{H} \xrightarrow[12\text{ yr}]{} {}^{3}_{2}\text{He} + {}^{\ 0}_{-1}\text{e}.$$

The short half-lives explain why these isotopes are rare in nature. They are artificial radio-isotopes produced in reactors.

The third radioactive emission is the only one which really is radiation in the usual sense. The gamma (γ) rays are electromagnetic waves, like radio waves, or light or X-rays. The special feature of gamma rays (and X-rays too) is that they have extremely short wavelengths and oscillate very fast, and this gives them their penetrating power. In emitting gamma radiation a nucleus of course gives up energy, but there is no change in the numbers of nucleons; it is a sort of settling-down process which often follows alpha or beta decay.

Other types of radioactivity are known but these three are the most common and – together with neutrons – are the main 'radiations' which give rise to concern about nuclear reactors.

Reactor products

As the fuel in a reactor burns, the inventory of radioactive nuclei which builds up has three main constituents:

- Unstable fission products, mainly beta- or gamma-active and contributing over three-quarters of the activity of the spent fuel. Their half-lives are mostly rather short: hours, days or a few years.
- Heavy isotopes resulting from non-fission interactions. Known as actinides, these are mainly alpha- and gamma-emitters with half-lives from decades to thousands of centuries, and account for about a quarter of the activity.

- Isotopes produced, mainly by neutron bombardment, in the fuel cladding, structures, etc. They vary widely, and contribute a few per cent of the total radioactivity.

An important distinction is that, if all goes well, the first two of these should remain *inside* the enclosed fuel elements whereas the third obviously does not. In discussing safety and the 'escape of radiation' we should be clear whether we mean the escape of the substances themselves or the penetration of containers by emissions from enclosed substances. For the latter the penetrating powers are obviously important, and these are very roughly as follows:

- Gamma rays and neutrons are the most penetrating, needing thick concrete or specially selected absorbing materials to stop them.
- Beta particles are stopped by, say, a thick metal sheet.
- Alpha particles are stopped by almost anything: a sheet of tissue-paper, or about an inch of air. Virtually the only way they can cause harm is if the source is either eaten or inhaled.

This last point brings out a major practical problem: that some reactor products are gases, and gases – unlike solids – will not just stay where you put them.

Containment

It is evident that if radioactive substances are potentially harmful the total containment system of a reactor must ensure that during normal operation the fuel cladding remains intact and the core is adequately shielded, and that in the event of an accident no radioactive substance – gas, liquid or solid – can escape into the surroundings. Containment of spent fuel is equally important during removal, transport, reprocessing and ultimate storage, until enough half-lives have passed for the danger to become negligible.

Reactors

The first problem for the designer of any nuclear reactor is to get the neutrons right. The composition of the fuel and the choice of other materials, the sizes and shapes of the component parts of the reactor, the methods used to extract heat energy, are all governed by the need to ensure that precisely one neutron from each fission event induces another fission. Too few neutrons and you don't have a power plant; too many and you have a catastrophe.

Table 8.5 Types of reactor

Reactor	Fuel	Moderator	Coolant
PWR	enriched uranium	ordinary water	ordinary water
BWR	enriched uranium	ordinary water	ordinary water
CANDU	natural uranium	heavy water	heavy water
MAGNOX	natural uranium	graphite	carbon dioxide
AGR	enriched uranium	graphite	carbon dioxide
LMFBR	highly enriched U/Pu mix	none	liquid sodium

Now *reducing* the number of neutrons is relatively straightforward. Certain materials are good neutron absorbers and will very effectively cut down the neutron flux when introduced into the reactor core. So control is a negative process: the reactor is designed to produce too many neutrons and the excess is absorbed. A common control system is a set of rods, any of which can be inserted or withdrawn as necessary.

Producing an excess of neutrons in the first place is a matter of putting in enough U-235 at sufficiently high concentration to allow for all losses and still have more than one spare neutron per fission. There is thus an inter-relationship between the composition and form of the fuel elements, the type of material used to thermalize the neutrons (the *moderator*) and the nature of the coolant which flows through the core and carries away the heat. Table 8.5 lists the main reactors currently in use for power generation, all but one of them thermal reactors. Their main differences come from the choice of moderator and coolant, a choice which is crucial in determining the entire structure of the reactor.

Moderators and coolants

The ideal moderator should reduce neutron energies as efficiently as possible. That is, quickly and with the minimum neutron loss. At first glance, hydrogen should be best because it has the lightest atoms, which means that a neutron gives up its energy in the fewest collisions. (A fast-moving billiard ball gives up most of its energy when it collides almost head-on with a much slower or stationary one. A fast-moving billiard ball colliding with a twenty-pound cannon ball bounces off with very little energy loss.) Ordinary water contains plenty of hydrogen, is cheap and has the further advantage that it can also act as coolant. Unfortunately however, it is a rather heavy absorber of neutrons. They combine with the hydrogen nuclei (protons) to form deuterons, nuclei of the hydrogen isotope deuterium. (Only in the case of hydrogen are the isotopes given names of their own. They are *deuterium*, written as either ${}^{2}_{1}D$ or ${}^{2}_{1}H$, and *tritium*, ${}^{3}_{1}H$.) The absorption of neutrons by water is so great that natural uranium is too dilute to sustain a chain reaction. So all reactors using ordinary water (light-water reactors or LWRs) must be fuelled with enriched uranium.

Deuterium itself is an excellent moderator. The neutrons need more collisions than with hydrogen but they will travel literally for miles in heavy water (D_2O) with very little loss. In consequence heavy-water reactors (HWRs) can use natural uranium. But the 500 te of heavy water needed as moderator and coolant in a large reactor must be obtained by separating out the small fraction of deuterium in natural hydrogen (about one atom in 7,000); a factor to be balanced against the saving on uranium enrichment.

The third common moderator is the one used in the first reactor ever: carbon in the form of graphite blocks. Although the atoms are relatively heavy, the neutron loss is only a quarter that in light water. Graphite cannot of course circulate, so something else must act as coolant. In British reactors this is carbon dioxide gas. The first generation (called the Magnox reactors, from the alloy with which their uranium metal fuel was clad) could use natural uranium, but the newer Advanced Gas-Cooled Reactor (AGR) needs slightly enriched fuel.

The LMFBR, being a fast reactor, has no moderator. It is sufficiently different from all the others that we shall discuss it separately.

Structures

Certain basic requirements determine the form of all the reactors discussed here. The fuel must obviously be distributed at the right density, in good contact with the coolant and surrounded by the moderator. Given the high level of radioactivity once the reactor has been running, replacement of spent fuel must be handled from outside. Hence the common use of tubes or 'pins' assembled into bundles, with the fuel (now normally uranium oxide, which can withstand higher temperatures than uranium metal) in the form of slugs or pellets inside a metal cladding (Fig. 8.2).

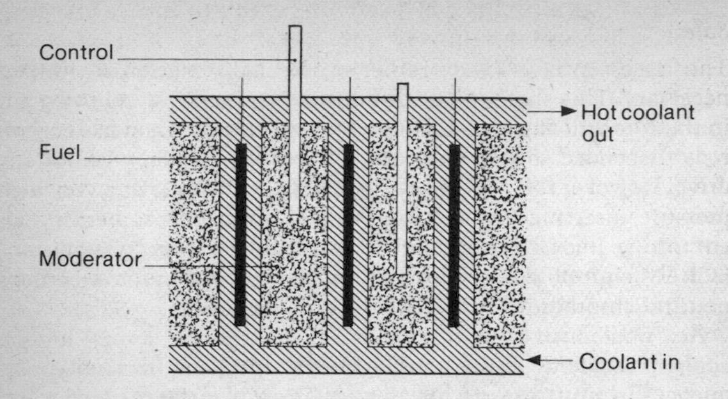

Fig. 8.2 Essential features of a thermal reactor core

Coolant must stream freely past the fuel, and for the most efficient power generation should leave at the highest possible temperature and pressure. This raises a problem, because only the HWR with its very low neutron loss in moderator and coolant can tolerate the additional loss in the material of individual pressure tubes carrying the coolant through the core, and still maintain its chain reaction. So in both the LWRs and the AGR the whole core must be submerged in coolant in a large pressure vessel.

Materials as well as people can be harmed by exposure to intense irradiation and this raises many new problems for engineers. With the need for all monitoring and control, inspection and maintenance to be handled remotely, the growth of the nuclear industry has meant the development of entirely new areas of instrumentation and materials science. Let us contemplate for a moment the interior of the core of a reactor. The coolant must not occupy too much space, so flow rates must be very fast to carry away the heat. Thus we have a hot fluid streaming at high speed, under pressure, through narrow gaps and channels past long thin fuel rods. It is not surprising that materials and measuring instruments, even when they survive the combined effects of temperature and pressure, chemical attack and bombardment by sub-nuclear particles, have sometimes shaken themselves to pieces. The designer must ensure that fuel rods do not distort and change the fuel configuration, that control rods can move freely without jamming – and that nothing is ever allowed to impede the coolant flow.

Safety

The first essential is a way of stopping the chain reaction in a hurry if necessary. This shouldn't be difficult because almost anything put in the core will absorb neutrons. One 'fail-safe' method has control rods suspended so that they can if necessary fall in, without any drive. However it is worth noting that if the inside is at a very high pressure, inserting *anything* from outside means that something has got to give. It is fairly common these days to have two systems for a SCRAM shut-down: control rods and also provision for injecting a neutron absorber into the coolant or moderator.

Any unplanned change in coolant flow, or worse, a total loss-of-coolant accident (LOCA) could have particularly serious consequences in a nuclear reactor. In a coal-fired plant, if the feed water were to fail, the mass of hot coal would be rather an embarrassment, but at least the generation of more heat could be prevented by stopping the air and fuel supplies. Not so in a nuclear reactor. Even after the chain reaction is brought to a halt the generation of heat continues. It comes from the energy released in the decay of the radioactive substances present, and immediately after shut-down can be almost a tenth of the full power output of the reactor. As we've seen, nothing can be done to stop this because you can't switch off radioactivity. Fortunately the half-lives of many of the products are very short and the rate of heating will fall quickly – to perhaps one per cent of full power within an hour. Nevertheless, without cooling, the heat produced in this time could melt the core and perhaps the floor of the reactor building as well: the scenario for the China Syndrome.

The likelihood or unlikelihood of this particular catastrophe, and the measures adopted to prevent it, must be a central concern when we ask about the safety of different types of reactor.

The fear that the core of a thermal reactor will somehow turn into an atomic bomb is not well-founded. Any explosion is the result of an energy-producing chain reaction (chemical or nuclear) which multiplies so fast that the energy hasn't time to escape by the usual 'peaceful' means, as heat or light. Instead, the energy density rises until the bonds holding the material together are ruptured. In the immortal words of the unfortunate apologist at Three Mile Island in 1979, it is indeed an 'energetic disassembly' – but we'll call it an explosion. Thus a nuclear explosion requires very rapid multiplication of neutrons and this in turn requires a high density of fissile material in a large enough lump to make the loss of neutrons through its surface unimportant. About 20 lb of pure U-235 will do it, or half this much Pu-239. If the fissile material is diluted, the critical mass is greater, and if the concentration is below about a tenth no runaway explosion is possible. So the only way for a thermal reactor to become a bomb would be for the U-235 or Pu-239 distributed throughout several tonnes of core to bring itself miraculously to one place.

This argument does not of course exclude an 'ordinary' chemical explosion. Whether this can occur depends here as in other industrial plant on the materials present and their condition. (An explosion of this kind did happen at Three Mile Island and was contained.)

Efficiency

A feature common to all present nuclear stations is that they use the conventional method for producing electric power: steam turbines driving generators. The coolant carries away the energy as heat which produces hot steam. Steam temperatures and pressures in nuclear plants have always lagged behind those of fossil fuel systems, partly because of the more severe conditions which materials must withstand. Consequently, thermal efficiencies are lower (34 per cent turbine efficiency is considered good) and the waste heat output is greater. Heat losses should be least if the coolant itself becomes the steam, and this is the method adopted in boiling water reactors (BWRs). The others all use two circuits, with the primary coolant giving up energy in a heat exchanger (steam generator) to a secondary flow of water.

The concept of 'overall efficiency' is not a very clear one for a nuclear plant. What do you take as starting-point? What is counted as energy input? At the mines (Fig. 8.3), natural uranium is extracted – or to be precise, an ore of natural uranium. This is treated to give an oxide (U_3O_8) known as *yellowcake*, which is in turn processed before use as fuel, even in a reactor using natural uranium. Increasing the proportion of U-235, if necessary, is an expensive process consuming a twentieth or more of the power output of the plant. The useful life of a fuel element ceases when the build-up of neutron-absorbing fission products starts to limit the chain reaction, so the spent fuel may have a fissile content equal to half that of a fresh element. On the other side of the account-book, as we've seen, some Pu-239 which wasn't even there to start with will have contributed to the output.

Thus any definition of efficiency must be rather arbitrary, and the best procedure is probably to compare different nuclear plants by

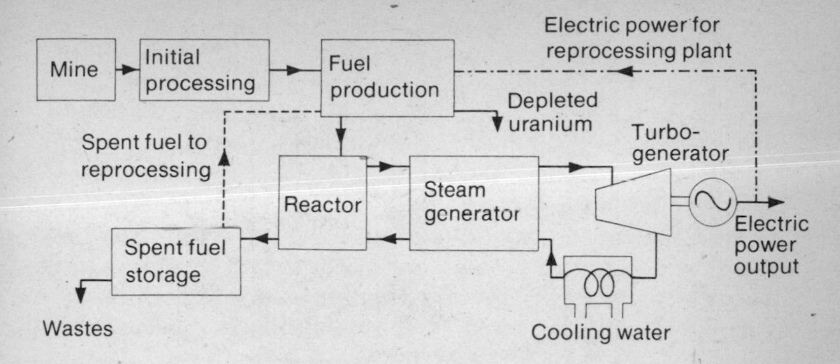

Fig. 8.3 Uranium fuel processes

asking how much uranium must be mined for each unit of useful electrical output, making it clear whether or not reprocessing of spent fuel is included in the calculation.

Four types of thermal reactor

We've seen how reactors work and to some extent why they are as they are. It remains to look at the actual structures. With an eye to later discussion this brief treatment concentrates mainly on the differences between the four and on the features which are important for their safety.

PWR

Over three-quarters of the world's nuclear power installations use light water reactors and more than half of these are PWRs. (Slightly surprising, for a system originally designed as a power plant for submarines!) The size we describe produces over 3,000 MW of thermal power, enough for two 550-MW turbo-generators. Its core consists of little more than the fuel elements, 50,000 long thin metal tubes (Fig. 8.4). Assembled into bundles, these are held in place between top and bottom plates, and the whole is submerged in water in a pressure vessel some forty feet high with walls eight inches thick (Fig. 8.5). The fully-loaded reactor holds about 100 te of uranium.

Water, acting as both moderator and coolant, circulates through the pressure vessel and through steam generators near it. High pressure (over 2,000 psi) prevents the water from boiling even at its maximum temperature of over 300°C, and maintaining this

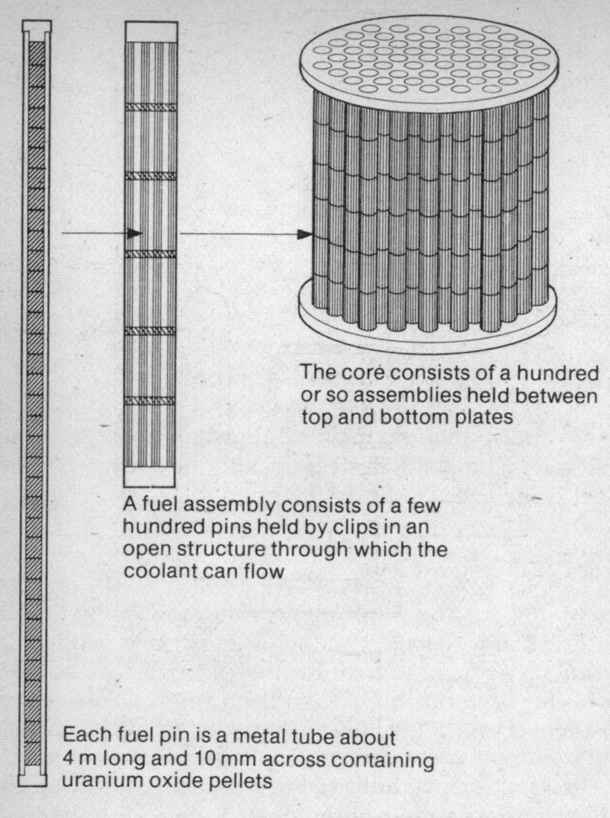

Fig. 8.4 Fuel arrangement in a pressurized-water reactor

pressure is very important, because cooling depends on the presence of water – not steam, which would carry away much less heat. The danger of a massive LOCA is taken very seriously because the coolant flows in a few large pipes and if one were to break, the water would flash to steam and be lost in seconds. This would leave the core with only itself to heat: a relatively small quantity of material to absorb a great deal of energy.

A point in favour of any LWR is that loss of coolant is also loss of moderator, so the chain reaction will in any case stop. Nevertheless, a fast and effective emergency core cooling system is essential, and a PWR of this type will have at least three. One consists of tanks holding water under pressure, and responds quickly by releasing this

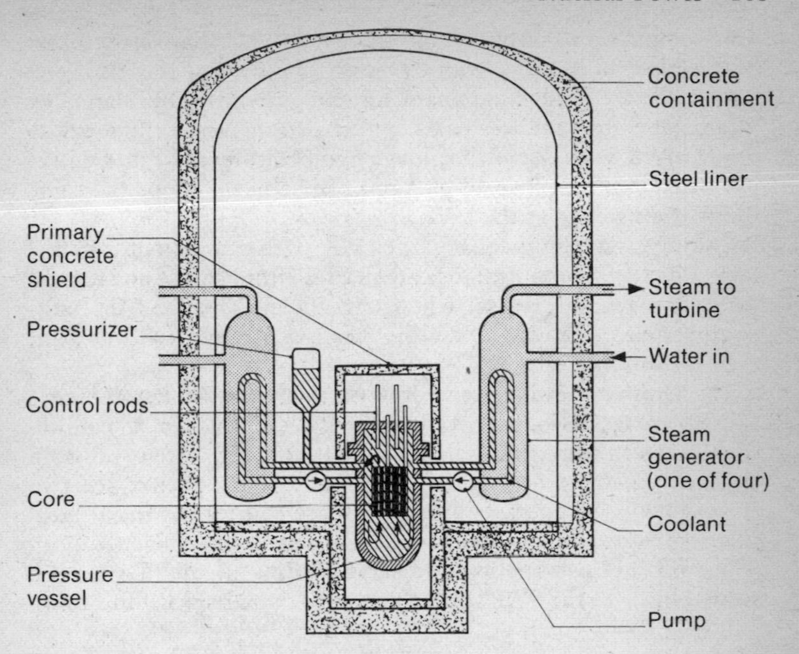

Fig. 8.5 A pressurized-water reactor

into the core. This should hold down the temperature until pumps can come into operation and provide a sufficient flow from other tanks. For lesser leaks a third system is available, injecting water at the full operating pressure of the reactor.

Refuelling means closing down for a couple of weeks – a procedure which strikes one as more appropriate for a submarine returning to port at long intervals than for a base load power-station. However, we shouldn't forget that any power-station is out of commission at times for maintenance, and also that a fossil fuel plant has the inverse disadvantage of short-term dependence on regular fuel deliveries. A more serious drawback of annual refuelling is the inefficient use of uranium. If the reactor is to remain critical as the fuel is increasingly poisoned, it must be designed for a large initial excess of neutrons, an excess which at the start of the fuel cycle must be absorbed by extra control: a waste of good neutrons and of uranium.

BWR

About a third of nuclear power plants in the USA and a number in other countries use BWRs. Like the PWRs these have light water moderators and coolants, with the essential difference that the water boils, directly producing steam for the turbines. Elimination of steam generators reduces costs, and it also increases efficiency so that a BWR with somewhat lower temperature and much lower pressure (1,000 psi instead of 2,000 psi) achieves about the same thermal efficiency as the PWR.

The core is not unlike that of the PWR, with similar long thin fuel rods (Fig. 8.6). The control rods are inserted from below, and remain inside the pressure vessel when withdrawn. This, and the large volume above the core where the steam separates, lead to a 70-ft high pressure vessel.

The fact that the radioactive primary water flows to the turbine is a disadvantage. Also, the action of the BWR depends on an equilibrium between water and steam, which is very sensitive to pressure changes and must be carefully controlled. The lower of the two cooling pools is designed automatically to control any major pressure rise by providing somewhere for the steam to condense. Both pools also act as supplies of reserve cooling. Refuelling is on a similar cycle to the PWR, and in general, comparisons of these two

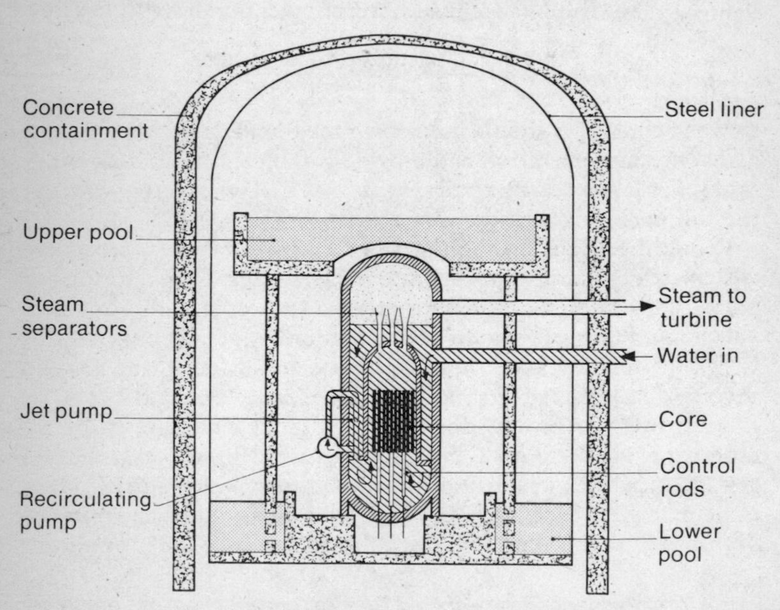

Fig. 8.6 A boiling-water reactor

light water reactors have tended to conclude that there is little to choose between them: in capital cost, fuel efficiency, and safety.

AGR

The AGR was adopted for Britain's programme in the mid-1960s and seven twin-reactor (2 × 660 MW) stations are now operating or under construction, joining the eleven Magnox stations. The structure of an AGR is of course completely different from an LWR. Instead of an open lattice of thin tubes, the core, about thirty feet across, is made of graphite, with some 300 vertical channels into which the fuel clusters slide (Fig. 8.7). There are 2,000 of these, short and chunky, not long as in the LWRs. The carbon dioxide coolant is pumped at about 600 psi through the core, with a flow pattern designed to keep the graphite as cool as possible. Gas temperatures up to 600°C were planned, but present AGRs are derated to 500°C because of materials problems.

The pressure vessel is not a steel dome but a massive reinforced concrete structure enclosing the complete system. Advocates of the AGR claim that this is less likely to fail catastrophically than the LWR steel vessels. Then the coolant is already a gas, so a sudden change in pressure or temperature should not have the same

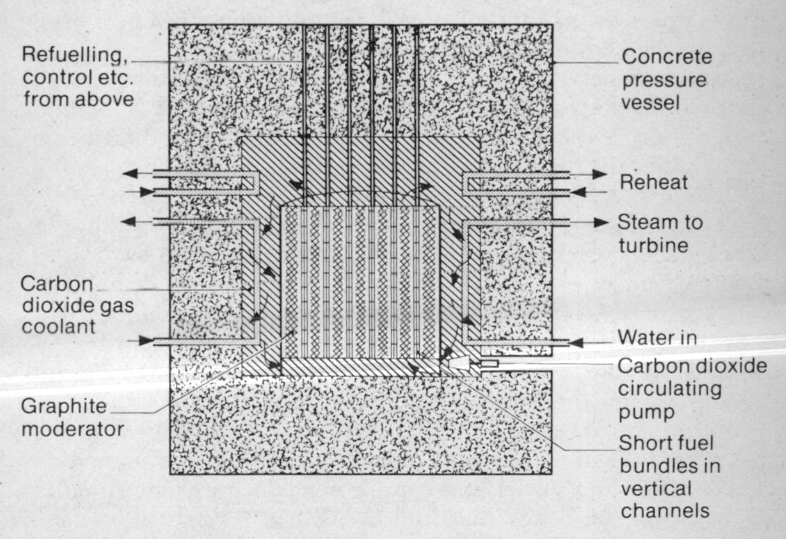

Fig. 8.7 An advanced gas-cooled reactor: core, boiler and pressure vessel

dramatic effect as with water cooling, and in any case the mass of graphite could absorb excess heat – enough, it is claimed, to survive for up to an hour without external cooling.

Unfortunately the rest of the case is not so strong. In principle the AGR should produce slightly more power per tonne of uranium than the LWRs, but the performance of the existing reactors has not yet established this. Nor is the planned on-line refuelling yet operating as intended. Not even their supporters any longer expect AGRs to compete on world markets, and Britain is now contemplating a change to PWRs.

CANDU

This is another case of one country committing itself to a reactor type which has not been much developed elsewhere. A dozen CANDU (Canadian Deuterium Uranium) reactors with heavy water as coolant and moderator provide power for Canada's nuclear power-stations, and this is also the only system to have made any dent in the domination of the international market by LWRs

The CANDU uses to advantage the low neutron absorption in heavy water. It has separate circuits for moderator and coolant (Fig. 8.8). The short fuel bundles lie in hundreds of horizontal channels and the coolant flows past them inside double-walled tubes. The moderator is thus kept cool enough that it need not be under pressure to avoid boiling. Refuelling is on-line.

The whole network of fuel channels and horizontal and vertical control rods is in a horizontal steel cylinder called the *calandria*, which for a 600-MW reactor is about 25 ft in each dimension. Surrounding the reactor is a concrete radiation shield and the entire system is inside a containment building some 200 ft high.

The use of many small pressure tubes makes sudden total loss of coolant unlikely (though a fracture could still lead to local overheating), and as in the AGR, the mass of moderator should hold the temperature even if the coolant pumps failed. The disadvantage that loss of coolant is not loss of moderator is recognized by a dual SCRAM system, with control rods to drop in and provision for injecting a neutron absorber.

With a coolant temperature only a little above 300°C the CANDU has poor thermal efficiency, but its efficient use of uranium compensates for this. Overall it is capable of producing about a quarter more power per tonne than the LWRs. The most striking fact, however, is what the CANDUs actually do produce. The capacity factor depends on demand as well as availability, so comparisons

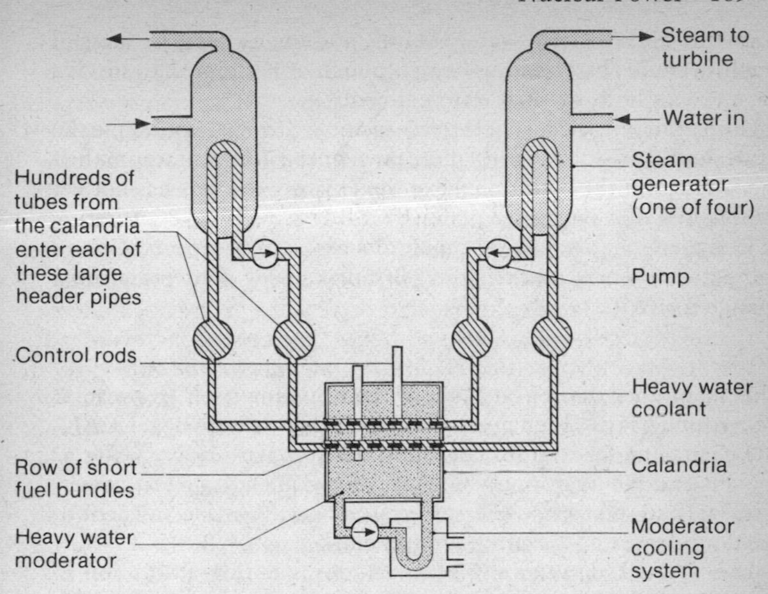

Fig. 8.8 A CANDU reactor: calandria and steam generators

need care. But the data in Table 8.6 show one reason at least why the CANDU has a quite surprising spectrum of admirers.

Costs

And finally, the other important factor. It is not difficult to find detailed analyses showing that the cheapest electricity comes from one rather than another type of plant. Britain's CEGB convinced themselves in 1965 that the AGR was cheaper than any LWR and in 1973 of the reverse, with as far as one can see little change in the hard facts. Current estimates of the capital costs of an American PWR are $850 per kW and also £850 per kW, which seems odd, as the normal exchange rate is two dollars to the pound. The anti-nukes draw to

Table 8.6 Capacity factors

The average capacity factor is the gross output to date from all commercial plants of the type, as a percentage of the maximum output which they could have produced in continuous operation.

Reactor type	**PWR**	**BWR**	**MAGNOX + AGR**	**CANDU**
Average capacity factor	57%	55%	46%	78%

our attention the sums spent on nuclear research and development, those still to be spent on waste disposal and decommissioning, and the supposedly hidden costs of enrichment. The industry replies with figures showing that (1) the 200-year development costs of coal are just as high; (2) waste disposal will not be expensive (and they show a picture of a hole in the ground to prove it); (3) decommissioning costs will also be small (with another picture of a little concrete box in a pretty landscape); and (4) enrichment is *not* subsidized.

But we mustn't allow the controversy over these peripherals to hide the central fact, which is that there is major disagreement over the relative costs of any two *particular* power-stations, even under reasonably well-specified conditions. Figures for the 'true cost' in pounds (or dollars) per kW can indeed differ by a factor of two, depending on assumptions about construction times, fuel and other prices, inflation and interest rates (or discount rates; see pp. 294–5), and capacity factors. Several recent independent studies surround their tables of costs with pages of qualifications and warnings; and the efforts of a Select Committee to establish the cost of Britain's proposed PWR are recounted in detail in their 1981 Report (see Further reading). The figures of the CEGB and of the utilities companies in other countries are published in their annual reports, and usually show nuclear power to be cheapest.

To attempt to adjudicate where committees, study groups and working parties of experienced economists and accountants have failed to agree would be presumptious, and to reproduce enough figures for a judgement would need far too much space, so we must leave the subject with the rather weak comment that the range of uncertainty due to our inability to predict the future seems to be of the same order as the difference between estimates of the future cost of power from the LWRs, the AGR, CANDU, the breeder – and coal.

Breeders

Breeder reactors are not new. The first ever nuclear generation of electric power was from an experimental breeder in the USA in the early 1950s; and Britain, France and the USSR each now have one plant with a few hundred MW output, while the two latter are building larger units. The fuel-saving attraction (see pp. 152–3) is obvious, but today the breeder is the subject of fierce debate. The two issues on which there is the most disagreement are whether a breeder programme is *necessary* and whether it is *safe*. The necessity or otherwise depends on estimates of demand and supply, to which we return later. The question of safety will largely govern our approach here. We shall not compare different systems because there are no 'different systems' to compare. All *existing* large breeder reactors are of one type, the liquid metal fast breeder reactor (LMFBR), so the immediate question is whether or not to have more of these.

We have seen how a breeder works in principle, and it is obvious that for the total number of fissile nuclei actually to increase there must be at least two 'useful' neutrons from each fission: one to carry on the chain reaction and one to interact with a fertile nucleus. Allowing for losses, well over two neutrons must be produced; and the reactor must then be designed so that the chain reaction can be controlled, losses are minimized and all possible spare neutrons interact with fertile nuclei. Let us see how the LMFBR achieves all this.

- The fissile material is mainly Pu-239, producing nearly three neutrons per fission – about half a neutron more than U-235, on average.
- It is a *fast neutron* reactor. Fast neutrons give slightly greater neutron yield per fission, with appreciably lower losses and better conversion of U-238 than thermal neutrons.
- Because fast neutrons are much less efficient in *inducing* fission a high concentration of fissile material is needed. The fuel is a mixture of uranium and plutonium oxides with 15–20 per cent fissile nuclei.
- The fertile material, U-238, surrounds the central fissile core in an arrangement known as a blanket.
- The coolant must carry away about ten times as much heat per cubic foot of core as in a thermal reactor (because of the high fuel concentration) and must not absorb neutrons. Nor must it act as a moderator. Sodium is used, a metal which is solid at room temperature but liquid at the 500°C operating temperature.

Most existing FBRs are of the pool type (Fig. 8.9(a)). All radioactive constituents are in one vessel, and the use of liquid sodium permits high temperatures without needing a high pressure. An intermediate sodium circuit ensures that under no circumstances can the radioactive primary sodium come into contact with water. (Sodium is chemically extremely active, igniting spontaneously on exposure to air and reacting violently with water.) The fuel assemblies are bundles of steel tubes loaded with pellets, and the required U-238

(a) Schematic picture of the main features

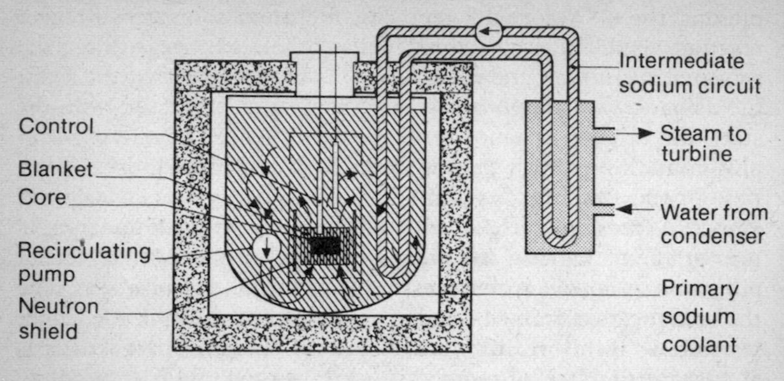

(b) Fuel element

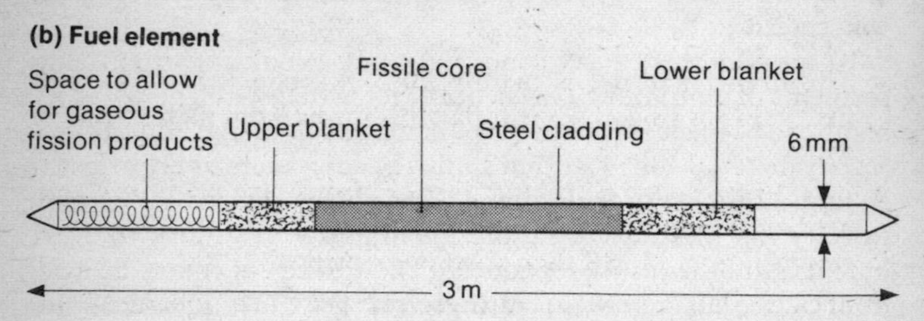

Fig. 8.9 The LMFBR

blanket is achieved by suitably distributing their contents (Fig. 8.9(b)).

The main concern about the FBR centres on its fuel, for three reasons:

- the concentration of the fissile material,
- the necessity for reprocessing,
- the central role of plutonium.

The first of these raises the question whether a runaway nuclear reaction could occur. A fully efficient atomic bomb is not possible, and even a 'nuclear fizzle' would require a large number of improbable conditions at the same moment. There are also reasons to believe that it would extinguish itself automatically. Nevertheless, there is some disagreement about probabilities, and the degree of unlikelihood does seem to be of a lesser order than the virtual impossibility in a thermal reactor.

Reprocessing is an option in a thermal reactor programme. At present, the USA does not reprocess fuel from commercial power reactors, the USSR issues fresh fuel to other countries against the return of spent fuel, and the CANDU is so designed that the spent fuel is hardly worth reprocessing. Britain has reprocessed from the start (the original Magnox design was influenced by the need for plutonium for military purposes) and Sellafield (ex-Windscale) carries out reprocessing for a number of countries, as does La Hague in France. At present, of the seventy or so countries with some form of reactor, only a handful have reprocessing facilities. With a breeder programme, however, reprocessing is essential, because it is how the bred fuel is extracted – and in any case plutonium is needed for start-up. So if the breeder were to become the standard reactor it is almost inevitable that reprocessing facilities would become more widespread.

To see why this causes concern, we must look at the particular problems of plutonium. One central fact is that with enough of it a bomb can be made. *Not*, it must be emphasized, easily. Plutonium is a nasty material and in any but skilled hands is more likely to lead to an unpleasant death than to unlimited power. (Indeed, its extreme toxicity and the consequent blackmail power of a threat to distribute it is another cause for concern.) Bombs can of course be made from U-235, but only by separating it from the U-238, and separation plants for isotopes are large, expensive and complex. Building one is not simple even for a body with the resources of a government. Plutonium does have diluting isotopes (also produced in the reactors) and 'weapons grade' Pu-239 must have less than 7 per cent Pu-240. But plutonium for breeders, containing 80–90 per cent Pu-239 (unlike the U-235 buried in many times its own mass of U-238), can still be used for a weapon, if a rather unreliable one. There can be little doubt that the best method of safeguarding plutonium against misuse is to design thermal reactors so that they burn up as much as possible, and then to leave the remainder in the highly radioactive spent fuel – which would mean no breeder programme.

This brief look at breeders has undoubtedly accentuated the negative aspects, but the justification must be that the advantages, if they do indeed work safely and efficiently, are too evident to need emphasis. Whether uranium utilization is increased twenty, fifty or a hundred-fold is not critical. What matters is that countries such as Britain and the USA have enough U-238 lying around as depleted uranium to keep them in electric power for centuries.

Table 8.7 Energy from fusion

Fusion reactions

Fusion of two deuterium nuclei leads to hydrogen and tritium nuclei $^2_1H + ^2_1H \rightarrow ^1_1H + ^3_1H$

Fusion of another deuteron with the triton produces a helium nucleus and a neutron $^2_1H + ^3_1H \rightarrow ^4_2He + n$

The neutron is unstable and decays, with a half-life of 16 minutes, into a proton, an electron and a neutrino $n \rightarrow ^1_1H + e + \nu$

The net effect is that three nuclei of deuterium have become two of hydrogen and one of helium.

Particle masses (u)

proton	1.0073
neutron	1.0086
deuteron	2.0136
triton	3.0155
helium-4	4.0015

Energy

The total nuclear mass has decreased by 0.0247 u, about one 240th of the total mass of the original three deuterons.

This 'lost mass' is the energy produced ($E=mc^2$), some 360 million megajoules per kilogram of deuterium consumed.

360 million megajoules will run a 600-MW power-station (30% efficiency) for roughly two days.

Sources of deuterium

Water is H_2O. As an oxygen atom is 16 times as heavy as a hydrogen atom, there are 2 kg of hydrogen in 18 kg of water.

1 in 7,000 hydrogen atoms is deuterium, so there are 2 kg of deuterium in 63,000 kg (about 60 tonnes) of ordinary water.

A 600-MW power-station therefore needs a net fuel input of 15 tonnes of ordinary water a day!

Footnote on fusion

Fusion is the coming together of two lighter nuclei to form one heavier one. As it is the reverse of fission we might expect it to consume rather than produce energy, and this would indeed be so if we tried to persuade strontium and xenon to fuse into uranium. However, if we start with two *very* light nuclei we find completely the opposite. The fusion of two deuterons, producing He-3 and a neutron, *releases* energy. The masses tell us so. Table 8.7 shows the figures, and reveals why billions of dollars, roubles, etc. have been spent on fusion research. Fusion, it seems, offers 'clean' energy from a fuel so plentiful that it will last for millions of years. No mining, no radioactive fission products, no reprocessing, no plutonium. Only one problem. We don't know how to do it. Not even in the laboratory. The only fusion power we have achieved so far is the hydrogen bomb.

It is easy to see in a general way why there is a difficulty. Two nuclei can only fuse if they come close enough together for the strong nuclear force of attraction to take over. Until this happy state is reached the repulsive electrical force dominates, pushing ever harder the closer they approach. Only if they can be given enough initial energy to overcome this enormous electrical barrier can they reach the promised land flowing with fusion energy.

We'll finish, then, with a very brief glance at one of several attacks on the problem. (The reaction being sought is not exactly the one shown in the table, but the idea is the same.) The *thermonuclear* method aims to give particles high energies by using the fact that hotter means faster. A temperature of some millions of degrees is needed, and a major technical problem is how to contain the mass of 'hot' particles. If they touch anything they will share their energy and cool down. One answer is a 'non-material' container, and a great deal of work has gone into the *magnetic bottle,* which holds the particles by surrounding them not with atoms at all but with a strong magnetic field. Designing this is difficult enough, but it is only the first step. There is the question of how to get the fuel in and the energy out. And how about side effects? Is it really going to be clean? Tritium is radioactive, and pollution of water by this hydrogen isotope is not an appealing prospect.

Optimists believe that we shall see controlled fusion within the next few decades, but no one expects fusion power-stations until well into the twenty-first century, and there are so many unknowns that it is quite impossible to incorporate this potential source realistically into present plans. A number of recent studies, while feeling that research should continue, have questioned the current scale of investment. Should we really try to fuse all our eggs in this one basket, which may prove either dirty or bottomless?

9 Energy from Water

Time is of the essence

We do tend to take for granted the flexibility of our energy supplies. In the early hours of the morning demand is low; then we all get up and put on the kettle and make the toast: millions of us, consuming several hundred joules a second each. And a little later still we are using hundreds of *thousands* of joules a second each to travel to work. And so on through the day, the week, and the seasons of the year. The relevance of this to the next few chapters is that they are concerned with those energy sources – tides and waves, sun and wind – which are predictably variable and unpredictably irregular. No matter how plentiful and environmentally beneficent it might be, a new source which provides energy between midnight and six a.m. on occasional Sundays in Lent, and provides it in a form which cannot easily be stored, will never solve our problems. So before moving on to the 'renewable' sources we might look at some possible ways of storing energy in reasonably large quantities.

Most of the new sources are seen as potential producers of *electric power* and this is a real problem, because it is extremely difficult to store electrical energy *as* electrical energy. It would mean separating large amounts of negative charge from equally large amounts of positive charge, holding the two apart until the energy was needed, and then letting a current flow. It can be done, but sufficient capacity to store even one *second's* output from a large power-station would probably cost a million pounds or so.

Table 9.1 shows some other possibilities, together with two natural sources of energy. It gives the total energy stored by 100 te of material, and if the First Law of Thermodynamics were all, this would be the available energy. But as we know, the Second Law is just as important, and column B reveals its effects. Notice that the energy stored mechanically is converted with little loss, but that in the other cases the efficiency is much less.

It is interesting to see how the storage capacity reflects the fundamental forces at work. Nuclear energy of course provides the most compact store. The following four all depend on the chemical energy

Table 9.1 Natural and man-made energy stores

A The energy stored in 100 tonnes of material.
B The time for which this 100-tonne store could provide, at present-day efficiencies, 600 MW of electric power.

Storage	A (MJ)	B
Natural uranium	60,000 million	one year
Fossil fuel	2 to 4 million	20 to 40 minutes
Steam at 540°C and 2,500 psi	350,000	4 minutes
Spinning flywheel	100,000	3 minutes
Rechargeable batteries	12,000	20 seconds
Concrete at 100°C	6,400	one second
Water at a head of 300 m (1,000 ft)	300	half a second

which binds atoms together: in the hydrocarbons, in the condensing of steam into water, and in the limits set for the flywheel by the forces which prevent it from flying apart. The battery also uses chemical energy, and its much lower storage capacity suggests that improvement should be possible – as indeed it is, but not cheaply yet.

The final two entries are not quite comparable with the others because they both depend on arbitrary choices: of the temperature of the concrete and the height of the water. Increase either and the stored energy will rise proportionately. We have already seen (Chapter 6) how to calculate gravitational energy. It is about ten joules for every kilogram raised through a metre:

$$\text{joules} \simeq 10 \times \text{mass in kilograms} \times \text{height in metres.}$$

There is no Second Law limit on the efficiency, and hydroelectric turbogenerators can convert over four-fifths of the stored energy into useful output, so it's not surprising that pumped storage is almost the only large-scale system in use for dealing with surplus electrical energy. Fig. 9.1 shows how it can work, and although the overall losses in transmission, pumping and generation may be as much as a third, it is still a more efficient way of dealing with the surplus from, say, a nuclear power-station than just throwing away heat at night. (One argument for Britain's new Dinorwic pumped installation, which will store almost 10 million kWh, is that nuclear capacity will in the future rise above the base-load level.) Of course,

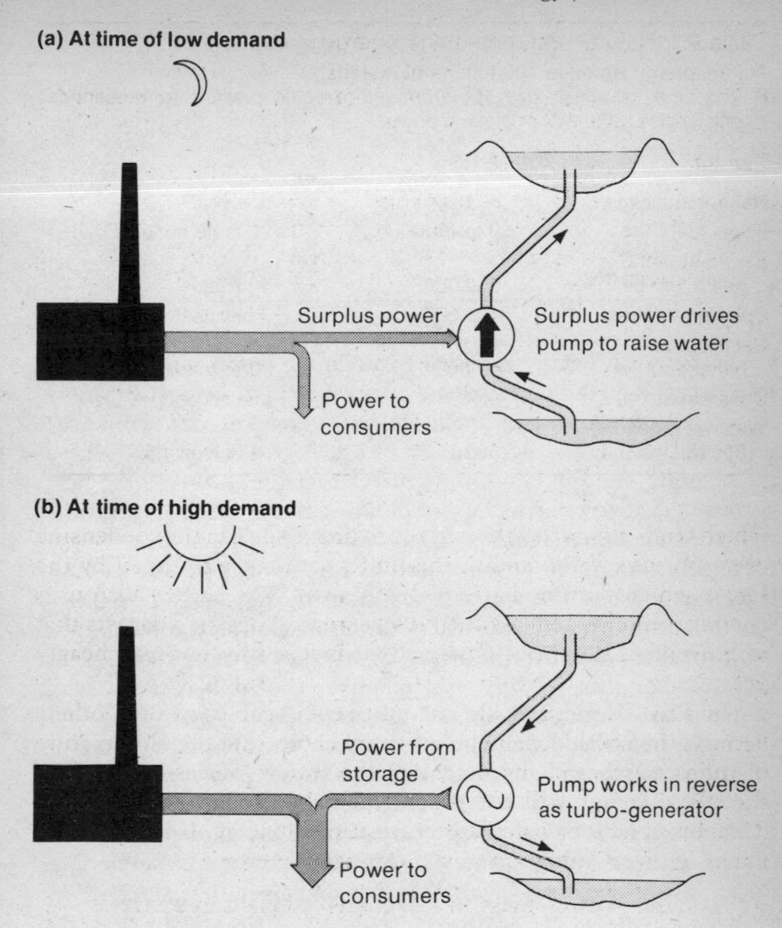

Fig. 9.1 Pumped storage

any hydroelectric system using water from a reservoir has the great advantage that it can respond very quickly to changing demand: no need to bring a large mass of material to a high temperature before there is any output.

Storing energy in hot concrete is not seriously proposed as a way of dealing with electric power, but it is a way of storing *heat*. We have night-store heaters, and then there is the potential as a store for solar energy. Again the calculations are quite simple. We know that things get hot when heated; or more precisely, that the temperature

Table 9.2 Specific heat capacities
A The energy needed to heat 1 kg by 1°C.
B The energy needed to heat 1 cu m by 1°C.

	A (J)	B (MJ)
Water	4,200	4.2
Concrete	800	2
Brick	1,000	2.3
Polystyrene	1,300	1.3
Cast iron	500	3.5

rises when heat energy is added. How much hotter for how much heat depends on the material and how much of it there is. Table 9.2 shows some specific heat capacities: the energies needed to raise the temperature of a specified amount by one degree. These can be used to calculate, for any chosen quantity of material, the stored heat corresponding to a particular temperature rise or vice versa. Insulation is of course the main problem. The average night-store heater used in Britain might be compared to a reservoir with such a large leak that most of the water flows away in the course of a day – whether or not you want to use it. Better insulation is possible, at a cost, but there remains the difficulty that the centre of a solid block will stay hot after the outer regions have cooled. It is not easy to design a large-scale, flexible and effective heat storage system.

For the man-made storage systems then, the present situation is that pumped storage works, provided you have the water and a high place to put it, but that all other methods suitable for electric power need considerable development. Compressed gases are the subject of several current studies, and one 300-MW power-station in Germany, using compressed air, has been in operation since 1978. Flywheels have their advocates, but when we realize that the equivalent of a large pumped system would need tens of thousands of tonnes of material spinning at tens of thousands of rev/min we must doubt whether they have a future in really large-scale systems. (On the other hand, a look at Table 9.1 suggests that if we propose to run cars by taking on electric energy overnight for use during the day, flywheel drive might be a better bet than present batteries.)

Highlands, lowlands, rivers and lakes

Hydroelectric power generation is simple enough in principle. Moving water forces a turbine to rotate and this drives a generator.

Mechanical energy is converted into electrical energy. There is no heat stage and thus no Second Law limit to the efficiency. The ultimate source of almost all our hydroelectric power is the sun. Solar radiation provides the energy to evaporate water from oceans or lakes and drives the convection currents and winds which deliver the vapour to hilltops or mountain-sides where it condenses and falls as rain or snow. It is not difficult to make a rough estimate of the total annual energy potentially available in any region. All we need is the annual precipitation and the height of the land above sea level. Allowance must be made for re-evaporation – important in large countries such as the USA or India where up to half the precipitation never reaches sea level. And of course there is the more difficult judgement of how much development is desirable. No one wants to put every trout river and mountain stream into a pipe or to fill every valley and canyon with a reservoir. Environmental considerations extend beyond aesthetics and leisure activities too: the Aswan Dam has brought benefits, but it has had deleterious effects on agriculture downstream, and even on fishing in the Mediterranean.

The present view of hydroelectric resources seems to be that Switzerland is close to saturation in use of her potential, that there would have to be very compelling reasons for any major increase in Britain and that the USA might at most double her present capacity. India could increase hers by a much larger factor, but at the cost of power transmission over great distances. On the world scene, two regions with large undeveloped capacity are Africa and South America, and the largest existing project, to produce over 10,000 MW, is being built on the border between Brazil and Paraguay.

All these comments on resources refer of course to what we might call 'conventional' hydroelectric systems, using the energy of existing rivers; but this by no means exhausts the potentialities of moving water, as we shall see in the remainder of the chapter.

Alternating currents from alternating currents

Surprisingly perhaps, the entire tidal flow of all the world's oceans dissipates only about a third as much energy each day as does its human population. Like many dissipation processes this one involves friction: the drag of the tidal flow over the sea-bed and along coastlines produces heat at a total rate of a few million megawatts. The effect of the drag is to slow down the Earth's rotation (by about a six-hundredth of a second a century), and any energy which we extract must accordingly come ultimately from this lost kinetic energy.

We have tides mainly because the Earth has a moon. Just as the gravitational pull of the Earth holds the Moon in its orbit, so the Moon pulls on every particle of the Earth. But not all particles are at the same distance from the Moon, so the strength of the pull varies from place to place. If we imagine a much simplified Earth, a sphere with a uniform layer of ocean, the result of these differences will be two bulges, one towards and one away from the Moon. If the bulges keep their orientation towards the Moon while the Earth rotates on its axis, we see that they will pass round and round the surface: a rise and fall of water level twice a day. The real situation is of course somewhat more complex. The movement of the Moon around the Earth means that slightly more than 12 hours elapse between successive tides; and the sun contributes an effect, seen in the regular variation between spring tides, when sun and Moon act together to give maximum tidal range, and neap tides, when they tend to cancel out. Then there is the detailed influence of the geographical features of the Earth itself.

The shapes of coastlines, the effects of islands and inlets, their directions and dimensions, are crucial in determining tides and therefore the energy which we might extract. Detailed flow calculations are beyond even the largest computer, but nevertheless a simple picture drawn from everyday experience will allow us to see why some features might lead to particularly large tidal ranges. Consider the waves you can set up in a bath or bowl by moving your hand to and fro at one end. If you paddle at just the right rate a big displacement builds up, with the water swinging from one end to the other. It is a sort of resonance effect, and the 'right rate' depends on the size of the container. Noting that places with large tidal range tend to be inlets, bays and estuaries – partially enclosed bodies of water – we can look for a similar explanation. The ocean tides could provide the regular 'push', at intervals of just over twelve hours, and if the size is right a large movement will be maintained. If correct, this analysis carries a warning: we must be careful that the barrages and breakwaters of a power installation don't alter the flows and reduce the very effect we propose to use.

Once the body of water is chosen, a rough calculation of its energy potential is possible. We'll take as an example the Rance estuary on the Brittany coast, site of one of the few existing tidal power-stations, operating for some 15 years. The tidal range is about 10 m and the area enclosed roughly 20 sq km, which means about 200 Mte of water stored at an average 5 m above low water level: 10 million MJ, or about 3 million kWh, of stored energy. The next question is how to extract it.

A simple method would be to capture the water at high tide, hold it until the next low tide and then let it out through turbines (Fig. 9.2 (a)–(c)). The pattern of the resulting power output reveals two obvious disadvantages of this simple scheme: the turbines must be designed to handle the very high peak power, and the power is delivered in very short pulses – at different times each day, of course. Is it possible to improve on this?

It is, but at the cost of reduced total output (Fig. 9.2 (e)). The pattern adopted at La Rance, using both rising and falling tides, running the turbines at less than the maximum possible head of water and incorporating an element of pumped storage, certainly gives a more uniform supply. However, the mean output per tide, about 760,000 kWh, is only a little over a quarter of the potential estimated above. To put it another way, the 24 10-MW turbines and generators produce a total average power of only 60 MW. Not, it would seem, a very cost-effective use of machinery.

The two-basin scheme is a possible way round the difficulty. In one simple version, Basin A (Fig. 9.3) is kept topped up at high tides and Basin B emptied at low tides, so that power can be obtained at any time by releasing water through the turbines between A and B. Additional turbines between each basin and the sea will increase the flexibility and the output. Twin basins have appeared at times in plans for two potentially important tidal sites: the Severn estuary in Britain and the Bay of Fundy on the border between Canada and the USA. These two, which must rank as the world's most planned power-stations, have been the subjects of studies for the past half century or so, and the most recent report on the Severn was published, after a two-year, £2.3 million investigation, in the summer of 1981.

The report, after considering proposals for a large single basin with 12-GW capacity and for a two-basin system, recommends that attention should be concentrated on a smaller one-basin scheme (Fig. 9.4). This would generate on the ebb tide only, but the installed capacity would be thirty times that of La Rance. The annual output of 13 TWh means a capacity factor only a little over 20 per cent, so a fair comparison would be with conventional plant of perhaps 2.5-GW capacity. At the estimated £5.6 billion for the scheme, this is equivalent to over £2,000 per kilowatt – more than twice the current cost of present power-stations. Fuel is of course free, but

(a) Water flows into the basin through open sluices as the tide rises

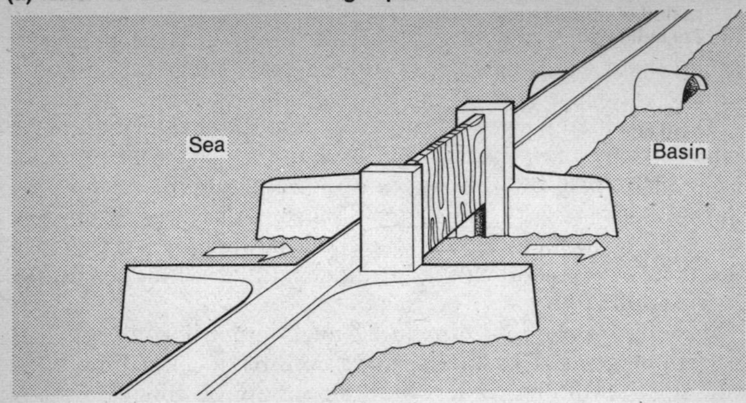

(b) The sluices are closed at high tide, so the basin level stays high as the sea falls

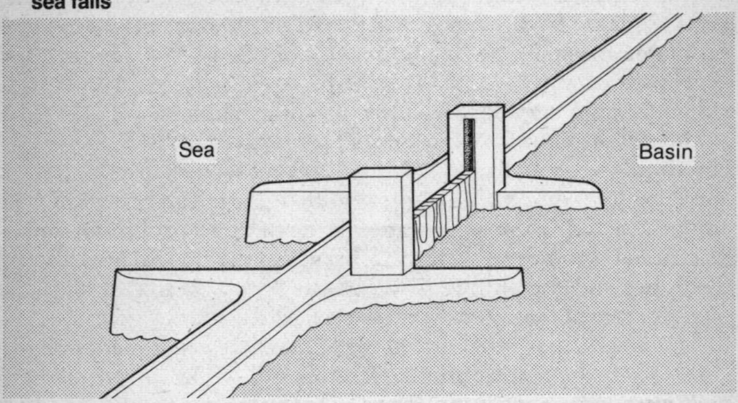

(c) At low water when the head is greatest, the turbines are opened and the water rushes out, generating power at a high rate

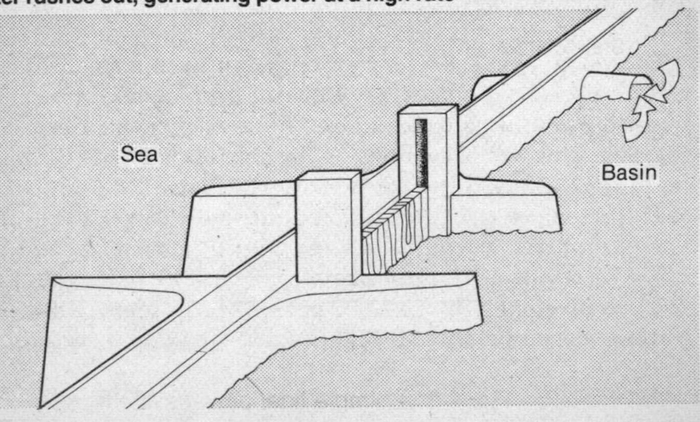

(d) This shows the changing level in the basin, and the power output, over a period of about twenty hours. (The times corresponding to (a), (b) and (c) are indicated)

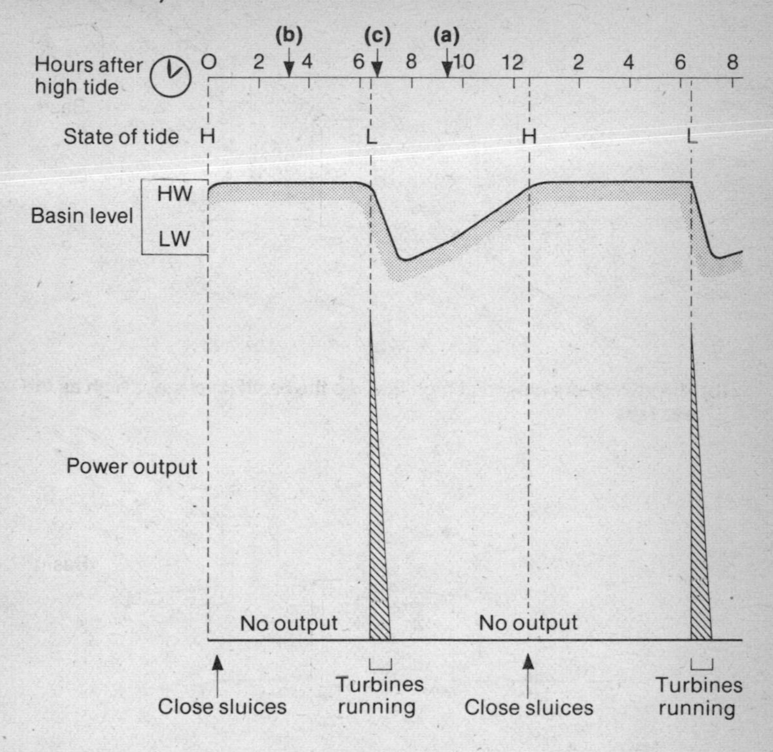

(e) A more uniform output is obtained by generating during the rise and fall of the tide

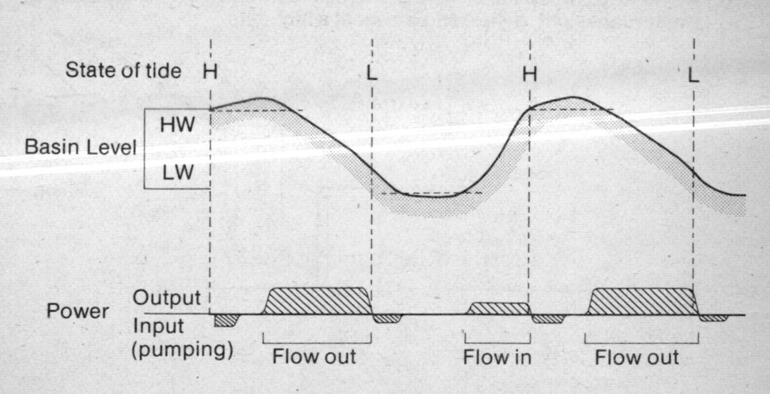

Fig. 9.2 A simple tidal power system

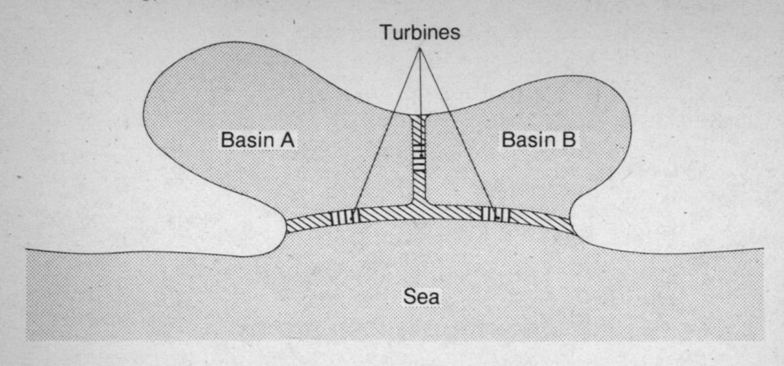

Fig. 9.3 A two-basin system

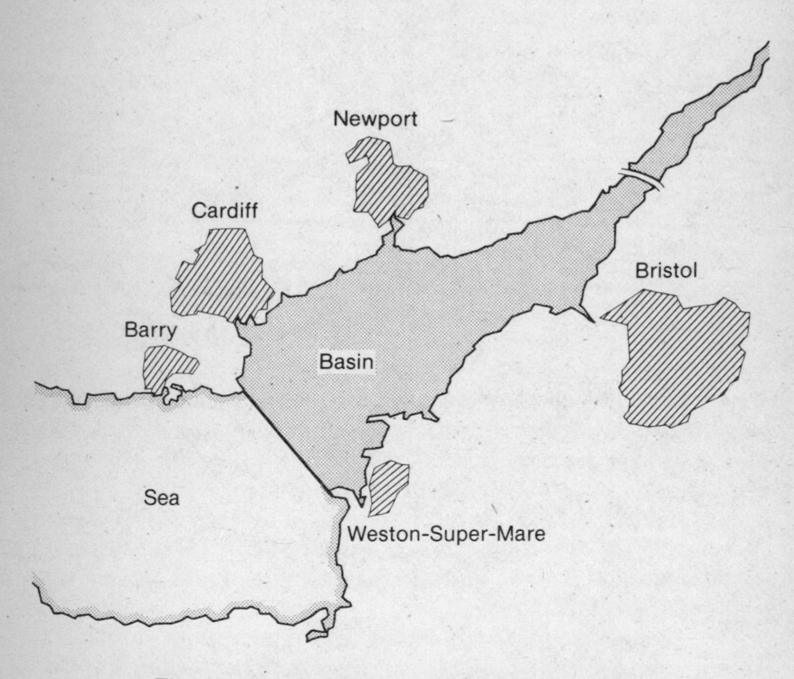

Fig. 9.4 Proposed Severn tidal power scheme

Single barrage, Lavernock Point to Brean Down.
Installed capacity: 7.2 GW from 160 turbines.
Estimated annual output: 13,000 million kWh.
Estimated cost (1981): £5,600 million.

maintenance is not, and whether the scheme will be competitive for the year 2000 depends on the future behaviour of coal prices and on developments in nuclear power. The Study Group regards the proposal as worth further consideration, and points out that it could be the first stage of a two-basin scheme which would extend the generating period to about 20 hours a day.

Little is yet known about the environmental effects. A barrage can change flow patterns in very complex ways; there are problems of land drainage and sewage, and anxieties have already been expressed about the effects on wild life. The more detailed study of these factors, and of the technical feasibility of the scheme, would probably take four years and cost over £20 million, and the minimum construction time would be nine years. The Government has 'welcomed' the report and 'looks forward to receiving comments'.

Wave power

It is not difficult to see that ocean waves are a potential source of energy. Just stand on the shore and look at the sea; or listen to the continuous roar of the surf, day and night, year in, year out, on an exposed beach. The power of the sea, built up by winds over thousands of miles of open ocean, is obvious. Obvious, but how much? What is 'the power of the sea'? How many kilowatts?

Let's start with a simple calculation which uses little more than the evidence of our eyes. Two features of a wave are clear from even a casual glance: the water moves incessantly and is lifted up intermittently. *Kinetic energy* and *gravitational energy* are both present. We'll make an estimate of the gravitational energy first. Consider a long slice of water one metre wide running out across the waves. In Fig. 9.5(a), the water lifted up is the shaded area. (Its volume would of course just fill the corresponding trough if there were no wave.) For the simplified squared-off wave (b), it's easy to see that the raised volume is 80 cu m, and as this is raised through one metre the energy stored (see page 178) is 800,000 J. The quantity will obviously be less for the curved wave, and more detailed calculations show that it is about half as much: 400,000 J.

This gives us a rough idea of the energy per metre width of wave. A complete analysis – rather too complex for inclusion here – shows that the kinetic energy of all the moving water in the same strip is *exactly the same* as the stored gravitational energy. But it also shows that on average only half this total energy flows past during a complete wave cycle. So if the time between the arrival of successive

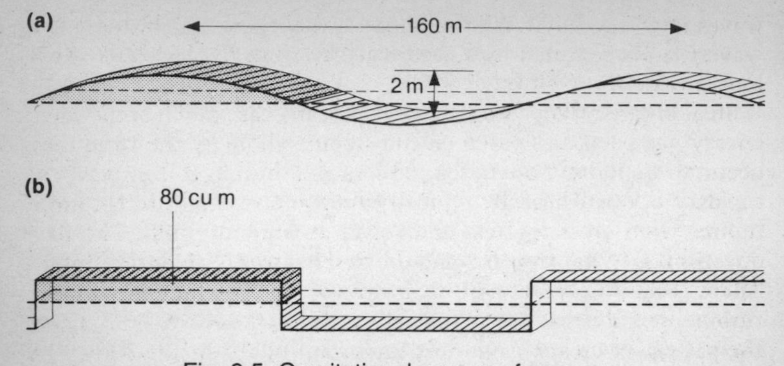

Fig. 9.5 Gravitational energy of a wave

crests is ten seconds (for the 160 m wave) the energy delivered per metre width is 400,000 J in ten seconds, an average power of 40 kW. We still need to ask how much of this could actually be collected, but if the figures are even approximately correct the power is just about large enough to be interesting. A really big power-station – a gigawatt or so – would stretch some 25 km (16 miles) across the waves, so it's not exactly a compact source. Nevertheless, it could mean a fuel-free supply of electric power, and one, moreover, which would normally be greatest at the times of the year when demand reaches its peak; so it does seem worth looking at.

Many people have indeed looked (and hundreds have filed patents) but until recent years few proposals moved off the drawing-board and into the sea. The Japanese have developed a bobbing buoy in which oscillating water compresses air to drive a small turbogenerator whose 100-W output gives enough power for a light, and several hundred of these are now in use. Then, during the 1970s, wave power became the main 'alternative' energy source receiving government support in Britain. Investment of some £10 million over several years has financed research into half a dozen or so devices, in some instances to the small-scale model test stage. The immediately striking feature of these is how very *varied* they are. So far, in our study of water power, whether the source was a fast-running stream or the slow massive movement of the tide, there was just one conversion system: the water forcing a turbine to rotate. Now, with waves, we have things bobbing and nodding, bending and wagging, a sort of tank with cat-flaps, and even a great squashy sack full of air (Fig. 9.6 (a)–(e)). And, significantly, no suggestion that energy might be collected by simply putting a turbine in the sea and letting the waves flow through it. All this suggests that there may be more to a wave than is revealed by a casual glance from the beach, so we'll look a little more closely.

Breakers are rather complicated and in any case much of the wave energy has been dissipated on the sloping shore by the time they occur, so we'll move out to sea and consider the long rolling waves of the deep ocean. These are typically a few metres high, 100 or more metres from crest to crest and travel at some 30 mph. The first question is, 'What travels?' and the first answer is, 'Not the water.' There is definitely not a 30-mph flow of water, which is why the turbine would be no use. What does travel is the *wave* – the shape, the pattern of crests. And most important of all, *energy travels*. A moving wave delivers energy from one place to another.

If it doesn't travel with the wave then, what does the water do? The answer is that it goes round and round in circles. That the water at any point goes up and down is obvious; but if you ask yourself where the water comes from to make a crest (or goes to out of a trough) it is equally obvious that there must be a to and fro movement, along the direction of the wave, as well. Fig. 9.7 shows how the motion of an individual particle of water fits into the pattern of the travelling wave, and we see that the water on a crest is moving forwards, in a trough backwards and on the front slope of a wave vertically upwards – facts well known to swimmers, surfers and sailors of small boats in large oceans.

The actual speed of a water particle is much less than the speed of the wave: it only travels once round its little circle in the time it takes one crest to replace another, so its speed is perhaps 3 or 4 mph rather than 30. The motions of particles below the surface are just like those at the surface except that the amount of movement decreases as you go down. This decay of the wave motion is quite rapid, falling to half the surface movement in little more than a tenth of a wavelength (a depth of ten metres for waves 100 m long).

Now to the problem of designing a wave-power device, and we might start with a couple of very general principles. The first is that, if energy is to be extracted at all, there must be *relative movement*: not only something which is moved by the waves but something else which isn't, or is moved less, or – as in the raft – is moved differently. And then, somewhere in the system there must be a 'one-way' element. Something must ensure that as the water moves up and down or to and fro the energy doesn't go in and out but in and in. Those 'cat-flaps' or louvres which allow water or air to pass must be one-way flaps, or valves.

(a) The rectifier

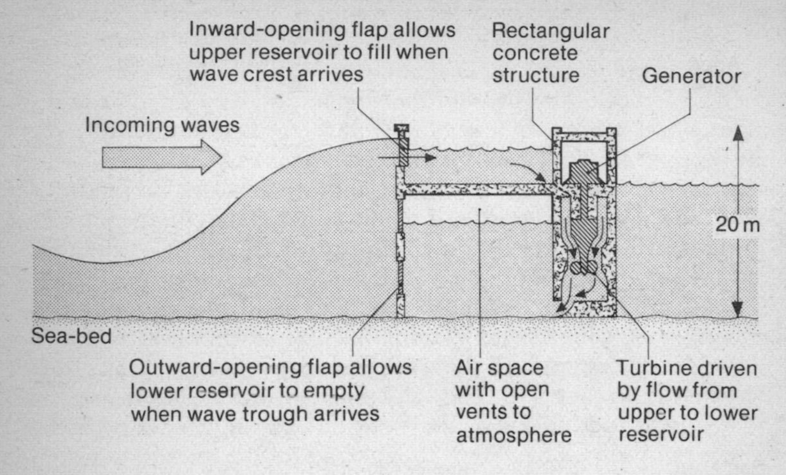

(b) The duct: an omnidirectional oscillating water-column system

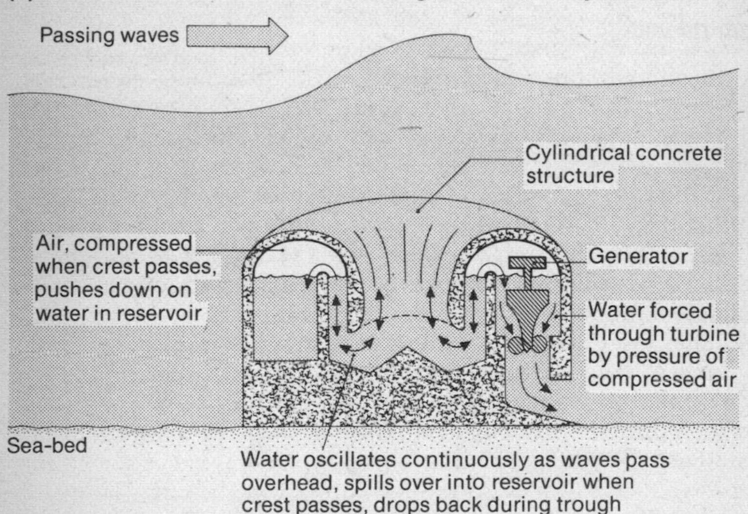

Fig. 9.6 Wave-power devices

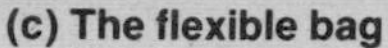

(c) The flexible bag

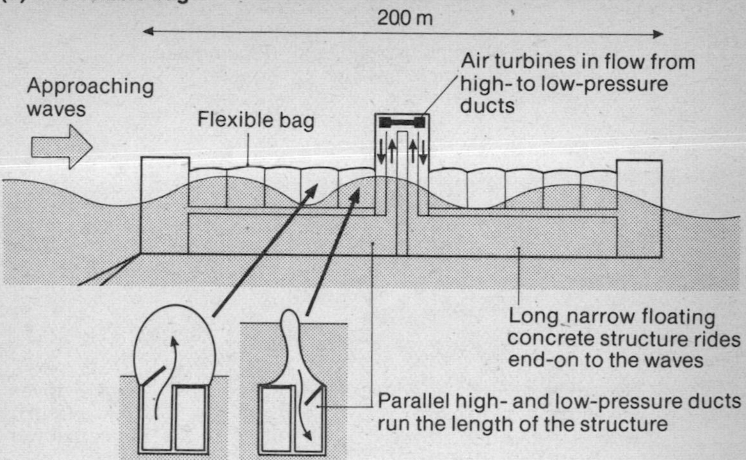

(d) The duck

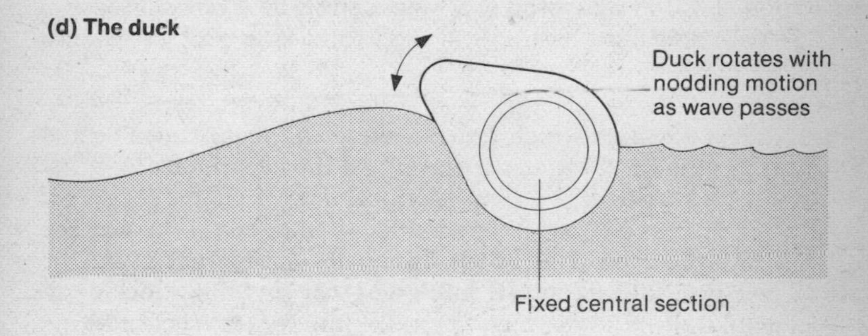

(e) The raft

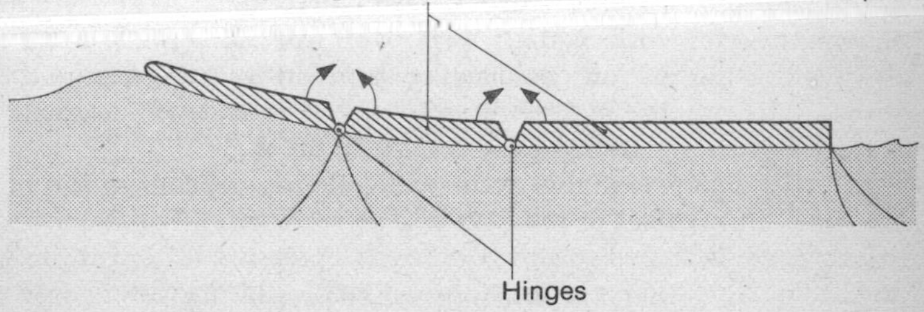

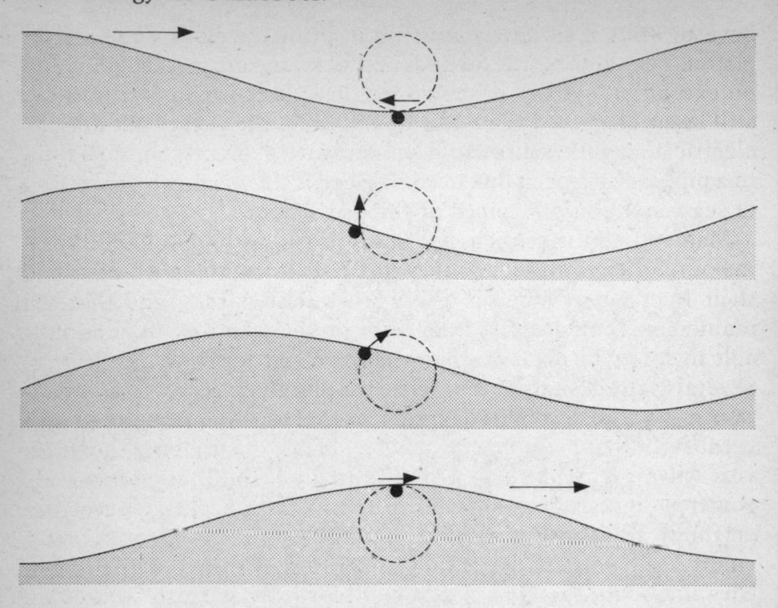

Fig. 9.7 Circular motion of a water particle as a wave passes
The wave height and therefore the circular motion is much exaggerated compared with the length of the wave.

Almost anything which satisfies these two requirements could extract some energy from the waves, but the next important question is the matter of *size*. This is critical not only in the sense that the bigger the collector the greater the output, but in the way the system is matched to the waves. This is obvious for the raft and the flexible bag, but it is equally important that anything which oscillates (like the duck, or any water-column system) should be designed so that its natural time of swing matches the time interval between wave crests. Another question is whether the device needs to respond equally well to waves from all directions. Of the systems shown, the duct works equally well for all wave directions, and the flexible bag and the raft could swing on moorings to point into the waves. The others have a preferred direction and will normally collect less energy from waves coming at an angle.

How the energy is finally extracted obviously depends in part on the basic mechanism of the device. The rectifier and the duct, for instance, are designed specifically for low-pressure water turbines, and the flexible bag for an air turbine; but oscillating water doesn't have to drive a turbine – it could in principle push some sort of piston. The same is true for nodding ducks and articulated rafts, and no extraction system is shown for these because possibilities are still being discussed. Then the final form of the energy need not be electric power. Pressure could be transmitted directly through fluid in a pipe, or hydrogen produced by electrical breakdown of the H_2O of sea water could be piped or shipped ashore.

Many possibilities then, but many problems too. Reliability is a major one. We must remember that these structures would spend their lives facing some of the world's nastier seas, and that the mainenance engineer, far from being on the premises, might not be able to get to the plant at all for weeks on end during winter storms. Several of the designs have reached the present stage of the competition not because they are elegant or extract the most energy but because they are simple and stand a chance of surviving the 'fifty-year wave'. Which, if any, will eventually be built as a large-scale power plant is still very uncertain. Britain has concentrated recently on the flexible bag, an oscillating cylinder, various types of water column and ways of extracting power from the duck. Other countries including the USA, Japan and Norway have also worked on some of these, and on others such as bobbing buoys. The technical problems are undoubtedly soluble – at a cost; but even with major investment it is hardly possible that a full-scale plant could be operating before the year 2000.

Cost estimates for wave power in Britain range from 5p to 20p per kWh: from twice to nearly ten times present quoted nuclear or coal costs. Some studies of possible environmental effects – on fish breeding grounds, for instance – are under way, and there is also debate about the needs for transmission lines across remote parts of the Scottish highlands. Then, even if all other problems are solved, the ultimate contribution is likely to be limited by the uncertainty of the waves. Periods of calm occur even in mid-winter, and heavy dependence on wave power would mean either a major back-up generating system or enormous storage capacity – the equivalent of a few hundred Dinorwics. It is possible to imagine a water-powered Britain, with fifty or more major wave-power stations and huge volumes of pumped storage using not only inland sites but bays, lochs and estuaries, but whether people would want it or be willing to pay for it is another matter. During the 1970s, government spending rose to about £3 million annually, but in Britain as in the USA the 'oil glut' accompanying the recession of the early 1980s has modified any official enthusiasm for alternative energy supplies, and current statements suggest that support will decline rather than grow in the coming years.

OTEC

Leaving the stormy waters of the north Atlantic for warmer tropical seas, we'll finish our survey of energy from water with a look at an entirely different source of power. *Ocean thermal energy conversion* is the name given to a scheme for producing electric power from the enormous reservoir of stored solar energy in the surface layers of the oceans. As we've seen in earlier chapters, the conversion of heat into mechanical or electrical energy is necessarily accompanied by the rejection of waste heat, and there must be somewhere for this to go: any heat engine needs a condenser as well as a boiler. The

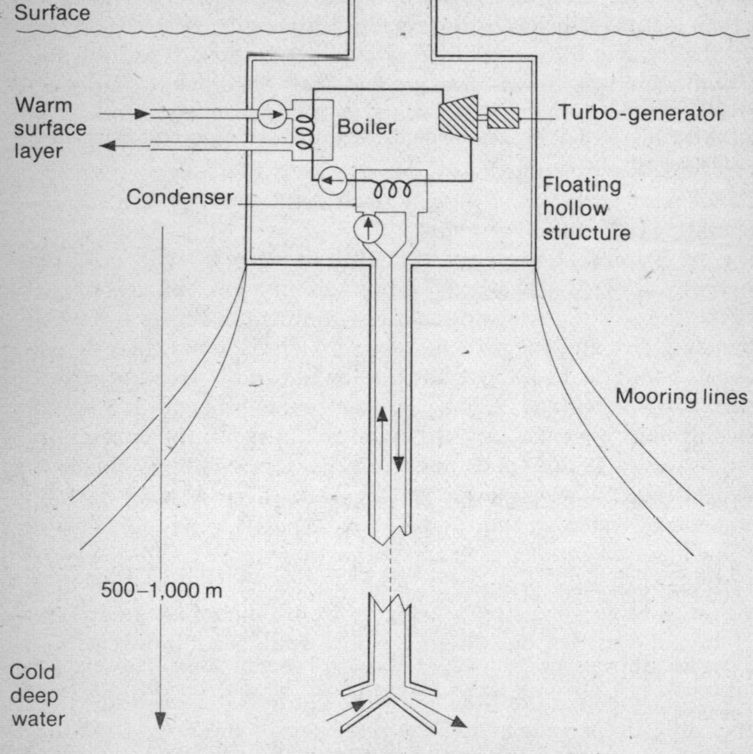

Fig. 9.8 An OTEC power-station

essential feature which makes OTEC possible, then, is that there is a temperature *difference* between the surface water and that at depths of a few hundred metres. In the tropics this can be as great as 20°C (40°F) and although the corresponding heat engine efficiency is low, the total available energy is so large that the scheme is considered to be worth further study.

Fig. 9.8 shows a possible OTEC system, and Table 9.3 gives data for a 600-MW plant. The temperatures are obviously too low to produce steam, so the turbine is driven by a fluid with very low boiling point; ammonia has been suggested, or freon – the liquid used in the heat pumps of domestic refrigerators. As the diagram shows, the system is otherwise just like the steam-driven turbo-generator of a conventional power-station. With suitable allowances for the necessary temperature differences in the heat exchangers (boiler and condenser) we see that for surface water at 25°C (80°F) the efficiency is unlikely to be greater than 3 per cent: less than a thirtieth of the input energy is converted into electric power.

Table 9.3 OTEC data

The diagram shows possible temperatures for the warm and cold water and the turbine fluid.

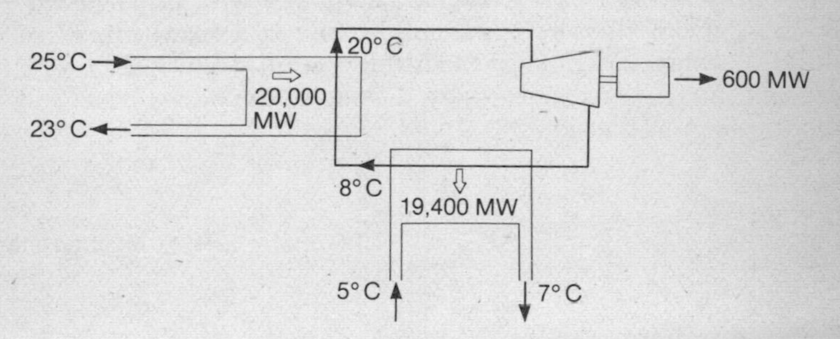

The maximum possible efficiency (heat engine rule) is

$$(20 - 8) \div (20 + 273) \times 100 = 4\%$$

If we assume a maximum achievable efficiency of 3%, the input needed for 600-MW output (600 MJ per second) is

$$600 \div 0.03 = 20,000 \text{ MJ per second.}$$

One tonne of water gives up 4.2 MJ in cooling through 1 °C, so if the warm water is cooled by 2 °C it gives up 8.4 MJ per tonne. Thus an input of 20,000 MJ per second needs

$$20,000 \div 8.4 = 2,400 \text{ tonnes of water per second.}$$

At present no one proposes a 600-MW OTEC plant, but it is interesting to compare the flows with those of other systems we've met. From Fig. 5.5 (p. 95) we can see that the cooling water flow for a conventional fossil fuel or nuclear plant is about thirty tonnes a second; and Table 9.1 shows that a 600-MW hydroelectric plant with a 1,000-ft head needs some 200 tonnes a second. So the OTEC scale is much larger, and even a 240-MW plant (the largest proposed so far) would need flow rates comparable with the world's biggest hydroelectric schemes. The point of this comparison is that flow rates determine size, and size to a great extent determines cost. American investment in OTEC is at present some $60 million annually, and a 'mini-OTEC' plant generating 50 kW has been under test for a few years off Hawaii. It is suggested that OTEC power-stations might be built at a few thousand dollars per kW output – five to ten times conventional power plant. But as with wave power, there are many unkowns still (not least, how to get the power ashore), and OTEC is certainly not going to provide a near-term solution to the energy problems of the USA or any other region.

Nevertheless it remains an attractive idea. Unlike almost any other known system, its input is absolutely guaranteed. It doesn't depend on short-term variations in solar energy because the top hundred metres or so of water act as a massive heat store. There will be seasonal variations but in suitable locations they can be small, so the OTEC plant can run day and night, year in, year out, following demand fluctuations if necessary, as long as the ocean exists and there's a sun in the sky.

10 Solar Energy

A very large source indeed

The attractions of solar energy are easy to see. Sunshine delivers energy as heat and light to the Earth at a rate equal to nearly 20,000 times our entire primary energy consumption; and this is only a tiny part – less than half a billionth – of the total power which the sun radiates continuously in all directions. Yet, on a human time-scale, this massive drain on its reserves has hardly any effect. The unimaginable 4×10^{26} W which the sun radiates means that its mass is decreasing by over 4 Mte in each second (see pp. 149–50) but even at this rate it would take several million years to lose one millionth of its present mass. For once we need not worry about diminishing resources: it seems that we have at last an energy supply which is enormous, continuous and free.

But, with solar energy as with other free-gift offers, you need to read the small print carefully to see exactly what you'll get. On your individual patch of the Earth, it may well turn out to be not all that enormous, it certainly won't be continuous, and it's free only if you forget to count capital and running costs. The first of these claims may seem surprising in view of the figures quoted above, so let's look more closely.

One problem is that solar energy is not very *concentrated*. Even in the sunniest parts of the world, in order to receive the solar equivalent of the output of a 600-MW power-station you would need to spread out an array of collectors reaching a mile or so in each direction – or ten times this area if you allow for present solar-to-electric conversion efficiencies. One difficulty is that even at its brightest and best the sun doesn't actually make things very hot. True, eggs *can* be fried on the streets of New York, and there's an old New England recipe for strawberry jam which tells you to put the fruit and sugar in a flat pan with a sheet of glass on top and leave it in the sun for two or three days. (The glass is very important, and we'll return to it.) But if you want to convert the heat into electrical energy the Second Law of Thermodynamics imposes its usual limits, and as we already know, you need temperatures of many hundreds of degrees to reach even 20–30 per cent conversion efficiency in practice.

But perhaps the point of solar energy is that we don't *need* huge power-stations? Let's try a smaller scale. The average British household uses a little over 700 therms a year of energy in all forms. The solar energy received by an average square foot of British flat surface is about *three* therms a year, so even if you could convert a third of this into useful energy you'd need 700 sq ft of collectors to provide your annual total. And a Californian, with twice the annual solar energy but twice the consumption too, would be in much the same position. Definitely not as compact as a gas boiler.

Another feature of solar power is that the sun doesn't always shine. It is missing at night, and in most parts of the world it is more visible at some times of the year than at others. Consider again the poor Britons, shivering on a grey winter's day. Only *one-sixth* of their annual solar energy arrives during the half year from October to March, while on an average December day that entire 700 sq ft of collectors would receive only half a therm. Things are better for the Californians because the spread is more uniform throughout the year; but in either case there remains the problem of the energy we need before sunrise or after sunset.

So there's a lot of it, but not necessarily where we want it, when we want it or how we want it. The contribution which solar energy can make in the future undoubtedly depends on how well it can be adapted to our needs. At present the technology which comes nearest is probably the *solar hot water system*, and it is certainly the most widely used. For this reason we take it as our main study in this chapter. *Space heating* is much more of a problem, chiefly because when we need the heat we often haven't got the sun. There have been many interesting experiments (and not only in sunny regions), but until the problem of storing large quantities of heat for long periods has been solved, the main lesson to be learned is the rather more general one that a well-designed building can run on far less energy than we use at present (we return to this in Chapter 14). Our second topic is a very different use of solar energy: the generation of electric power. This is attracting great interest, as a technology which may see major developments in the near future and which really could change world patterns of energy consumption.

However, we'll start as usual by looking at the energy itself. What exactly is this gift from the sky?

Electromagnetic radiation

Almost all the solar energy which reaches us over the intervening 92 million miles of empty space comes in the form of *electromagnetic radiation*. (There is a relatively tiny contribution from high-speed particles, mainly the almost undetectable neutrinos. As they are undetectable precisely because they shoot straight through almost everything, including the Earth, we shall not consider them further.) Electromagnetic radiation carries energy in the form of travelling waves – *electromagnetic waves*. We've already met in Chapter 9 the idea that a wave can carry energy; and we've also met the idea of electric and magnetic effects produced by charged particles and electric currents. Now, you can 'generate' radiating circles of water waves by paddling something (your toe, for instance) up and down or to and fro. Similarly with electromagnetic waves. If an electric charge wobbles up and down at some point, a spreading electric and magnetic effect – the electromagnetic wave – travels outwards from it, carrying away energy.

The solar energy sequence is thought to be roughly as follows. Nuclear fusion reactions (see pp. 175–6) release huge quantities of energy in the sun's interior, maintaining its temperature at millions of degrees. The energy travels outwards and is eventually radiated as electromagnetic waves by the rapidly oscillating charged particles of the hot surface region. When the electromagnetic radiation reaches the Earth, the charged particles which make up atoms or molecules are pushed around by the electric effect, picking up energy from the waves. And so you sit, warming yourself in the sunshine.

Everyday experience helps us to add further details. Think what happens when you gradually heat something – a piece of metal for instance. Before there is any visible radiation to see, you can feel that it starts to radiate *heat*. This heat radiation consists of *infra-red* electromagnetic waves, 'infra-red' meaning *below* the red: the oscillations producing the waves, and correspondingly the wave vibrations themselves, are slower than the slowest of visible vibrations, those of red light. As you continue to heat the piece of metal, its particles vibrate with more energy and it begins to glow, first red and then yellow. By the time it is white-hot, it is sending out all the colours of the rainbow, all the visible radiations from red to violet (and a lot of infra-red still, as well). Finally, at extreme temperatures like the 6,000°C of the solar surface, a few per cent of the energy is *ultra-violet* radiation, waves vibrating faster even than violet light (Fig. 10.1(a)).

(a) Colour spectrum of radiation

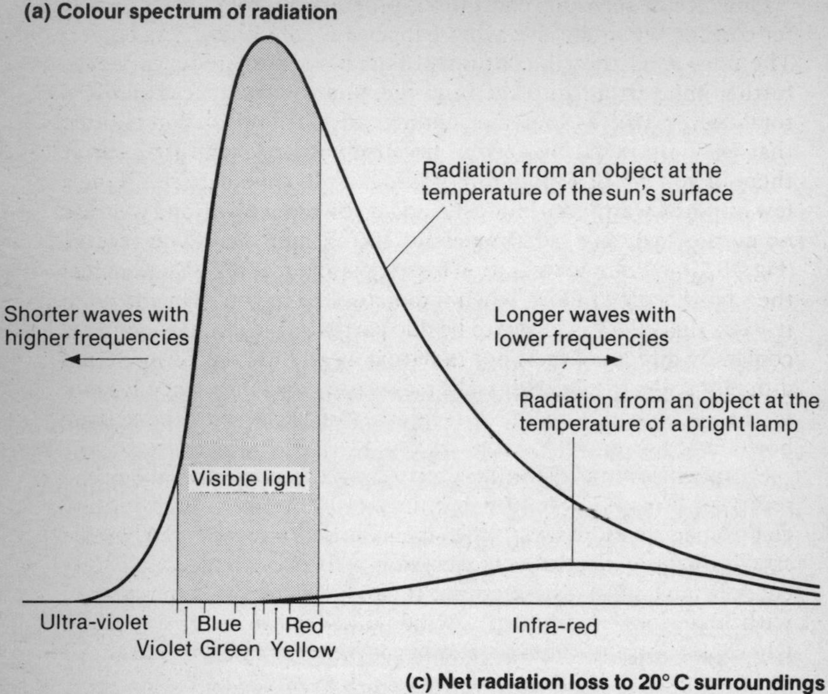

(b) Total power radiated per sq m of surface

Temperature	Power
6,000°C	90 MW
2,200°C	2 MW
100°C	1 kW
37°C	500 W
20°C	400 W

(c) Net radiation loss to 20° C surroundings

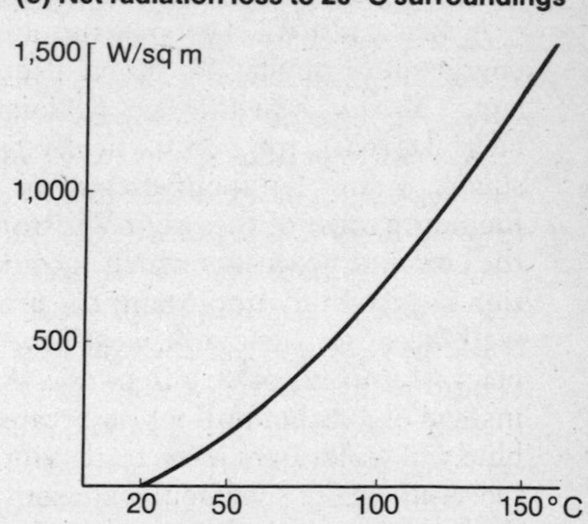

Fig. 10.1 Radiation from hot objects

1. All the data are for 'ideal' surfaces. Real objects will radiate somewhat less power at the given temperatures.
2. The radiation curves for objects in the range 20–100°C would be far in the infra-red and only a thousandth as high as those in (a).

Once we've seen this continuous progression from hotter to hotter objects, we might ask what happens as something gets *cooler*. The answer is that it continues to radiate, but produces waves further and further into the infra-red, and sends out less and less total energy. Unless your surroundings are very hot you don't notice that they are radiating at you, because you are radiating back at them. If you are in a room full of things at 20°C you are receiving a few hundred watts continuously; and if *you* are at 37°C (and wearing no clothes) you are radiating about 100 W more than you receive (Fig. 10.1(b)). Your net radiation loss, (c), is about 100 W. You can feel the 'coolth' of a window, when it radiates less to you than you do to it – but this effect is likely to be due partly to the circulation of air cooled by direct contact: not radiation at all but convection. On a different scale, the Earth must be radiating energy into space at a rate to match the solar energy it receives. Otherwise we'd be getting hotter and hotter.

So much for the radiation emitted by objects. How about the receiving process? One of four things can happen to light (or any electromagnetic radiation) when it encounters matter. It can bounce straight back, *reflected* as in a mirror. It can be *scattered*, with its energy going off in all directions. It can be *transmitted*, passing on with little loss, as through a window. Or it can be *absorbed*, in which case the absorbing substance gains energy and gets hotter. All these happen to sunlight in its interactions with the Earth.

A spacecraft just outside the atmosphere would receive solar energy at a rate of 1,350 W on a square-metre collector facing the sun. The very best that can be done at sea-level, on an extremely clear day with the sun vertically overhead, is about 1,000 W per square metre. So about a third of the original energy has gone, including most of the original ultra-violet. We need not discuss all the complex processes which account for this loss, but one interaction is extremely important: the atmosphere *scatters* light. Just as well, because if it didn't we'd have a blazing sun with an almost black sky in all other directions. And the sky is a beautiful blue instead of a rather dull white because the tiny particles scatter the blue and violet light – the faster vibrations – more than they do the red. And we see spectacular sunsets because the light coming to us through a greater length of atmosphere has lost more of its violet and blue by scatter, leaving the yellows and reds. Clouds, on the other hand, are white because their relatively large water drops scatter all colours equally well.

All this does have some relevance to the practical uses of solar

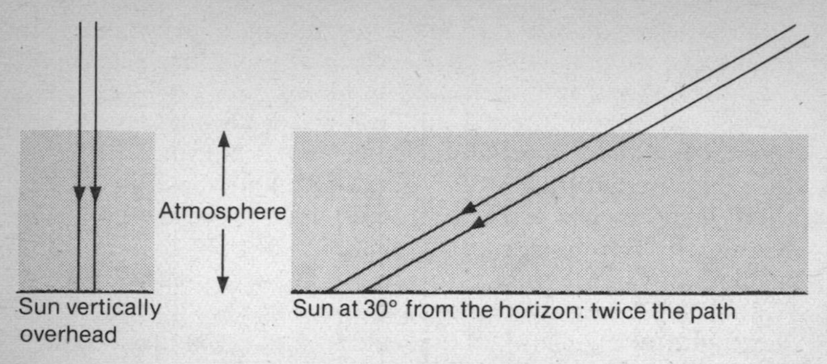

Fig. 10.2 The angle of the sun

energy. There is the obvious point that the lower the sun in the sky, the longer the path of the rays through the atmosphere and the less energy reaches us directly. At 30° above the horizon (Fig. 10.2) the amount of atmosphere between the sun and you is double that when it is vertically overhead, and the energy reaching a square metre of collector is less than 800 W at best. This is why the intensity rises and falls in the course of a day, and also why it falls as you travel to higher latitudes. But if you travel to higher *altitudes* it increases – by some 4 per cent every 1,000 ft at European latitudes. Finally, the fact that clouds scatter light means that even on overcast days we still receive energy from the sky. This scattered light, often called the *diffuse* contribution, can be very important. Even on a 'clear' day in a European city it can provide a third or more of the energy reaching a collector.

Hot water

We shall deal mainly with Britain because it represents the marginal case, and if solar heating is worthwhile for a British household it is worthwhile almost anywhere, unless fuel is extremely cheap – or the population so poor that they can't afford a hot water system of any kind. At the start of the chapter we saw some rather gloomy figures, so we'll begin by looking at these in more detail. Instead of total household consumption, let us just consider the energy needed for hot water: allowing for tank and pipe losses, a little under half a therm a day (Table 3.8, p. 54).

Now let's look again at the solar energy. We'll take the daily average month by month, and assume that flat collectors are used, facing more or less towards the south. Day-to-day variations are of

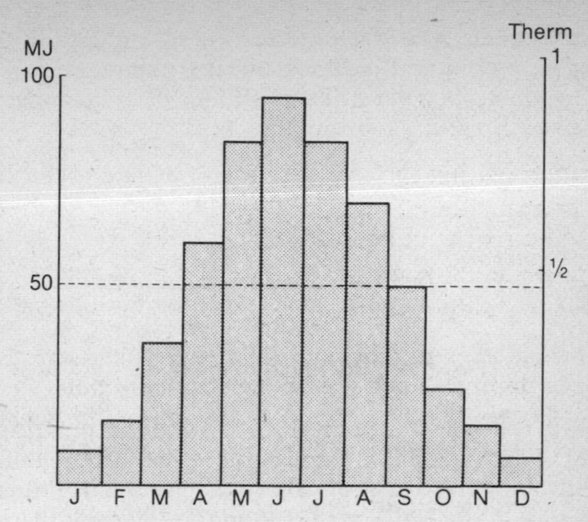

Fig. 10.3 Daily solar energy, southern England
The columns show the average daily solar energy received by a 5-sq m array of collectors tilted to the south, for each month of the year.

course large, and the figures also depend on the environment of the collector, so they are bound to be approximate. Fig. 10.3 shows the data for a fairly modest 5 sq m of collector – about 50 sq ft, which should fit on most roofs. (To avoid awkward sub-multiples, we'll use megajoules, roughly hundredths of a therm, for most of the discussion.) We see that the energy is by no means negligible compared with the required 50 MJ. For the summer months it is more than enough, and even in spring and autumn hovers around a half. Winter looks pretty poor, with probably less than a quarter of the daily demand from mid-November to mid-February. However, the annual total of nearly 17,000 MJ, or about 160 therms, is large enough to be interesting.

There is of course a further question. These figures are for the incoming energy, the primary supply; and if we've learnt one thing it is that there's many a slip 'twixt supply and consumer. How many of those therms will reach the hot water tank?

The answer depends on the collector and the hot water system. Yards of black hosepipe coiled on the roof may have provided the hot water for rural Californians sixty years ago, but on a fresh spring day in England the heat losses could equal the solar gain before the water was even lukewarm. The old method has one thing right, however.

Black surfaces are indeed good absorbers. But it does little about reducing losses and that is almost equally important.

Moving to a modern but still fairly simple flat-plate collector (Fig. 10.4) we see that it has four main parts:

- an outward-facing absorbing surface – the face of the collector plate,
- pipes to carry the heated fluid,
- insulation on the sides and back,
- a sheet of glass on the front.

The principle is pretty obvious. Solar energy is absorbed by and warms the absorber which warms the circulating fluid. Good thermal contact between these two is clearly important, and although the fluid can be either a liquid or a gas, liquids are generally thought to be more efficient. We'll return to the fluid, but first let's study the ways the collector can lose heat once it is warm.

There is *conduction* through any surrounding material, and this is why insulation is important. Then there is *convection*, when air warmed in contact with the outer surface carries away heat. (Wind strength has a marked effect on this.) Finally there is *radiation* from the collector surface.

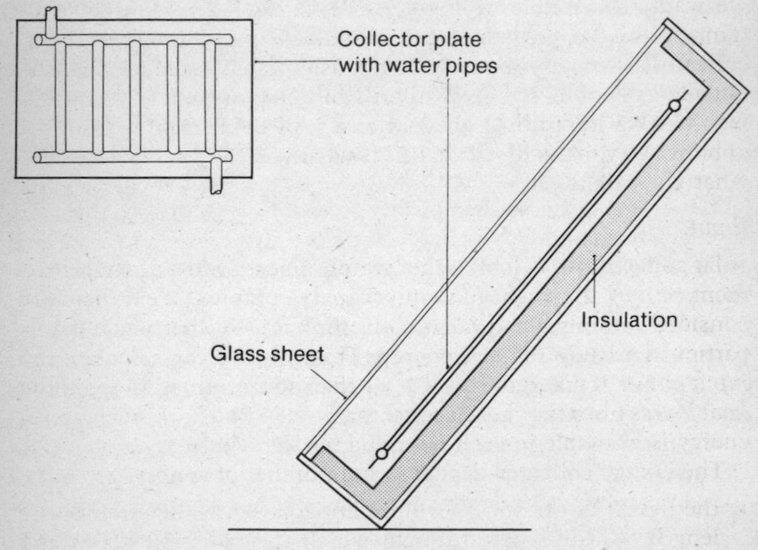

Fig. 10.4 Flat-plate collector

Now we can see the reason for the sheet of glass. First it cuts down convection losses because its outer surface will be cooler than the surface of the absorber itself. Any material would do this, of course, but glass has the well-known and very useful property of being transparent to visible radiation: it lets the sunlight in.Moreover, it has another equally useful if less well-known property. It is *opaque* to infra-red radiation (except for a short region of waves just outside the visible). Remembering the discussion of radiation from heated objects, we know that the absorber, being warm but nowhere near red-hot, will radiate its energy almost entirely as infra-red waves. So a glass sheet will allow through most of the incoming solar energy but will trap nearly all the out-going radiation from the hot absorber. (This is called 'the greenhouse effect'.)

Of course the outer glass itself will radiate to the surroundings, but much less than the absorber would because it is cooler. If we mentally cancel the sun for a moment (behind a thick cloud perhaps), and consider only the exchange of radiation between the collector and the environment – like you and the room in our earlier example – the net radiation loss from a surface at 35°C to surroundings at 10°C will be less than half that from a surface at 60°C.

The not very surprising conclusion of all this is that heat losses will be less the cooler the outside surface of the collector. Naturally we want the circulating fluid *inside* to reach a reasonably high temperature, so the whole art – or science – of designing solar collectors is to get the inside as hot as possible while keeping the outside as cool as possible. Apart from ensuring that the system won't leak water through the roof, rust to pieces in eighteen months or blow away in the first gale, this is what you pay for. Now let's see what you might get.

Inputs

Solar collectors may look rather simple objects, but a general treatment of how they behave is surprisingly complex. We'll therefore consider not only a particular collector but also a particular day: a particular pattern of solar energy. This way we can calculate the *input power* at different times. If we then add information about the *heat losses* from the chosen collector, we can find how much useful energy is available in each hour and for the whole day.

The energy collected depends on a number of factors:

- the height of the sun above the horizon, which determines the length of atmosphere through which the radiation passes and therefore its strength when it reaches the collector,

- the state of the atmosphere: clear, hazy, cloudy etc.,
- the angle which the collector surface makes with the direction of the sun,
- the fraction of energy transmitted by the glass,
- the fraction of energy absorbed by the absorber.

Fig. 10.5 shows the sad decline of the kilowatts – the gradual reduction as we take into account each of these factors. The picture is for a summer's day in the south of England, and all the curves show rates of arrival of energy on one square metre of collector during the course of the day. It is quite important to distinguish between these in discussing useful solar energy, so we'll take them in turn.

The outermost curve (A), with the highest intensity at all times, gives the *maximum possible* energy which could be received, through a very clear atmosphere and with the collector *rotating* so that it always points towards the sun. (This is called a *fully-tracking* collector.) Unfortunately we don't have *very* clear skies in Europe, even on sunny days, so the next curve (B) gives a more realistic account, still for a fully-tracking collector. Of course, if scatter does reduce the direct radiation, there will be a diffuse contribution, from regions of the sky not in line with the sun. This is shown in the low dashed line (B′) and is included in the total shown by B.

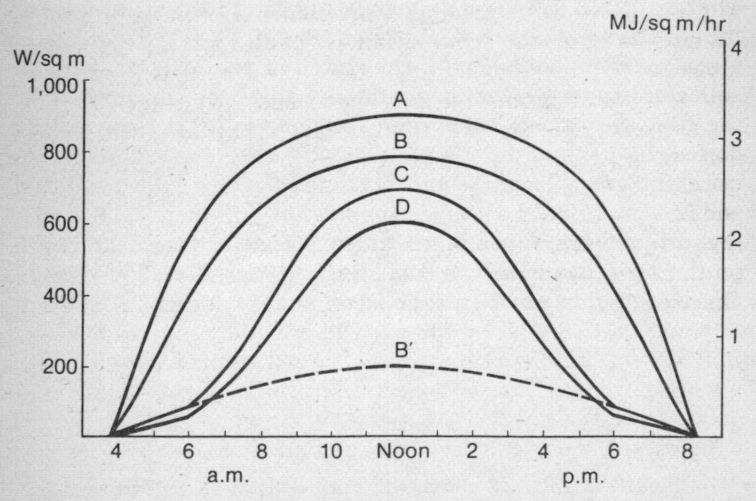

Fig. 10.5 Solar intensity during a summer day
See text for explanation

Tracking systems are complex and expensive, so we must see what a fixed collector can do. Curve C shows this, for a collector tilted at 51½°. (This is the latitude, which means that our collector would face the noon sun in spring and autumn, at the equinoxes.) At midsummer noon the sun is high in the sky and its rays come a little slantwise on to our collector, delivering slightly less than the full 780 W/sq m. And of course when the sun is towards the east or the west there is even more slant until, at 6 a.m. or 6 p.m. for this tilt, the direct illumination is cut off altogether although the sun has not yet set. (A *horizontal* collector would receive more energy in the course of a long summer day, but very much less in winter.)

The final reduction is for the energy which reaches the collector but does not heat the absorber. Some is reflected away and some absorbed by the glass, with total loss of perhaps a sixth if it's a good collector. The remainder is curve D, and we see that the energy input rises to a maximum of just over 600 W/sq m at midday. The total input for the day, however, is only about 5 kWh, so our array (5 sq m) would receive just under one therm.

The sad object in Fig. 10.6 is the mid-winter version of 'D', giving a total daily delivery of about a quarter of a therm. And we haven't yet taken all the losses into account.

Heat losses

The rate of heat loss depends on the temperature difference between the collector and its surroundings, and Fig. 10.7 gives the picture for a reasonably good collector, showing how the losses rise as the collector temperature goes up or the air temperature down. Without the glass sheet these losses would be about twice as high, and the dotted curve shows the effect of a strong wind. As an example of the use of this graph, let us assume that fluid enters the collector at 55°C and leaves at 65°C. Its average temperature will then be 60°C, so if the surroundings are at 18°C the difference will be 42°C. Finding this on the main curve, we can read off the losses: about 250 W/sq m. One thing is then obvious. If the solar input is less than 250 W/sq m,

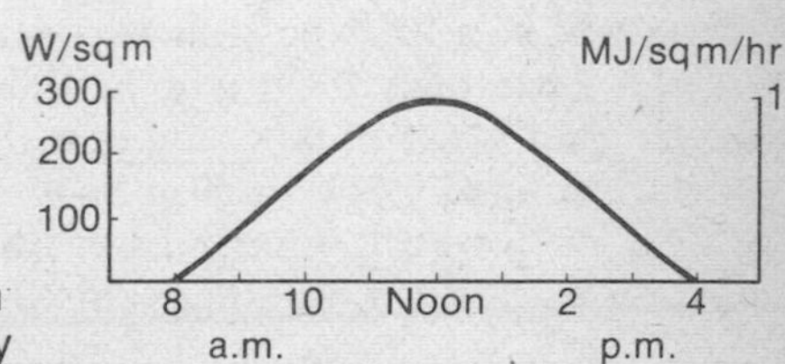

Fig. 10.6 Energy collection rate during a mid-winter day

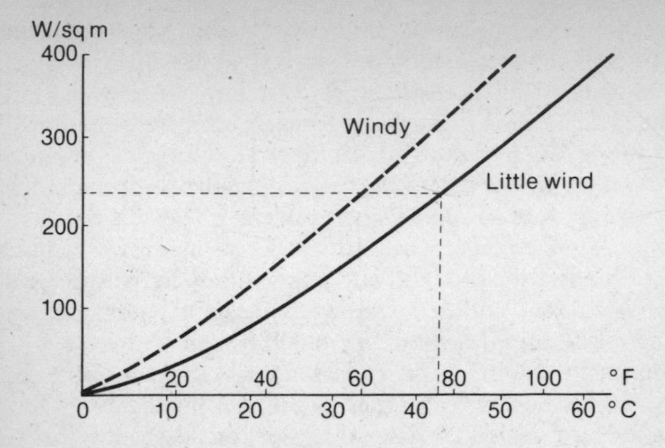

Fig. 10.7 Heat losses from a collector

The curves show how the total losses depend on the temperature *difference* between the collector and its surroundings.

it won't even maintain the collector at this temperature. The losses will be greater than the supply and the temperature will fall.

Daily cycles

In order to continue, we need to make another decision. Is our system *directly* or *indirectly* heated? In the direct system the water to be used circulates through the collector. This has the advantage of less waste heat, but it has two drawbacks. First, the continual supply of new water deposits its 'fur' or other odds and ends in the collector unless measures are taken to prevent this. Secondly, the collector must be drained if there is danger of freezing – unless you don't mind bathing regularly in anti-freeze. The indirect or closed-circuit system avoids these problems, but at a cost in lost heat. There are losses in the heat exchanger where the primary fluid heats the water (Fig. 10.8); and then the fluid needs to be hotter than the final water, which means greater losses from the collector than with a direct system. We shall assume indirect heating, which means that our results will be a little pessimistic compared with direct systems.

To see what might happen in the course of a day let's assume that we want hot water at 60°C and that the fluid therefore leaves the collector at 65°C. If it returns at 55°C, its average temperature will be 60°C. We'll take a summer day, with surroundings at 18°C, and a collector with 85 per cent input efficiency and heat losses as in Fig. 10.7.

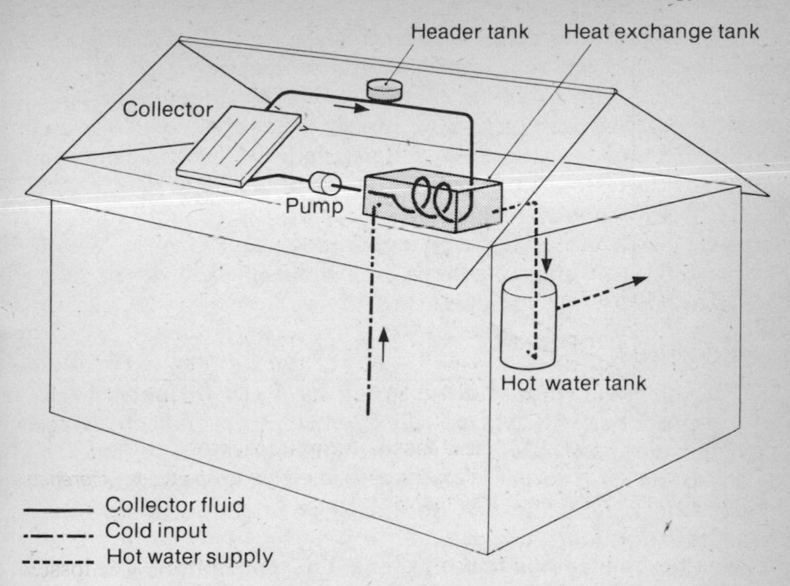

Fig. 10.8 An indirectly heated hot water system

The remaining factor is that it takes energy to heat the collector, which we assume to have cooled overnight. How much depends on how massive it is, and we take a figure of one-hundredth of a megajoule per degree for each square metre. Using the summer day input curve, we find that it takes until a little after 8 a.m. to warm up the collector with no fluid circulation (Fig. 10.9). 18°C is a rather optimistic outside temperature for this hour, but we'll keep it constant to avoid further complication.

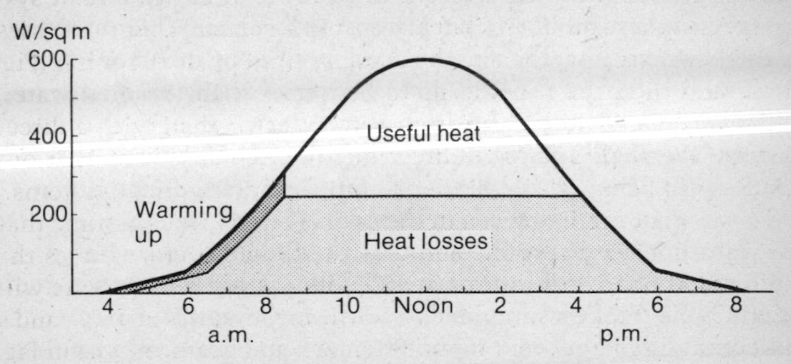

Fig. 10.9 One day's useful heat

When the system is hot we start circulating the fluid, at a rate to carry away the excess energy and keep the temperature at the chosen 60°C average. The losses stay constant at 250 W/sq m, and the extra above this is the useful input. At midday this is about 370 W/sq m, and it then falls until at about half past four the input is only just replacing the losses. After that, except for a little heat to be gained as the system cools, it makes sense to stop the fluid circulation because it will be carrying heat *from* the hot water *to* the collector.

The total heat gain, calculated from the graph, is a little over 2 kWh, or about 8 MJ, for one square metre.

Is it worth it?

The 50 sq ft would provide about two-fifths of a therm in the course of a summer's day. Its average 'collection efficiency' is about 40 per cent. But one lesson to be learned from this exercise is that 'efficiency' is not a very clear term for a solar collector, unless you know the conditions under which it is working. Two brief comparisons will bring this out more clearly.

Another look at Fig. 10.9 reveals that the useful heat would be more if the 'losses' line were lower down. And we know how to lower it: reduce the temperature. If we were willing to accept an average fluid temperature of 50°C instead of 60°C, the losses would drop by a third, giving two-fifths more useful heat. The 'efficiency' has suddenly become 55 per cent!

On the other hand, the above analysis carried out for the winter input (Fig. 10.6) gives no output at all if we assume an outside temperature of 10°C and demand a fluid temperature of 60°C; and virtually no output for 50°C. The same collector, but the efficiency is now zero.

So our rather arbitrarily chosen 50 sq ft of collector just fails to provide enough water for the average household in mid summer, and then gets worse. But perhaps we are asking it to do the wrong thing. As we've seen, cooler is better, and rather than trying to squeeze out a gallon or so of very hot water at very low efficiency, a more effective way would be to use the system to *pre-heat* water to 35°C (100°F) or so. This should halve the bill for other heating fuel.

Is it worth it, then? Some people evidently think so, because there are some 10,000 sq m of collector in use in Britain already – although much of this has been installed by public authorities rather than private householders. In parts of southern California, solar water heating is mandatory in new houses, and householders installing systems in existing houses receive income tax credits. (Very useful if you want to cut down on fuel costs for heating your pool.) Without such incentives the general view at present in Britain is that, provided you have a suitable south-facing place to mount collectors, and don't live in the shadow of buildings or in a particularly cloudy part of the country, or on a very cold, exposed hillside, solar water heating is a worthwhile investment if you are paying for full-price electricity to heat all your water. As fuel costs rise (which they surely will) and system costs fall (which they should with increased production) the balance is likely to shift, and it is no longer only the sun freaks who foresee 10 or even 20 per cent solar contribution to Britain's hot water by the end of the century.

Solar power

Let us design a solar power-station. We start with an idealized system to see if it will work at all, and Fig. 10.10 shows the general idea. Solar energy is absorbed, heating a fluid, and the hot fluid drives some form of turbogenerator. The maximum possible efficiency is determined by the usual heat engine rule (p. 75). It is equal to the temperature difference between the 'boiler' (the solar collector) and the condenser, divided by the temperature of the collector measured from absolute zero. Which means, of course, that it increases with increasing collector temperature. Unfortunately, so do the heat losses from the collector, and it's easy to see that we have the perfect Catch 22 situation.

Suppose that the solar input is a good healthy 1,000 W/sq m. Fig. 10.1(c) shows that the net radiation loss is just this much if the (unshielded) collector surface is at 120°C. We therefore have a maximum possible temperature – but not a very useful one because

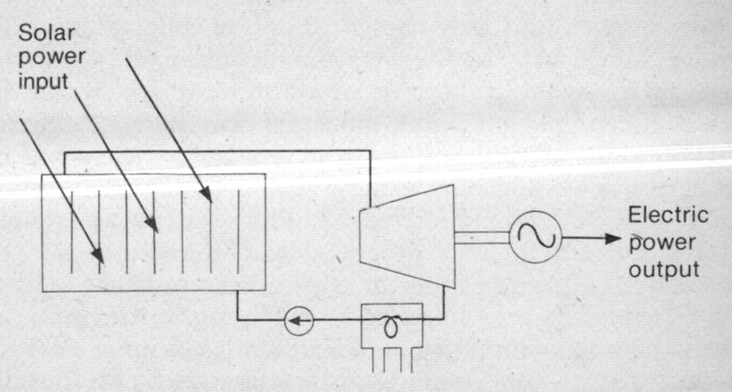

Fig. 10.10 A solar power-station

there is no energy left for the turbine! So we must reduce the collector temperature. But that reduces the turbine efficiency, and going too far in this direction leaves us with virtually no output from the turbo-generator. There is an optimum between the two extremes, but it is easy to calculate that, for a collector with the losses shown, the best possible efficiency is less than 10 per cent (Fig. 10.11). And this for an input corresponding to high noon in the Arizona desert. If the solar input is less, the efficiency will be even lower. It seems that we must look for some improvements.

There's nothing we can do about the Second Law of Thermodynamics; and as we've seen elsewhere, not a lot more to improve turbines. So we'll look at the input end of the system, and there are two possibilities: reduce the losses or increase the input. We've already discussed the reduction of losses in a simple domestic collector, and there is much more which can be done in this direction – at a cost. Special selective coatings on the absorber and the outer windows, and removing the air from the region around the absorber to reduce conduction and convection losses: these can bring the total losses to a tenth or less of their original value. But the major improvement in collector temperature comes by increasing the input, and the method is very simple. Instead of pointing the collec-

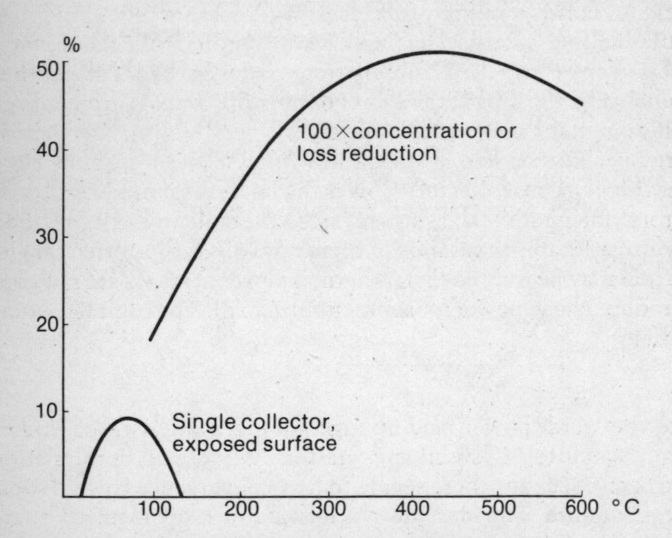

Fig. 10.11 Solar system inefficiencies

The curves show how the overall efficiencies of idealized solar power plants depend on the temperature of the fluid in the collector.

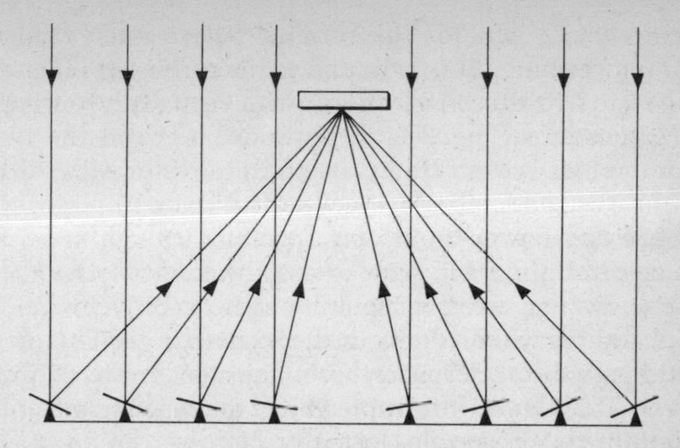

Fig. 10.12 Concentrating the power with a mirror array

tor at the sun, you use a large array of mirrors and reflect the energy which falls on them up on to it (Fig. 10.12). Instead of 1,000 W, each square metre of collector receives ten or even a hundred times this. It is important to see that this is not the same as spreading out ten or a hundred times the number of collectors. By *concentrating* the power, the operating temperature is raised, and the conversion efficiency correspondingly increases. Fig. 10.11 includes a curve for a hundred-fold increase in input (or a hundred-fold reduction in losses, or any equivalent combination), and we see that at last there is a chance of working at the sort of temperatures that turbines need.

There remain many problems. It's all very well to say that the energy is absorbed and heats a fluid which drives a turbine. How? Does it boil in the collector? How is the heat transferred? And those mirrors which reflect the sun's rays into the collector. All day? Then they must rotate; thousands of them, and all at the correct angles. The fuel may be free, but solar thermal power-stations are not going to produce cheap power for some time, if at all. The capital costs are too high.

SPS

Take the word 'power' and arrange around it any two or three of 'solar', 'satellite', 'system' and 'station'. Almost all combinations have been used, but there seems to be a convergence towards Solar Power Satellite. The idea was put forward in 1968, inspired by two well-known facts: the intensity of solar radiation is at least a third greater outside the atmosphere, and the sun need not set on a satellite. The satellite would be in *synchronous orbit*, circling vertically above the equator at a distance such that its time for one orbit is 24 hours, keeping it far out but always over the same point on the surface. Using 10–20 square miles of collector (Fig. 10.13) the SPS would gather enough power to generate more than 5 GW, with either a thermal system, or more probably photovoltaic cells (see p. 216).

This electric power would be converted on the satellite into *microwave radiation* and beamed to Earth in this form. The point about microwaves, electromagnetic waves in the radar region (vibrating nearly a million times more slowly than light), is that they travel through the atmosphere, including clouds or haze, with very little loss. They would be received by an antenna array eight or ten miles long by six miles wide – usually called the *rectenna* – and then converted into electric power at normal voltage and frequency.

The scheme has been supported by the space industries, and a $20 million study in the USA in the late 1970s found no insuperable technical problems, although the cost per kilowatt might be ten times that of conventional power-stations. It does seem that there are still a few outstanding difficulties however. The system using the power must be able to tolerate the sudden loss of 5,000 MW if

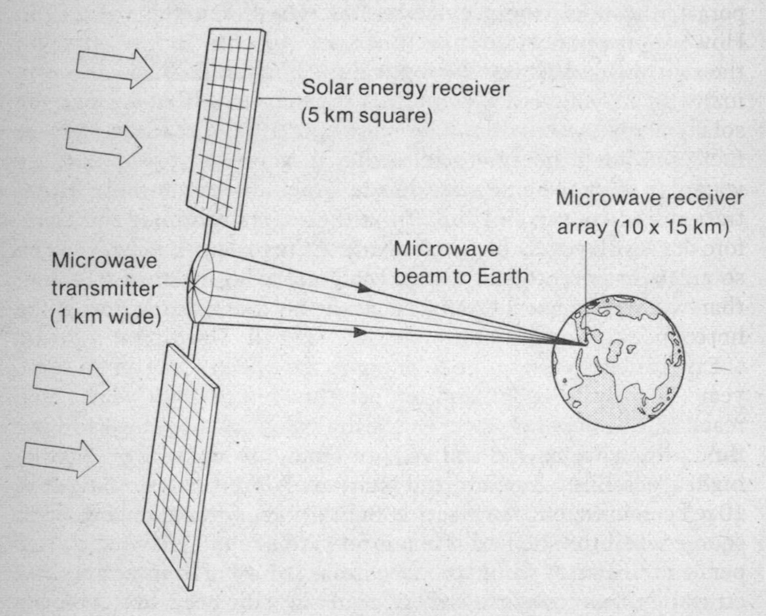

Fig. 10.13 The solar power satellite

something goes wrong. (In fact the power will predictably vanish for up to an hour every night for a few weeks in spring and autumn when the Earth eclipses the satellite.) Then it seems that when there is power we'd better use it. Otherwise the rectenna reradiates energy, and no one seems quite sure what will happen to that. Another main concern is what happens if, as one American scientist put it, 'the beam twitches and zaps LA'. The problem of finding sixty square miles of land well clear of people certainly exercises the minds of European supporters of SPS. At present, however, the system seems to be losing rather than gaining support, and we shall risk leaving it without further discussion, as a potential rather than probable future supply.

Saline solar ponds

One way around the difficulty which we've seen with simple flat collectors, that they have low efficiencies and therefore need large areas, is to look for large areas. The idea of using sheets of water as collectors has been considered for a long time, but the problem is that the hottest water, being the lightest, will rise to the top, and the heat losses from the surface will therefore be very high, with evaporation making them even greater than for a solid collector. However, it was noticed some thirty years ago that in very salt lakes the stirring-up does not happen and the temperature consequently *increases* as you go down towards the bottom where most of the solar energy is absorbed. The reason for this is that the density increases with the concentration of dissolved salt (ordinary sea water is a few per cent denser than fresh water). So a natural concentration gradient can build up, with the most concentrated and therefore densest layer at the bottom; and the density difference may be so great that even when heated the bottom layer remains heavier than the water above. So it stays at the bottom, getting hotter and hotter.

Experiments with saline solar ponds have been going on for some years, mainly in Israel, and temperatures up to 100°C have been reached. The very hot water can be drawn off and used to vaporize a fluid which would then drive a turbine. Obviously 100°C is the highest possible temperature, which makes it difficult to reach even 10 per cent efficiency in practice. But the heat input to a pond 300 m square when the solar radiation is 500 W/sq m is 45 MW, so even 2 per cent efficiency would produce nearly 1 MW of output. Very low efficiency does not necessarily rule out a system, provided it is cheap enough. A number of analyses have suggested that, for large-scale power generation, it is not the power tower or the SPS but the solar pond which is the one to beat. There are still problems to solve, and it is clearly a system for places where space is available, and enough water to make up the evaporation losses; but a programme aiming at a few megawatts of electric power is now underway, and it will be interesting to see the results.

Photovoltaics

From the massive power tower of the solar thermal installation to a tiny silicon flake a few millimetres wide, generating a voltage when light falls on it. From the bludgeon to the rapier. If our technological instincts are right this simple, *direct* conversion from radiation to electric power must surely be superior to boilers, turbines, generators and tonnes of hot steam. Sadly, they are wrong. The rapier is no more efficient than the bludgeon. The millimetre solar cell generates only microwatts, and for megawatts we need hundreds of square metres, just as before. Yet of all the new energy systems the solar cell has probably the greatest potential for really wide-scale use. A few square metres per kilowatt is not necessarily a serious problem, and countries still developing full electricity supplies could avoid all the paraphernalia of large power-stations, transmission networks, transformers and the rest by installing clusters of solar cells supplying power as and where it is needed, for a town or village, a factory or even a single household.

Naturally there is a catch, or rather, there are several. The first is cost. At present, solar cells are about ten times as expensive as nuclear power-stations. There is, as we have seen, a great deal of debate about costs, but most figures hover around one pound a watt. Against this, the present cost of solar cells is about five pounds a *peak* watt, where 'peak' often means the output corresponding to a very high input – 1 kW/sq m or so. Now even the sunniest regions accumulate only some 2,200 kWh/sq m in a year, which means an 'availability factor' for the cell of about 25 per cent (2,200 hours of a possible 8,760). Thus a fair comparison should at least double the cost-per-watt figure.

The £50 an ounce you must pay for solar cells is certainly not due to scarcity of raw material. Silicon is the second most common element in the Earth's crust, and reserves tend to be listed as 'unlimited'. The cost rises in two steps. The first, taking it to about one pound an ounce, is the production of extremely pure silicon – one 'wrong' atom in a million would be regarded as very poor. Then this must be melted and re-formed very slowly and carefully so that its atoms are in an almost perfect crystalline array. The need for large, high-quality single crystals is the real difficulty, and much effort is going into attempts to produce efficient photovoltaic cells with less demanding structures.

To explain exactly why crystalline perfection is so important would require a more detailed picture than can be developed in a brief paragraph or two, but a much simpler outline of the processes in a solar cell will be enough to show why the efficiency is bound to be poor. The name is a good starting-point: *photovoltaic* because when light falls on it a voltage is generated. Like a battery, it will send a current round any circuit connected to its terminals (Fig. 10.14). In the earlier discussion of electric currents we saw that a voltage supply does two things. It maintains a voltage difference so that electrons passing through it are 'lifted up', and it continuously replenishes the energy which is used or dissipated as the electrons travel round the circuit. In the photovoltaic cell this energy comes from the light, and the basic process is that it frees an electron from its state of atomic bondage, allowing it to move off through the crystalline lattice. Solids which need just a little energy to free their

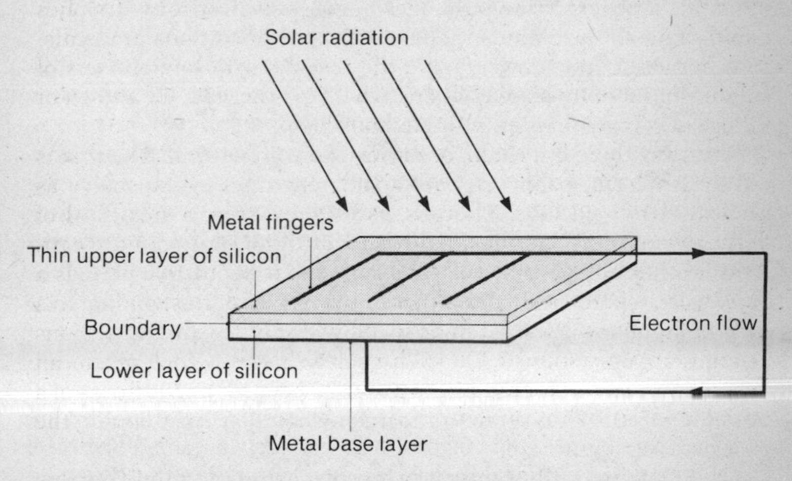

Fig. 10.14 A photovoltaic cell

Electrons are released near the boundary by solar radiation which penetrates the thin upper layer of silicon. Both silicon layers are single crystals but they incorporate very small amounts of different 'impurity' elements and this leads to a voltage across the boundary, which maintains a flow of electrons around any external circuit.

electrons in this way are the *semiconductors*: germanium, silicon, and various compounds such as lead telluride, gallium arsenide, etc.

Unfortunately, there's a very basic reason why photocells have low conversion efficiencies for solar radiation. It has to do with the well-defined quantity of energy needed to free an electron. If the amount offered is less than this the electron stays bound, and if it is more the excess is wasted, dissipated as heat. And now, in the middle of an account of a technological device, we meet one of the more revolutionary ideas of modern physics. Light, it seems, is not quite the smooth flow of radiated energy which has been pictured. There is a great deal of evidence (including the type of photocell behaviour we are studying) that it delivers its energy bit by bit, in very precisely defined chunks; and these amounts of energy, the *quanta*, are related in a very simple way to the particular frequency of the oscillating wave. A wave vibrating at a certain number of cycles per second delivers its energy in quanta of a certain number of joules, never more and never less. Double the frequency (by going, say, from infra-red to ultra-violet) and you double the quantum of energy.

The consequence for the photocell follows immediately. There will be just one type of radiation, just one frequency, which provides exactly the energy the electron needs. A lower frequency will be useless and a higher one wasteful. But solar energy includes a whole range from far infra-red to ultra-violet, so there is bound to be some loss.* The 'ideal' radiation for silicon is infra-red, just outside the visible region. About a fifth of solar energy at the Earth's surface comes at lower frequencies and is therefore no use at all, and nearly three-quarters of the rest becomes wasted excess energy. A silicon cell can never convert more than 21 per cent of solar power into electric power. Indeed, if we subtract losses for light which is reflected off the surface and for electrons which never reach the outside circuit, we reach a limit of about 18 per cent in the best possible case.

And this best possible case means an almost perfect single crystal. Any impurities or imperfections increase the electron loss and further reduce the efficiency, so there is clearly a trade-off between cost and useful output. Much of the development in the 1960s was directed towards the use of solar cells on spacecraft where cost does not seem to have been a major consideration, and efficiencies close

* The fact that radiation from a hot object *must* cover a range of frequencies follows from the Second Law of Thermodynamics, which is therefore making yet another appearance in its usual role of limiting the thermal efficiency.

to the 18 per cent limit were achieved. To see what this means, take for instance a single cell half an inch across, an area of 1.6 sq cm. With a solar intensity of about 600 W/sq m this would receive 0.1 W and give an output, at 18 per cent efficiency, of just under a fiftieth of a watt.

Returning to Earth, what happens when a cloud moves across the sun? The power falls, of course – instantly; which introduces another problem. Suppose you have a battery of 600 half-inch cells, connected as a 6-V supply, giving 12-W output when the solar intensity is 600 W/sq m, and you are running equipment which needs just this: it carries a current of 2 A when connected to a 6-V supply. Then the sun goes in. The power drops and the cell can no longer supply 2 A at 6 V. The result is like trying to start a car on a flattish battery – the voltage falls. It is obvious that any system, household or larger, which relies on solar cells for electric power needs an instantly available alternative generator or adequate storage capacity in rechargeable batteries. Either will of course add further to the cost.

Despite all this, there are plenty who see photovoltaic power as having great potential. The problem of small-scale fluctuations in power output can to some extent be overcome with suitable control circuits, and these become cheaper every day with the rapid development of microelectronics. New methods of production are bringing down the cost of single crystal cells, and another approach reasons that if the cells are cheap enough, even efficiencies below 10 per cent might not rule them out. One possibility is an amorphous (not crystalline) layer of semiconductor deposited on plastic film, which could be really cheap. Solar cell cost has fallen by more than a factor of ten in the past decade (which can hardly be said of any other power system) and there does seem to be justification for the view that it may be below one dollar a watt before 1990. Perhaps we won't see a third of America's power from roof-top photocells by the end of the century as some predict, but it is true that even at present costs more and more houses in remote areas, from Alpine chalets to Alaskan homesteads, are blossoming with little patches of silver-grey disks. A solar cell is, after all, about the simplest power supply in existence. It consumes no chemicals and has no moving parts, needs no maintenance, and provided you don't drop a brick on it or cook it, will quite possibly outlive you.

11 Wind Power

The winds that blow

The energy of the winds is stored solar energy. A few per cent of the radiation reaching the Earth is absorbed in the atmosphere, and the resulting uneven heating of the air leads to large-scale circulation patterns. During most of the year the tropics receive appreciably more energy per square metre than the polar regions, so the redistribution of this over the Earth's surface before it is radiated back into space means a massive energy flow, carried partly by the great ocean currents but mainly as the kinetic energy of moving air – the winds. This energy is gradually dissipated by friction with the ground, but in low latitudes the total annual flow is a few hundred thousand exajoules a year, several hundred times the present world total energy consumption.

Not all this is available, of course. For a start, we don't (yet?) know how to build mile-high windmills, so we can only count as useful the one per cent or so within 1,000 ft of the ground. And if we further discount the flows far out at sea, and in other inaccessible regions, we reach a figure which is about equal to our present energy use. The calculation is very rough, but similar results come from more detailed studies; of the USA, for instance, and of Britain (whose smaller land area per unit of energy consumed is partly compensated by greater average wind strength). So once again we have a renewable source which is just about large enough to be interesting. And once again we must ask how, when and where the energy is available, how much of it we can extract, and what it will cost in financial and environmental terms.

Calculating the energy delivered by the wind is easy. It is *kinetic energy*, and we know (p. 104) how to find this for any mass moving at a known speed. Given that the mass of a cubic metre of air is a little over a kilogram and that a mile an hour is a little under half a metre per second, you find that each cubic metre of air in a 20-mph wind carries slightly more than 50 J of energy (Table 11.1). Now suppose we put up a 4 ft-diameter windmill – an area of about one square metre facing the wind (Fig. 11.1). The calculation in Table 11.1

Table 11.1 Power from a 20-mph wind

Data: wind speed = 20 mph ≃ 9 m/s
mass of air in one cubic metre = 1.3 kg
diameter of windmill = 4 ft ≃ 1.2 m

kinetic energy of one cubic metre of air: $\frac{1}{2} \times 1.3 \times 9 \times 9 = 53$ J

area of windmill blades: $\pi \times 0.6 \times 0.6 = 1.1$ sq m

volume of air reaching blades in each second: $1.1 \times 9 = 10$ cu m

energy reaching blades in each second: $10 \times 53 = 530$ J

530 J/s = 530 W of input power

reveals that the wind power input is roughly 500 W. But Fig. 11.2 brings out the extremely important fact that the delivered power rises *very steeply indeed* as the wind speed increases. It's not difficult to see why. The kinetic energy of a given mass of air depends on the square of its speed, and then the mass arriving each second depends on the speed again. So doubling the speed multiplies the power by eight; tripling it, by 27; and so on.

The next question is how much of this power can be extracted. It obviously can't be the lot, because if you stop the wind altogether, where does the air go? On the other hand, there must be *some* reduction in speed if energy is to be extracted at all. Between these extremes there must be an optimum, and the condition for this

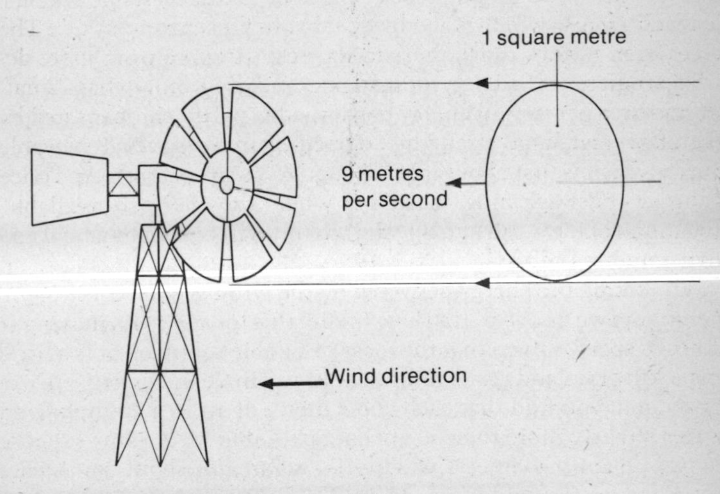

Fig. 11.1 A 4-ft diameter windmill

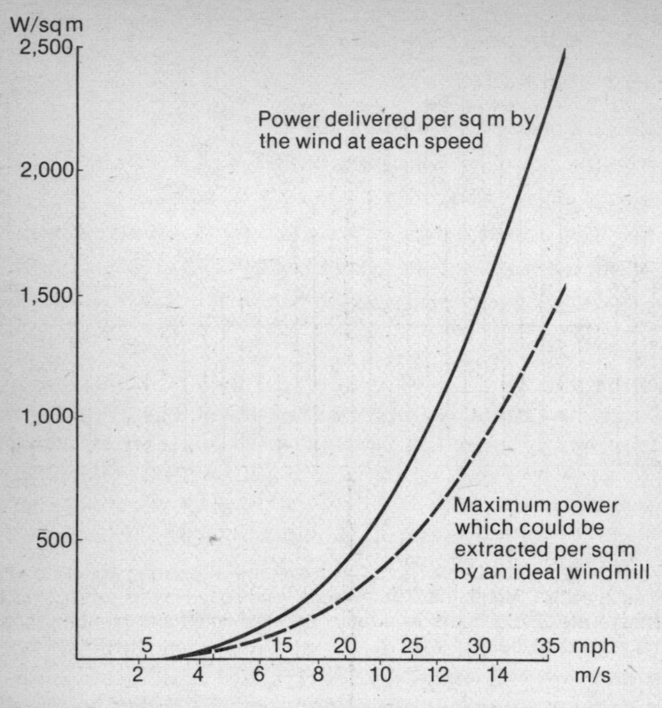

Fig. 11.2 Wind speed and wind power

turns out to be very simple, and valid for a wide range of wind machines and wind speeds. It is that the wind speed be reduced to *one third* of its initial value. A little arithmetic shows that in this case the wind gives up just under two-thirds of its energy, so this is the maximum which can be extracted by an ideal machine. The lower curve in Fig. 11.2 corresponds to this ideal situation, and can thus be used to assess the wind potential of any site.

The assessment, however, involves much more than looking up the average wind speed (assuming that such a figure is available) and using Fig. 11.2 to find the corresponding power. Wind speeds are at least as variable as solar intensities or the heights of ocean waves, and the effect of variations is considerably greater, so the *patterns* of winds can be just as important as the total power. Winds are never completely predictable of course, but there may be regularities. Is there a windy season? Does an afternoon wind blow from the sea on most days? Are the strongest winds the 'wild North-Easters', or is

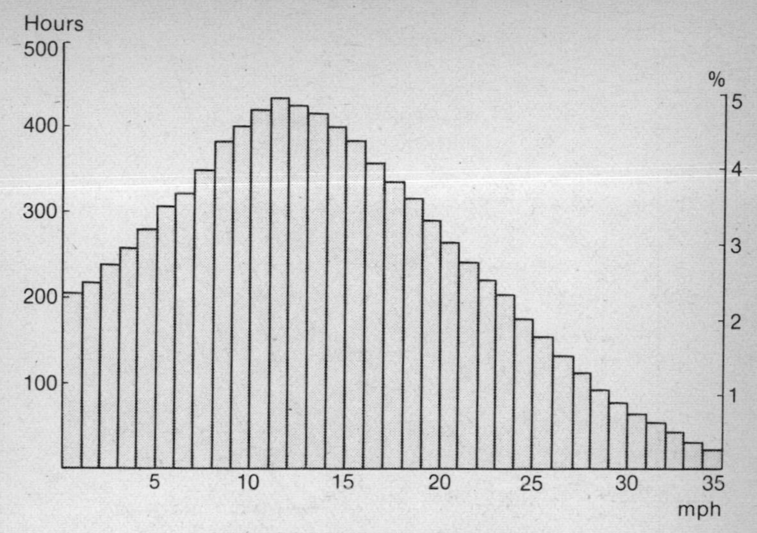

Fig. 11.3 How the winds blow

The blocks show the numbers of hours in an average year during which the wind speed lies in each 1-mph band. The average wind speed in this example is 15 mph. (There are 8,760 hours in a year, and the right-hand numbers show percentages of the year.)

there useful energy in balmy southern breezes too? For how many hours in a typical year is the wind speed less than, say, half the average for the particular site? (Fig. 11.3.) How fierce are the strongest gusts, and how long do they last? Such factors can be crucial in deciding the siting and the type of wind machine – or indeed whether it's worth having one at all.

Wind turbines

Windmills come in many shapes and sizes (Fig. 11.4) but the general idea is always the same. Forces induced by the air flow keep the blades or sails moving, causing an axle to rotate, and the axle serves to transmit a twisting force, a *torque*, to the millstone, pump, generator, or other machinery driven by the mill. Energy taken from the air is ultimately converted into frictional heat in the millstones, gravitational energy of raised water, or electric power. Our subject here is the first stage: how is 'energy taken from the air'? Or, the same question, what is the origin of the force which keeps the blade moving?

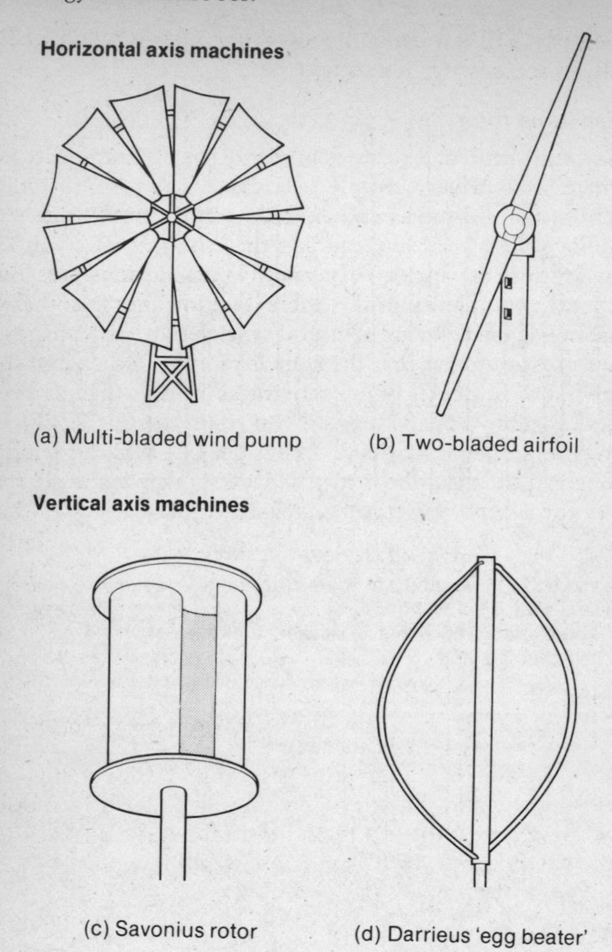

Fig. 11.4 Four types of wind turbine

The first answer is a pretty obvious one. It is the force of the wind hitting the blade which drives it. We know that air has mass, so it is just as though a group of people stood throwing stones at the blade. Naturally it is pushed away. Unfortunately it is very difficult to explain on this theory why (a) and (c), with their broad sails, actually extract *less* power than (b) or (d). Surely most of the air must miss those thin airfoils? And then, what makes a system like (d) go round one way rather than the other? The stone-throwing theory of windmills clearly will not explain everything. Nevertheless we'll start with it, to see how far it can take us.

The Savonius rotor

The Savonius mill, or S-rotor, is probably the best candidate for this treatment. It is a very simple vertical-axis machine (sometimes made from an oil-drum sawn in half down the centre!) and, viewing from above (Fig. 11.5), we can see that the impact of air on the concave surface of vane A will certainly push it backwards. But this is not yet the complete picture. There is for instance the other side of A, pushing against the air behind it, which naturally pushes back. And we mustn't forget that the vane is moving. If the wind speed is 20 mph but the outer tip of A is moving at 15 mph, the *relative* wind speed, the wind you experience if you sit on the tip, is only 5 mph. Meanwhile, someone on the tip of B is heading into a 35-mph wind. Obviously, if the machine is to produce useful power at all, the drag force of the 5-mph wind flowing past A must be greater than the

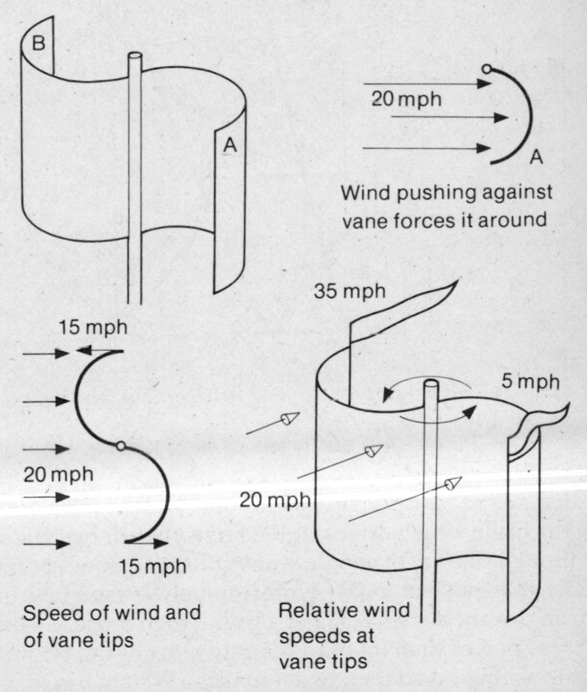

Fig. 11.5 Savonius rotor

reverse drag force of the 35-mph wind against B. Hence the need for the curved vanes.

Drag is a topic we've already discussed; in an entirely different context, of course, but the phenomenon is the same, and we can specify the *coefficient of drag* for a windmill vane just as for a car. Indeed the data (Fig. 6.2, p. 106) used to calculate the energy *dissipated* by a vehicle moving *against* the drag could equally well be used to find the energy *extracted* by a vane moving *with* the drag. The units in Fig. 6.2 (MJ/100 mi) are hardly appropriate to the present case, so Fig. 11.6 shows a more useful version. Notice that if the value of the drag tells us the joules per metre moved, we have only to multiply it by the speed of the vane (m/sec) to find the power input in watts. It is however the *net* power that matters, which means that the difference between the drag forces on vanes A and B must be used in any calculation, and it is not difficult to see that C_D

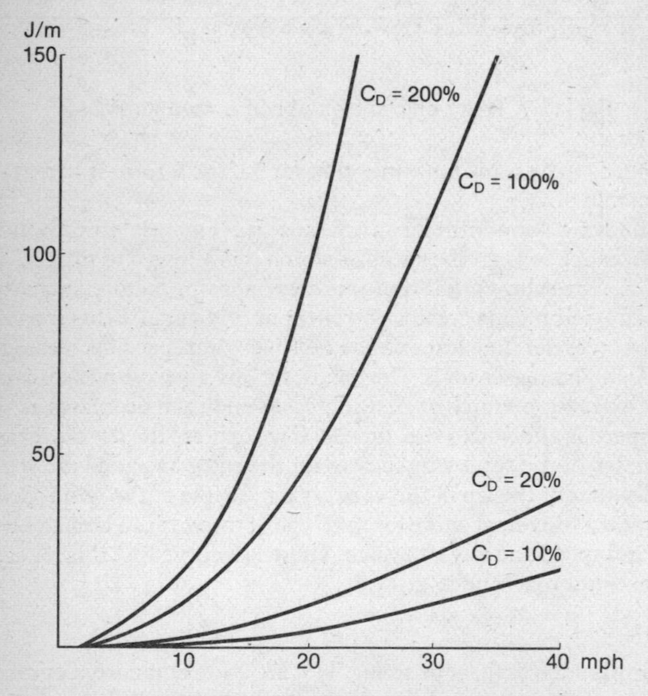

Fig. 11.6 Drag force and relative wind speed
The curves show how the drag force on 1-sq m area of vane depends on the relative wind speed over the vane and the coefficient of drag, C_D.

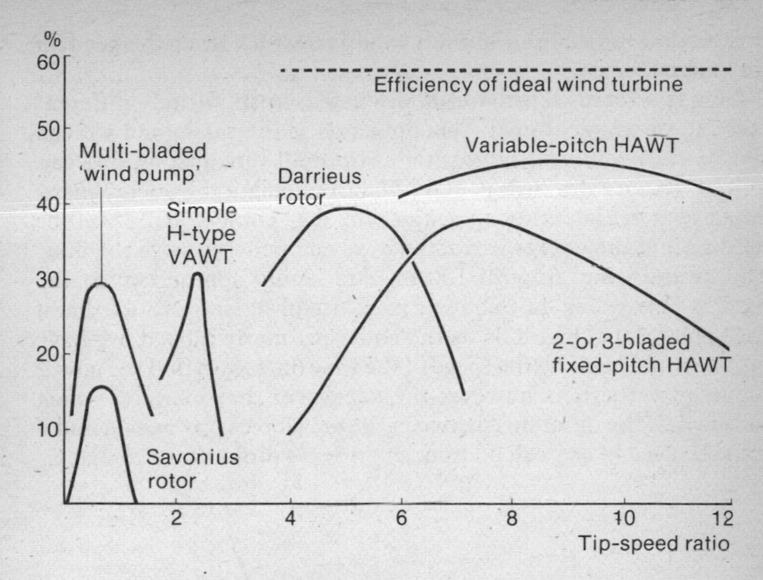

Fig. 11.7 Rotor efficiencies of some wind turbines

for A needs to be some ten times that for B if the S-rotor is to extract appreciable power.

In choosing a windmill for any purpose it is essential to know how the power depends on the speed at which it is allowed to rotate. The analysis above shows that there are two opposing factors: increased speed in itself means greater power input, but against this must be set the lower net drag force as the relative wind speed decreases for vane A and increases for B. The result, for any given wind speed, is a power extraction which at first increases and then decreases as the rotor speed is allowed to rise. Indeed, if we express the net power as a percentage of the total wind power for the rotor area, and the speed (usually that of the tip of the vane) as a multiple of the wind speed, we obtain a universal curve for this type of rotor. It can be applied to any wind speed and any vane size. There is a curve like this for each type of wind machine (Fig. 11.7).

Airfoils

Windmills such as (b) or (d) in Fig. 11.4 are quite evidently not much like a rotating oil-drum, and the first step in understanding how they work is to realize that they are in fact *flying machines*. Those thin blades are shaped like the wings (or perhaps the propellers) of aircraft

– and for very similar reasons. So we'll start by looking at the effect of a flow of air over such an *airfoil* shape, and then see how the ideas are applied in two current designs of windmill.

The important thing we learn from diagram (a) in Fig. 11.8 is that the air above the airfoil moves faster than the air below it. The crowding-up of the streamlines tells us this. (Think of water flowing

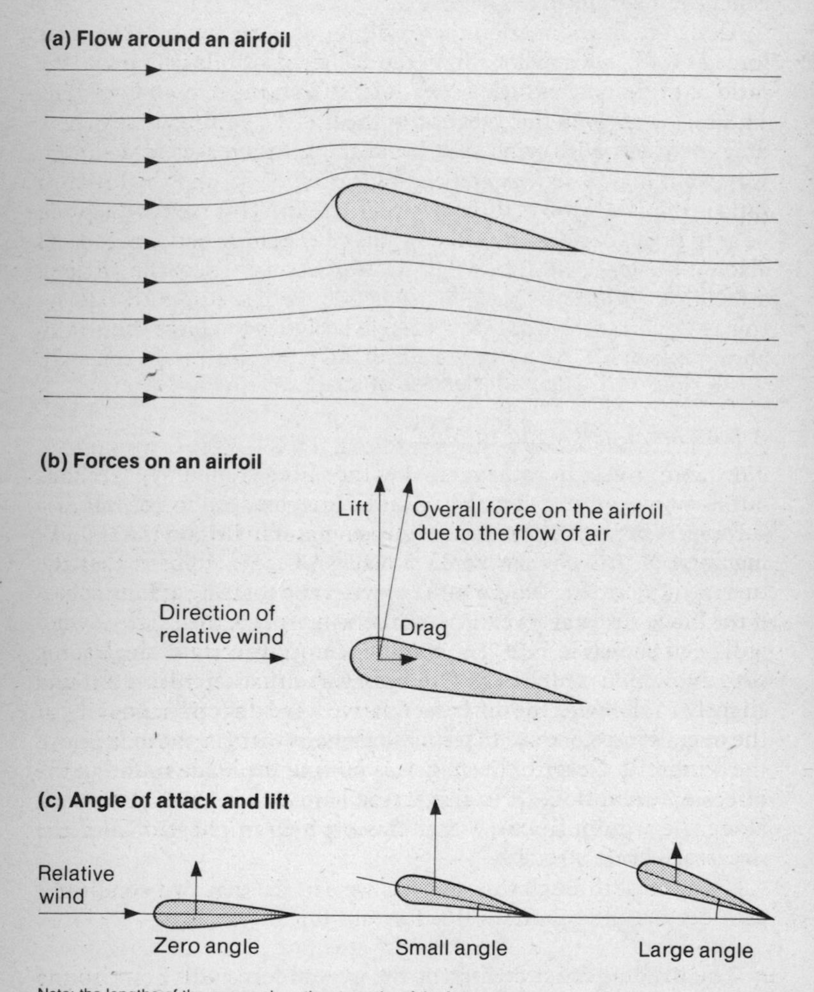

Fig. 11.8 Airfoils

through a narrowing pipe.) And the important consequence of this speed difference is that the air *pressure* is less above the foil than below. A lower pressure above than below means a net upwards force on the airfoil, which is of course what keeps an aircraft airborne. Naturally, if the foil is vertical, as in a windmill, rather than horizontal, this force is not upwards but sideways, but it is still generally referred to as the *lift*.

Along the flow direction there will be, as with any object, a drag force as well, and a major aim in the design of airfoils is to make the ratio of lift to drag as high as possible, so that the overall force (Fig. 11.8(b)) is almost in the direction of the lift. We've already seen how drag increases with wind speed, and the lift increases in a similar way. So it makes sense to talk of a *lift coefficient*, and the lift–drag ratio is then the ratio of these two coefficients. This ratio, which can be as high as a hundred to one, depends on the shape of the airfoil and also on the *angle of attack* (Fig. 11.8(c)). As we'll see, the angle of attack for a windmill blade depends on how fast it is moving, and this is important, because if the angle becomes too large there is an abrupt loss of lift. An airfoil windmill, just like an aircraft, can stall.

A horizontal-axis wind turbine

Now we must see how the lift makes the sails go round. We'll take as our example a fairly straightforward, three-bladed, horizontal-axis turbine. (The largest operating wind generator in Britain is a 100-kW machine of this type, with 25-ft blades.) Let us suppose that the turbine is pointing straight into the wind and rotating at high speed. If the blade speed is several times the wind speed, the *relative* wind will be as shown in Fig. 11.9, coming almost directly at the leading edge. Now let us redraw Fig. 11.8(b) for this situation, tilting it round slightly to allow for the different relative wind direction. Looking at the overall force, we see that it is slightly *forward* of the axis line of the windmill. Despite the drag, it is pulling the blade round in the direction of motion. (It is also trying hard to bend the blade back along the windmill axis, which has obvious implications for the necessary blade strength.)

Going back through this analysis we can see that two conditions are necessary to maintain this forward force:

- The lift-drag must be high or the overall force will point to the wrong side of the axis, slowing down the blade instead of pulling it round.

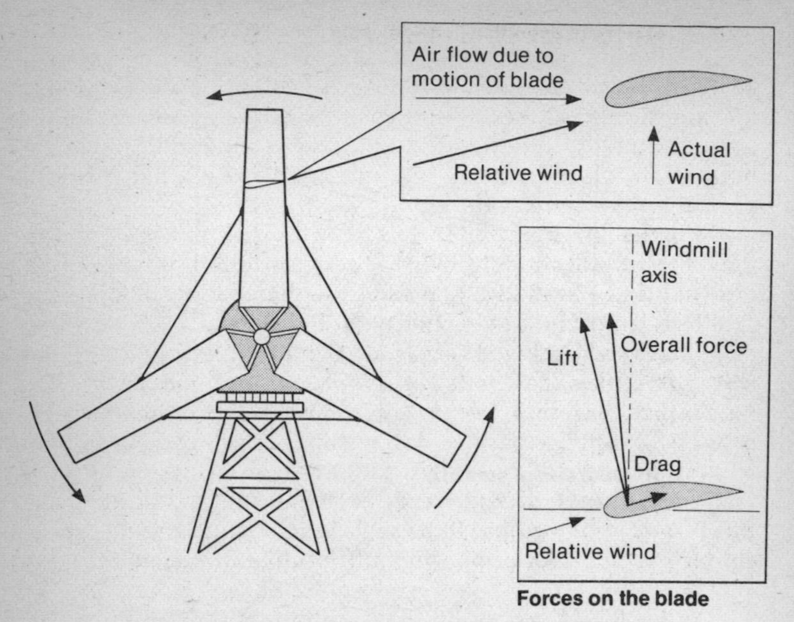

Fig. 11.9 Action of a horizontal-axis wind turbine

- To keep the angle of attack small so that the blade doesn't stall, the blade speed must always be large compared with the wind speed.

For any airfoil there is a 'best' angle of attack which gives the greatest lift–drag ratio, and the extraction of wind energy will be most efficient at tip-speed ratios which produce approximately this angle. The performance curve will be like that shown in Fig. 11.7.

There are many other factors affecting design and performance that we've not considered at all so far. The blade speed, for instance, is very different at the tip and at the inner end, so the angle of attack will decrease along the blade. Introducing a *twist* – making it more like a propeller than a set of wings – can improve efficiency. And then there is the fact that according to our theory the windmill will never start; there is no forward lift until the blades are moving quite fast. In practice some machines of this type *are* self-starting, but the torque is always low until the speed builds up.

A vertical-axis wind turbine

The machine in Fig. 11.10 is an obvious cousin of the 'egg-beater'

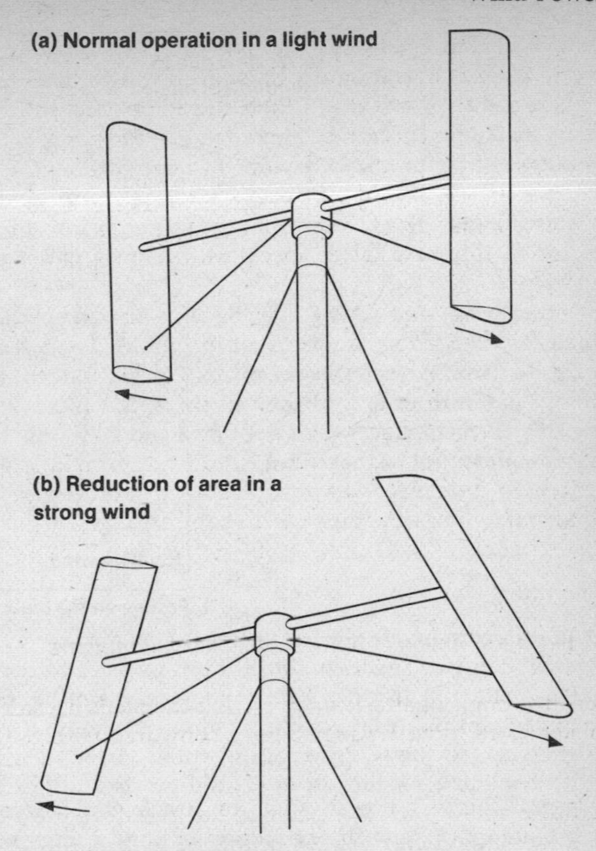

Fig. 11.10 A vertical-axis wind turbine

(11.4(d)), and their general principle is similar. These vertical-axis turbines share two advantages: they are panemonic, responding equally well to winds from all directions, whereas a horizontal-axis machine needs a tail vane or auxiliary rotor to turn it into the wind; and then the vertical axis means that there is no need to mount all the machinery – generator, gears, etc. – at the top of the tower, which saves construction costs.

A significant difference between the S-rotor and this airfoil machine is that here there is a forward force on *both* blades most of the time, similar to that shown for the blade in Fig. 11.9. As the rotor spins, however, the relative wind shifts and with it the direction of this force, so the structure must be able to withstand not only the downwind force but an oscillating sideways one as well. The efficiency in normal operation depends on the angle between the vertical blade and the rotor arm, which affects the angle of attack, and in some machines this can be altered to suit the wind. Fig. 11.7 includes curves showing the behaviour of a fairly simple H-type vertical-axis machine and of a Darrieus rotor. As with all airfoil systems, starting may be a problem and the output is low until the tip-speed (or in this case, blade-speed) ratio comes closer to its optimum value.

On the other hand, one VAWT developed in Britain approaches the problem of very *strong* winds in an ingenious manner. If the arms are fixed a little above the blade centres (Fig. 11.10(b)) the lower ends will fly out further and further as the speed increases. By flattening the 'H' this reduces the swept area and keeps the forces within acceptable limits, so that the machine can continue to generate power where other machines must be stalled or otherwise 'switched off' for safety, wasting the power of the strongest winds.

Wind generators

Using windmills to generate electric power is hardly a new idea. Of the estimated 6 million small windmills built in the USA over the past hundred years, the majority were of course the familiar multi-bladed water pumps (Fig. 11.4(a)), but hundreds of thousands were two- or three-bladed propellers running small generators. With the advent of reliable supplies from central power-stations these rather variable sources fell from favour, but recent interest in all renewable forms of energy has led people to look at them again. Unfortunately the wind is no more constant now than it was fifty years ago, and whether we think of small machines generating about a kilowatt, or huge multi-megawatt arrays, there are still a few difficulties to be sorted out before wind can be regarded as a serious competitor with our present sources of electric power.

The three main problems with wind generators are in a sense all to do with *matching*:

- The turbine must match the winds.
- The power output must match demand.
- The voltage generated must match our equipment (or the rest of the system, for large-scale generation).

To see what this means we'll put together some of the different pieces of information from the previous section. The 'hours' data from Fig. 11.3 are reproduced in 11.11 (as a smooth curve for simplicity), together with another curve showing how many kilowatt-hours of energy are delivered by winds with each speed. A 10-mph wind, for instance, blows for about 400 hours and carries a power of 50 W/sq m (Fig. 11.2); so winds at this speed contribute 20 kWh/sq m in an average year. Now any wind generator must be designed for a particular range of wind speeds, and the first problem arises when we try to decide which range is best for a site with, say, the wind pattern shown.

The 'hours' curve suggests that a system designed for 8–20 mph would be working for half the total time, which seems good; but the 'energy' curve shows that it would entirely miss *two-thirds* of the total annual energy. On the other hand, a machine working between 20–40 mph, while catching all but a third of the energy, would be running for only about a quarter of the time. Do we want maximum energy per machine but all delivered in the equivalent of three months, or should we sacrifice the extra energy for a better spread?

The third matching problem adds yet another limitation. In Chapter 5 we saw that electric power is generated by rotating a coil inside a magnet (or vice versa). Now the *speed* of rotation determines the *frequency* of the alternating voltage: the number of cycles per second, or hertz; and all our present mains-operated equipment is

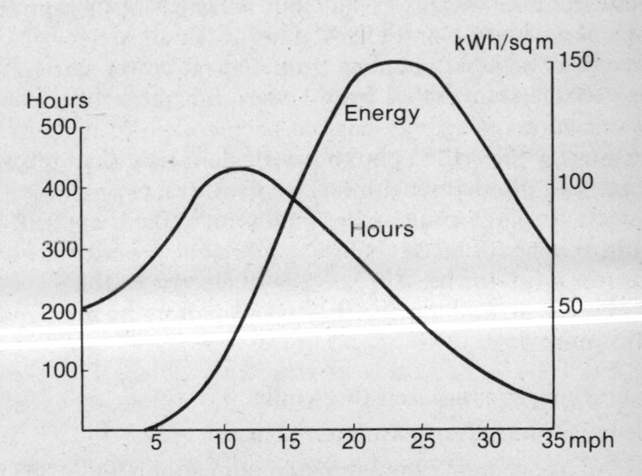

Fig. 11.11 How the winds deliver energy

For an explanation of the hours curve, see Fig. 11.3. The energy curve shows the total annual energy delivered by winds of each speed.

designed for a 50-Hz supply (60-Hz in the USA). But the efficiency curve for a wind turbine peaks at a certain tip-speed ratio, which means a different 'best' turbine speed for each wind speed. So there is another compromise to be made. Either the turbine runs at a fixed speed, in which case it cannot always be at its peak efficiency, or it runs at the most efficient speed for the wind that's blowing, but we have the cost, complexity and losses of the electrical equipment needed to convert the frequency.

The largest wind machine being built in Britain, for instance, has a 200-ft horizontal-axis turbine driving a synchronous generator. This means that it must maintain almost constant rotor speed over its entire operating range of wind speeds, from 15 to 60 mph. It is designed to be most efficient at 30 mph (with a tip-speed ratio of about seven), so at wind speeds below or above this the efficiency falls off considerably. The rated output of the turbine – the maximum power it can deliver to the generator – is just under 4 MW in a 50 mph wind, and a little calculation shows that this represents an efficiency of less than 20 per cent at this speed. The *average* wind speed will of course be much lower and even if it is as high as 30 mph for the site chosen (on one of the Orkney Islands), the energy extracted in a year will be perhaps a sixth of the total which the winds deliver. In a sense this doesn't matter, because they do after all deliver it free, and the remaining five-sixths is not some harmful waste product or unwanted heat but just a breeze. However, there is the capital cost; and as with wave power or pumped storage, installing expensive generating capacity and then using it for half the time and at a fraction of its rated output will need careful justification in terms of fuel saved.

The turbine described above has fixed blades, but one way to overcome the problem of the varying tip-speed ratio with synchronous generators is to have *variable pitch*. We've already seen that the blades of horizontal-axis turbines are often twisted like propeller blades, and it is also possible to design them so that the whole blade can be turned, varying the angle of attack to suit the wind. Turbines like this have been developed more in the USA – with the aim of using relatively light winds as efficiently as possible – than in Britain, where the idea seems to be to build simple, robust systems which will work reliably in fairly strong winds.

None of the above problems necessarily puts wind power out of court, but they do mean that careful studies of wind patterns, of the suitability of different machines, and of the relationship between their output and the existing power systems will be necessary before there can be major development on a national scale. It's one thing to put up a small generator to charge a few batteries if you live on a windy prairie and have no other power, but quite a different matter to abandon the security of known systems when you are responsible for a nation's entire supply.

Environmental effects

Like the sunlight from which it derives its power, the wind is not a very concentrated energy source. Fig. 11.12 shows the rotor diameters needed in various winds to produce a specified electric power at a fairly optimistic 40 per cent overall efficiency.

To take one example, we see that even at a site which could rely on a 30-mph wind for an appreciable fraction of the year, an output of 2 MW would need a 200-ft rotor. With the generally accepted spacing of about ten diameters, adjacent towers would need to be about a

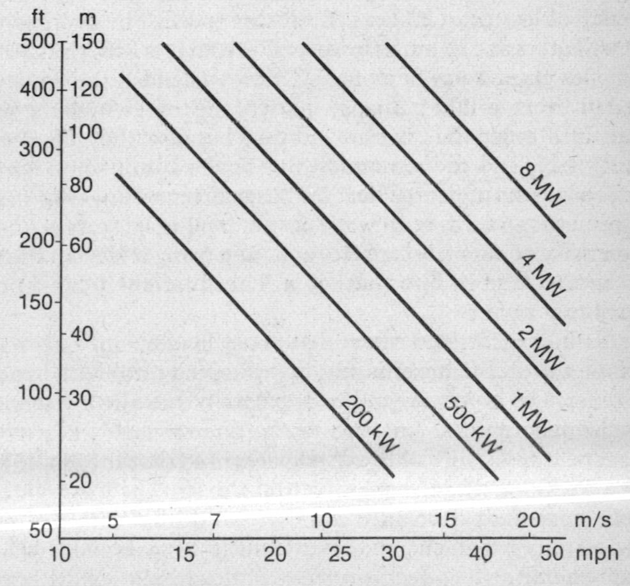

Fig. 11.12 Wind speed, rotor diameter and power

1. The lines show how power output depends on rotor diameter and wind speed, for a system with 40 per cent overall efficiency.
2. The output is for a steady, constant wind of the speed shown. If the speeds are annual averages the generator rating needs to be about twice the value given.

third of a mile apart; and a 600-MW power-station would be an array of three hundred of these, each about 300 ft high, occupying a stretch of land roughly ten miles by three. Or a stretch of sea, perhaps. In a crowded country like Britain, inland siting is likely to be a problem, especially if areas of natural beauty such as the Welsh or Scottish mountains or the remaining unspoilt coastline are excluded. Environmentalists in the USA are expressing concern about such regions too; but with ten times the land per person, the opportunities for acceptable siting are obviously greater. (And it should be added that by no means everyone finds these big turbines ugly.)

There are a few other unresolved questions. Noise is one. Just how loud will the 'high whirring sound' be, when it is produced by two 100-ft blades cutting the air at over 100 mph? And there is the interesting question of what happens if a blade breaks. Although the blades of a megawatt generator may weigh only a few tonnes, their outward pull as they spin in a fresh wind can be the equivalent of hundreds of tonnes, with a downwind bending force equivalent to tens of tonnes on each blade. (Turbines with long thin blades are often designed to run with the rotor downwind of the tower, so that the blades bend away from it.) The thought of this 100-ft object launched from a 200-ft tower at a speed of 100 mph or so, and whirling through the air on a no doubt interesting but probably unpredictable trajectory, suggests that people may be a little wary of living in the line of fire – where ever that is. However, the wings and propellers of aircraft very rarely fly off, and it is probably for this reason rather than an assessment of our priorities that the environmental effect most discussed is a very different one: TV interference.

Compared with the environmental problems of other power systems, however, the hazards and undesirable effects of wind power seem likely to be small, and it is generally regarded as one of the more benign sources. An array of the type described above might extract perhaps a third of the wind energy over a ten-mile front, and would obviously have some effect on the local climate. But a relatively small reduction in wind strength does not seem likely to bring serious problems; indeed, it might even be regarded as an improvement.

Realities

The construction of fifty 1,000-MW clusters of wind towers, sited some forty miles apart in shallow waters or on flat coastal land, might be the most acceptable way of satisfying Britain's future demand for electric power. But would it? It certainly couldn't do so alone because of the variability of the winds; so if acceptability depends on cost, and the total cost depends on the amount of variation the system must allow for, the fluctuations in wind strength are going to be very important. It isn't surprising, then, that a great deal of effort is currently being devoted to the study of these variations and their effects on individual turbines, large arrays, or whole generating systems.

The general line of argument isn't too difficult to follow, and starts with the observation that wind speeds vary in a number of different ways. On a very short time-scale, we have *gusts*, sudden changes within seconds or minutes. Then there are lulls and freshenings over a period of an hour or a few hours: the sort we experience as successive rain storms pass on a blustery day. There are the very important *daily* changes such as the drop in wind strength at sunset. And finally, there is the weather, windy or calm over periods of days or weeks; and the overall *annual* variation. Looked at with a wind power system in mind, these mere differences in timing suddenly become very important. Gusting, for instance, serious for a single small generator, can be virtually smoothed out by adding the outputs from the several hundred turbines of a large array. And variations over a few hours are at least partially smoothed if you feed power to a national grid from several installations 100 miles or so apart. But sunset is sunset within about twenty minutes for a country the size of England; and calm or windy weather, or the annual variations, are likely to affect most sites in rather similar ways.

Thus it seems that a system relying mainly on wind power would need at least some reserve capacity capable of producing power at a couple of hours' notice; and then, for longer periods of calm, a virtual duplication of the entire system. The obvious candidates for the first category are pumped storage and gas turbines, or possibly coal-fired stations kept running as 'spinning reserve'; whilst almost any type could satisfy the second requirement (except nuclear power, which makes no sense as an occasional supply). Weighing the greater capital cost of this dual system against the fuel savings, the minimal environmental harm and the security of an inalienable supply isn't easy. No one has yet tried off-shore windmills, and no one has built an array of any size, so the technical uncertainties are many. If there *is* to be a major contribution from wind power, it surely won't come in this century, and cost comparisons therefore depend on assumptions not only about the reliability of the winds but about future coal prices and developments in nuclear power. Estimates of the cost per kilowatt-hour in thirty years' time, from a system with a major wind component, range – not surprisingly – from a little below those for conventional systems to several times greater.

Which is not to say that nothing is known. World-wide, the past twenty years have seen a great deal of work on wind turbines, and on the problems of integrating wind power into existing systems. The US government has invested half a billion dollars in wind energy research and development, and the first few generators in the megawatt range are now operating. (Plans were for another billion dollars over the next five years, but the change of administration in 1980 has meant major reductions in programmes for renewable resources.) Britain's scale is somewhat different, but spending by the Department of Energy has grown and is now approaching a million pounds a year, while a government grant is supporting construction of the 3.7-MW machine described on p. 234, due for completion in 1984 at a cost of £5.6 million.

Small-scale, local wind generators have received much less attention from governments. They have their supporters, as we'll see in Chapter 15, but, to draw a comparison with another renewable source, there must be hundreds if not thousands of houses equipped with solar panels to every one with its own windmill. For a number of reasons it seems improbable that this relative position will change in the near future, and if wind energy does become important, it is much more likely to be in large-scale installations. Indeed, if we look only at electric power production from renewable sources, it is arguable that of all the alternative systems – solar power-stations, wave energy converters, tidal barrages, and others we'll meet later – the one with the fewest technical and environmental obstacles to overcome before it can make a really major contribution to national supplies is that somewhat surprising example of high technology, the windmill.

Part III The Future

12 Penalties

Introduction

Moving on to discuss the future doesn't mean leaving behind all the technologies studied in the central part of the book. On the contrary, we must take them with us, together with all their potentialities and problems. They *are* the future, or at least the only future we can plan for. Planning, with its necessary forerunner, prediction, is the main subject of the chapters which follow. How does anyone know how much energy we'll need, and what we'll have, in the year 2000? One chapter is devoted to the methods used in predicting future demand and an assessment of resources; and another presents some of the resulting forecasts – conventional scenarios and radical alternatives. We also take another look at conservation, for some writers the best 'energy resource' of all. And we'll finish, as promised in Chapter 1, not with a grand solution but with some hard questions.

But to start, a gloomy look at just a few of the penalties of our energy systems, present and future.

. . . and statistics

One thing soon becomes obvious when we start to look at the health and environmental effects of energy systems. The available information, the factual background, is often very different from the facts about, say, the efficiencies of generators or car engines. The difference is that more often than not the data are essentially *statistical*. We've met statistical data before, of course, but usually in the form of simple averages: the average solar energy falling on a collector in a year, for instance. Here, they play a much more essential role – or rather two different roles.

First, there is the question of the *effects* of the by-products of our energy systems. These effects, or at least the ill-effects, commonly appear at low levels, or are diffuse, or distant in space or time from their original causes. (Foolish we may be, but we usually have enough sense not to use energy systems which are obviously and instantly lethal.) Quite often our only way to link an effect with its cause is to look carefully for correlations in the past variations of the two. It is believed, for instance, that the bad smog in London in December 1952 caused 4,000 deaths in five days. Not because that number of otherwise healthy people fell gasping in the street and expired on the spot, but because there were 4,000 more deaths from respiratory and related diseases than would have been normal for that population in that place during that period. It could of course have been a random fluctuation, or the effect of a wicked fairy or a conjunction of planets; but because there was other statistical evidence relating air pollution and respiratory illness, and at least the beginnings of an understanding of why they might be related, the smog was held responsible.

Now 4,000 extra deaths in London in a few days is a pretty obvious effect; but what happens when the scale is much smaller? You may feel that something should be done about emissions from power-stations or internal combustion engines even if they 'only' cause 100 extra deaths a year in the whole country, but detecting such a subtle effect would be like searching for a hundred extra pieces of hay in a haystack. We must ask whether statistics will allow us to find them – reliably.

Then we have the statistical problem of *risk*. Suppose that we do satisfy ourselves that none of our energy systems is detectably harmful *in normal use*. We must still assess the chances that something goes wrong, and the probable harm if this happens. The assessment of risk is not a simple matter, but it plays such an important role in the comparison of different energy systems that we must attempt to look at the issues involved, the difficulties, and some of the predictions. You need only think of the everyday phrase, 'the *chances* of an accident' to see that statistical arguments will be central to this discussion.

The four topics of the present chapter, then: air pollution, radioactivity, the carbon dioxide problem, and the risk of accidents, are chosen mainly because they are important penalties – or potential penalties – of our energy systems, but also because they exemplify in different ways the particular sorts of problem arising when we try to assess the side-effects of any technology.

Air pollution

We know that burning fossil fuel produces toxic substances, and we also know roughly how much from any given process. It is therefore possible to give estimates for the total emissions from a power-station (page 126) or a country (Table 12.1). However, to assess the effects of these we need at least two further sorts of information. We need to know the resulting *concentrations* of pollutants: how much SO_2 in each cubic metre of air, for instance; and we need to know how these concentrations affect the health of people (or other life-forms) exposed to them. Unfortunately we still have too little information about either.

Systematic measurement of pollution is a relatively recent development. Reliable records go back only a few decades, and are chiefly for cities or industrial areas. Even in the developed countries there remain regions about which little is known, and record-keeping outside the industrialized world is still in its infancy. Overall, the few data firm enough to allow us to relate health effects to their causes do little more than support the evidence of our eyes and noses.

Undoubtedly the best-known case history is that of London during the twenty years from 1950. In the great smog of 1952 (see page 242) the levels of particulates – smoke – and sulphur dioxide reached several thousand micrograms per cubic metre. (Pure air weighs a little over one *kilogram* per cubic metre, so 4,000 μg/cu m would be about three parts per million by weight.) In the early 1950s the annual average levels were a few hundred μg/cu m, but the catastrophic consequences of the 1952 smog finally led in 1956 to a Clean Air Act. The burning of untreated coal was prohibited in London, and the mean annual smoke concentration fell in twenty years to under 50 μg/cu m – less than a fifth of its earlier level and only a hundredth of the 1952 peak. It has been argued that the change would in any case have taken place with the increasing use of central heating, but whatever the cause, the effect has been striking. Even the house martins have noticed and returned to central London!

The example is well-known precisely because it is not typical.

Table 12.1 Emissions from fossil fuel combustion in the USA

Source	**Emissions per year, millions of tonnes**			
	SO_2	**NO_x**	**Particulates**	**Hydrocarbons**
Power-stations	18	7	3.5	—
Other furnaces	4.5	6	1.5	1.5
Other industry	4	1	6	15
Vehicles	1	9	1	11
Total	**27**	**23**	**12**	**28**

The results are usually much less clear-cut. Nonetheless, it is an interesting case, in part because it exemplifies some differences between the approach to air pollution in Britain and the USA. The London air was improved by tackling the smoke, the particulates. The concentration of SO_2, the other major pollutant from coal, has also fallen, but this was not the main target. A similar concern has meant that new coal-fired stations in England are equipped with electrostatic precipitators (Fig. 7.4, p. 128) but are not required to reduce SO_2 emissions. In the USA, although particulates are also controlled, the emphasis has been on stringent regulations enforcing 'flue gas desulfurization' – even for users of low-sulphur coal. Another aspect of the American approach is seen in the precisely specified limits for permitted atmospheric concentrations of many individual pollutants (Table 12.2).

Unfortunately it is virtually impossible to compare the effectiveness of the two approaches. Some data are available: that, for instance, ground-level SO_2 concentrations near Britain's modern power-stations, with their tall chimneys, are on average no greater than elsewhere in the country; that total SO_2 emissions in the USA fell slightly during the 1970s while electric power production from coal rose by some 50 per cent. But comparisons are made difficult by a number of factors. Britain still has older power-stations without modern flue gas cleaning. The regulations in the USA were honoured more in the breach than in the observance in the late 1970s when the need to conserve oil was allowed to justify the burning of coal without SO_2 removal. And of course there are many other sources of air pollution than power-stations.

The case of the acid lakes points up yet another problem: the distances which can separate cause and effect. Recent years have seen more reports of increasing acidity, sometimes to the extent that all life is killed, in lakes so remote from industrial activity that

Table 12.2 Permitted concentrations of pollutants, USA 1979

A Maximum concentration over a 24-hour period.
B Maximum annual average concentration.

	Concentration in µg/cu m	
	A	**B**
SO_2	365	80
NO_x	—	100
Particulates	260	75

the cause can hardly be the 'usual' one of direct pollution by effluents. And the reports come not only from regions such as northern Germany or the north-east United States, known to be directly downwind of major industrial areas, but also from Norway and the Colorado Rockies, hundreds of miles from large-scale sources.

The problem is currently attracting the attention of an interestingly diverse array of scientists. Chemists and biologists are trying to establish the precise nature of the acidity. Does it come ultimately from SO_2 or from some other emission product; and in either case, by what processes? Geophysical techniques have produced the interesting information that successive layers of ice in Greenland's glaciers show rising acidity over the past century or so. Meteorologists, making use of the largest single atmospheric pollutant of recent years, have followed the emissions from Mount St. Helens for thousands of miles, along tracks which curve and bend, rise and fall, split, and even loop back on themselves. And a transnational study, initiated by the OECD after complaints by the Scandinavians, is producing tentative data on the emitters and receivers of pollutants in Western Europe (Table 12.3).

At which stage it may be worth noting that all this effort is dealing with just one effect of (probably) one pollutant from (mainly) one type of source.

Table 12.3 Emitters and receivers of SO_2, Western Europe 1974

All figures are in thousands of tonnes a year. They are regarded as uncertain within plus or minus 50%.

Receiving country	**Emitting country**							**Total received**
	Denmark	France	W Germany	Norway	Sweden	UK + Ireland	Other	
Belgium	—	60	20	—	—	60	200	**400**
Denmark	100	5	10	—	5	20	50	**200**
France	2	1,200	100	—	—	200	600	**2,000**
W Germany	15	200	1,400	—	5	200	500	**2,500**
Norway	15	20	20	60	20	100	250	**500**
Sweden	60	20	20	10	200	100	450	**1,000**
UK + Ireland	5	40	40	—	—	1,500	250	**2,000**
Other	200	500	900	20	150	1,500		
Total emitted	**400**	**2,000**	**2,500**	**80**	**400**	**3,500**		

Emissions from motor vehicles present if anything an even more confused picture. Los Angeles, with its unfortunate combination of the world's highest vehicle ownership and a location and climate which exacerbates pollution, is the extreme case. Its particular eye-stinging brand of smog is thought to be the result of chemical reactions between nitrogen oxides and unburnt hydrocarbons. The reactions are promoted by the sunlight, and the concentration of the smog is increased by the frequent presence of an atmospheric inversion: a layer of cooler and thus denser air sitting like a lid over the Los Angeles basin. The 1970 Clean Air Act imposed limitations on the emissions from internal combustion engines, and any vehicle must now have a clean bill of health before it may be bought or sold. The Act also set detailed air pollution limits which were to be met by 1975. As they would have required the elimination of about three-quarters of the Los Angeles traffic they were naturally not met. Whether the emission controls have appreciably reduced the smog is not entirely clear; but then neither is there clear evidence that the smog is in fact a long-term danger to health.

Which brings us to another question: the *effects* of pollution. We know that oxides of sulphur and nitrogen are toxic at high concentrations, as are smoke and carbon monoxide. And there is general (if not universal) agreement on the broad trend as concentrations get lower. It seems, and this is obviously important, that there is no *threshold* level below which these pollutants cause no harm at all. Even if there are effectively safe levels for an individual healthy person over a short time, we are concerned with effects over long periods, and on people who are old or already suffer from respiratory illness or heart disease. For a total population like this, the available evidence suggests a continuous proportionality between mortality and exposure to pollutants. In this sense there is no 'safe level'.

So how do we proceed? One possibility is as follows. Let us first find out how many extra deaths result from different levels of a particular pollutant. Then we can set a limit to its permitted level such that the expected number of deaths due to this cause lies in the statistically undetectable region: it is less than the normal fluctuations in death rates. You may or may not like the approach, but at least it gives a recipe – provided we have the basic data. How are these to be obtained? Almost certainly from epidemiological studies.

We must look at changes in death rates or other health effects in the population and try to see whether these are correlated with changes in pollution levels. The data, where they don't already exist, should not be too difficult to obtain. Unfortunately they are by no means easy to interpret. Consider just a few of the problems. Levels of SO_2 and of particulates both rise if more coal is burned, so how do we know which is responsible for any observed effect? The death rate rises during a cold spell. Is this due to the weather or to the extra fuel burned? How many excess deaths should be subtracted for an influenza epidemic? What about social or 'life-style' effects? (People who live in the more polluted parts of cities may be less healthy for other reasons.) How do we count the sick whose deaths are accelerated rather than 'caused' by air pollution? Long-term effects or short-term effects? And what about effects other than mortality?

Part of the difficulty for epidemiological studies is precisely that pollution levels *are* now rather low, and the effects are difficult to discern against the background of statistical fluctuations. Fig. 12.1 shows one set of data for a single effect at low SO_2 concentrations. The line is the relationship which was deduced from the data: no effect up to a threshhold of about 7 μg/cu m and then a rising incidence of attacks. It certainly requires the eye of the believer to decide that this particular line follows from this information; and it

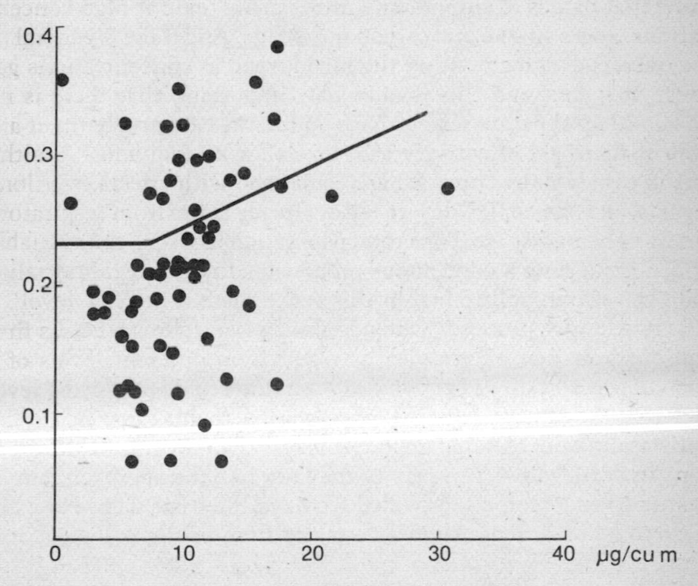

Fig. 12.1 An effect of air pollution?

Each point shows the rate of asthma attacks in a group of sufferers and the concentration of SO_2 at the time. See the text for discussion of the graph.

isn't surprising that a number of recent surveys have concluded that *no figure at all* can be given with any confidence for the death rate resulting from a specific pollutant level.

This conclusion seems unnecessarily defeatist. After all, we know that we don't want a return to pre-1950 London, and there are *some* data, even if they show no more than general trends. Fig. 12.2, bringing together results from different times and places, will serve as an example. It doesn't have well-defined single points representing precise values, but shaded areas within which the measurements lie. To see what it tells us, consider the two lines at roughly the upper and lower limits of the data. Suppose that the present average smoke concentration in a city is 50 μg/cu m. Line A predicts 0.5 per cent and line B 5 per cent excess deaths. In Western countries the death rate is roughly 10,000 a year per million people, so the excess for that population would be between 50 and 500 deaths a year.

There's a further point worth noting, however, before we make use of these or similar numbers. Fig. 12.2 shows death rates rising with smoke levels, but it cannot show that death rates rise *because*

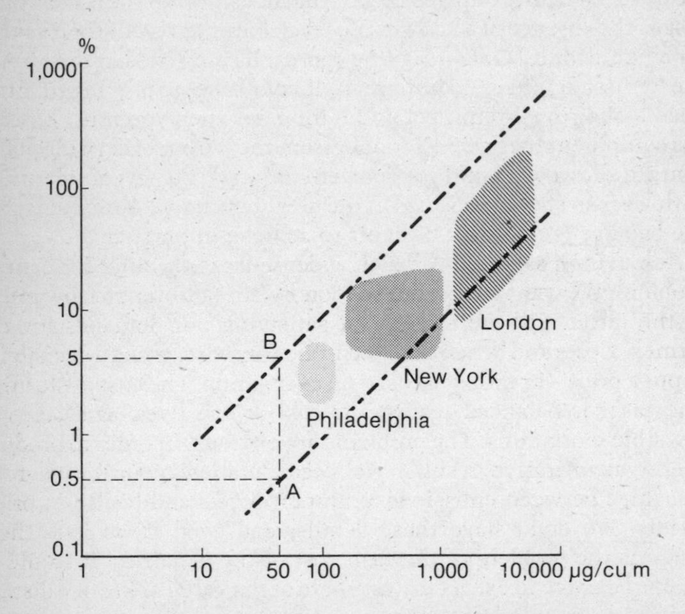

Fig. 12.2 Percentage excess deaths and concentration of particulates
See text for explanation

of smoke. A similar pattern would appear if we used SO_2 instead, and there is not yet enough reliable information to separate the two. It is *probably* the case that burning fossil fuel is indeed the cause, but we can't be certain. This is clearly another field where more studies are needed.

Of course any study is likely to provide answers only to the questions we choose to ask. There may be pollutants not mentioned here which are health hazards. And there are undoubtedly other environmental effects. In Western Europe, as in many industrial regions, we live our lives under a dirty blanket. Even on days when we think we see a clear blue sky, aircraft passengers could tell us otherwise, as they descend into the grey-brown layer. But there are still places in the world where it is normal to see mountains a hundred miles away in crystal-clear detail. How much do we care if they too sink into the haze?

Present methods of dealing with proposals to burn more coal, to build a new power-station or a conversion plant, vary widely, ranging from, 'build it and see', to a demand that the proposer should calculate in precise detail every last consequence. (The latter is as we have seen quite unrealistic, but that doesn't stop the legislators – just as the absence of a known safe level doesn't prevent their asking for a figure for it.) One intriguing approach being tried in the USA is the 'emission off-set' requirement. If your scheme is going to introduce X additional tonnes of pollutant a day then you must provide the equipment to *remove* X tonnes from the output of existing plant. You are allowed to build your power-station if you pay for treatment of flue gas in six old factories in the neighbourhood. A nice idea, but the balance is no doubt difficult to achieve in practice.

Achieving a balance in a much wider sense is of course the central problem. We want energy but we don't want pollutants. How much of the latter will we tolerate in satisfying our demands for the former? Costs and benefits. Indeed, one way of reaching a decision is to put a price – or rather, a cost – on everything. The cost of cleaning flue gases is balanced against the cost of lost lives, acid lakes or invisible mountains. The problems are obvious. In order to produce useful quantitative results, we need detailed quantitative relationships between emissions, pollutant levels and health or other effects. We don't have these details; and even if we had, there remains the problem of the cash cost of the penalties. It would of course be naïve to expect *my* estimate of the value of life or a distant view to be particularly relevant, and I'm sure that all our energy planners are honourable men. Nevertheless . . .

In conclusion, it seems that the main lesson we learn is a negative one: to be wary of statements like, 'Each megawatt of coal-fired electric power causes X deaths a year from air pollution.' We've seen what a long and complex analysis and how many doubtful data lie behind such a beguilingly quotable claim. We've seen that there can be no *simple* formula for calculating pollutant concentrations from individual emissions, and we've seen some of the difficulties of measuring health effects. These two, which we might call the *meteorological problem* and the *epidemiological problem* are just as important for other pollutants such as radioactivity, to which we now turn our attention.

Biological effects of radioactivity

To assess the environmental effects of a nuclear plant we need to know what its emissions are, how these are distributed in the atmosphere, how much is received by individuals or the population, and what are the health effects of these amounts. It does all sound rather familiar, and it is partly in order not to cover the same ground twice that the emphasis of this section differs from that of the previous one. But there are other important reasons for concentrating here on the last stage, the health effects. There is the fact that, unlike SO_2, radioactivity is not a single substance but a whole range of phenomena, and it is important – particularly in such a controversial area – to be clear exactly what effects we are talking about. Then there is another significant difference between the emissions from fossil fuel and nuclear plants. It lies in the range which separates 'routine emissions' from 'worst possible emissions'.

The routine discharges from a properly-run nuclear reactor are, as we shall see, very low. That is, they add only a small fraction to our normal exposure to radioactivity. It is arguable (if not universally agreed) that they present less hazard than emissions from the equivalent coal-fired plant. However, the total radioactive content of a reactor is a million or more times its normal annual discharge. Concern therefore centres less on routine emissions than on the potential for accidents. We return to this later, but here we shall accept without much comment the 'routine' data while looking in some detail at health effects over a wide range of exposures.

In order to discuss its effects at all we obviously need to know how radioactivity is measured. The 'radiations' were introduced in Chapter 8, and one way of characterizing a radioactive source is to give the

Table 12.4 Half-lives and activities

Isotope	**U-238**	**U-235**	**Pu-239**	**Sr-90**	**I-131**
Type of particle	α	α	α	β	β
Half-life	4.5×10^9 years	7×10^8 years	24,000 years	28 years	8 days
Activity of 1 gramme*	12,000 Bq	80,000 Bq	2,000 MBq	5 TBq	5,000 TBq
Mass for 10,000 Bq	1 g	0.1 g	5 μg	0.002 μg	2 μμg†

* These are the activities of the pure samples, before they have had time to produce 'daughters' which would change the sample.

† To appreciate this extremely small mass, we can note that 50 parts per million, million, million of I-131 in air gives an activity of 400 Bq per cubic metre (the maximum permitted concentration for workers).

type of particle and the number it emits per second: its strength in *becquerels.** A piece of anything with an activity of 1MBq emits a million particles a second and so on. Table 12.4 shows the half-lives and activities of samples of a number of isotopes of interest. To put these into perspective, we can observe that one gramme is the mass of half a tea-bag (or half a dozen coffee beans); that a microgram (μg) is a millionth of this – a speck of dust; and that the encapsulated radioactive sources regarded as safe for school laboratories have strengths of about 10,000 Bq.

Does this mean then that 10,000 Bq is 'a safe level of radioactivity'? Absolutely not! First, it isn't a level but the strength of a particular source. And secondly, safety depends on *what* the source is and *where* it is. An encapsulated 10,000 Bq source used occasionally in the laboratory adds a negligible amount to the radioactivity we all receive continuously, but 10,000 Bq of plutonium in soluble form entering the bloodstream through a scratch is serious enough to justify cutting away flesh to prevent its distribution. To assess safety – or harm – then, we evidently need to measure something other than strengths of sources.

A relevant quantity which *can* usefully be determined is energy: the energy deposited in living matter by the high-speed particle or penetrating radiation. When either of these passes through matter it

* The becquerel has replaced the earlier unit for the strength of a radioactive source, the curie. A one-curie source would emit 3.7×10^{10} particles a second, so the conversion is approximately

$$1 \text{ Ci} \simeq 40{,}000 \text{ MBq}.$$

tears apart the atoms or molecules, and one measure of the damage is the energy it uses in doing this. So we have a new unit:

- One *gray* is the quantity of radiation which deposits one joule of energy in each kilogram of material.

Notice that we have moved away from the source and its strength to the recipient and the effect, and we can see at once why a 10,000 Bq source sitting on a laboratory bench is a very different matter from a 10,000 Bq source in your bloodstream or lungs. If it is an alpha source, the radiation in the first situation may not even reach you whilst in the second it will be absorbed continuously in a rather small quantity of tissue; very different numbers of grays. (Of course doses *can* come from external sources. Gamma rays or X-rays, energetic beta particles and neutrons will all penetrate living matter from outside.)

The dose in grays is still not the end of the story, because not all grays are equal. The biological damage resulting from a particular dose in grays is roughly ten times greater if this dose comes by means of heavy particles (alphas or neutrons) than if it is due to betas, gammas or X-rays. Strictly, the damage depends on the type of particle and its energy, the type of tissue, and also the *rate* at which the energy is delivered, but we shall adopt a usual practice in simplifying to the ten-to-one ratio. To distinguish, then, between the energy dose in grays and the biologically effective dose, we introduce a third unit, the *sievert* (Sv).

- We take the dose in sieverts to be equal to the dose in grays for beta particles, gamma rays and X-rays.
- We take the dose in sieverts to be ten times the dose in grays for alpha particles and neutrons.

It is this dose in sieverts then, which must be measured or calculated in order to assess health effects.*

How large a dose is one sievert? What effect does it have? Data come from three main sources: laboratory experiments on tissue samples or on animals, studies of those unfortunate groups of people who have been exposed to high radiation levels, and epidemiological studies of larger populations at low doses. They show that radiation in large single doses can produce acute effects with symptoms which appear within hours or days, but that in addition – and over a

* It is still fairly common to find data given in the two older units, the *rad* and the *rem*, which are now superseded by the gray and the sievert respectively. Conversion is easy, because 1 gray = 100 rads and 1 Sv = 100 rems.

very wide range of doses and exposure times – it can also produce long-term effects. It can induce cancers and it can cause mutations: genetic changes which appear usually as deleterious effects in later generations.

To summarize the results of many years of research in a few lines may mean dangerous over-simplification, and the figures which follow omit such important matters as the specific effects of radiation on particular organs in the body. Nevertheless, they do give some idea of magnitudes, if only on a rather broad scale. The list starts with massive doses and works down to the very small amounts, in each case for whole-body doses.

- A dose of 10 Sv or more in a short period almost certainly means death within hours or days.
- Doses of 1–10 Sv lead to radiation sickness and disability for weeks or months, and can be fatal.
- Below 1 Sv the symptoms decrease, until at about 0.1 Sv there may be no immediately obvious effects.
- Over a wide range of doses, whether received by a few people or a large population, in a single dose or over a long period of exposure, the long-term rate of induced cancers appears to lie between *one and two per 100 person-sieverts of total exposure.*
- The total number of serious genetic defects down to about the tenth generation is thought to be of the same order as the number of induced cancers. (This figure, coming mainly from studies on mice, is recognized as being very tentative and subject to correction by at least a factor of five each way.)
- The *annual* radiation dose to which we are all continuously subjected lies in the range 1–2 *milli*sieverts. (Table 12.5 shows some details of this.)

The figure for the rate of induced cancer plays such an important role in discussions of nuclear power that it may be worth spelling out its meaning in more detail. It implies that, in a group of people all receiving doses of 5 Sv (roughly the maximum of Hiroshima survivors) between one in ten and one in twenty will eventually develop cancers; and at the other extreme, that the 1 mSv a year natural background will lead to 10–20 cancers for each year of exposure in a population of a million people. (The annual death rate from all cancers would be a little under 2,000.) However, we should note that these interpretations say nothing about *when* the cancer will appear: another element of chance. They also assume that the

Table 12.5 Annual radiation dose, Britain

The figures are population averages. The dose to a particular individual could be greater or less than that shown in all cases.
The figures are 'body averages'. The dose to a particular organ could be greater or less than that shown in all cases.

Annual dose to an average person	**Microsieverts**	
From natural sources		
cosmic radiation	300	
environmental radio-isotopes	450	
radio-isotopes in the body	250	
Total, natural background		**1,000**
Man-made contributions, medical		
X-ray diagnosis	300	
radiotherapy	120	
radio-isotopes	20	
Total, medical		**450**
Man-made contributions, non-medical		
weapons tests, fall-out	60	
miscellaneous (industrial radio-isotopes etc.)	10	
nuclear power programme	3	
Total, non-medical		**70**
Total		**1,500**

effect is proportional to the dose, no matter how small; that in the sense discussed in the previous section there is no threshold. By no means all experts agree that this is so.

It is on the basis of figures like these that international standards for acceptable exposures have been set (Table 12.6). These standards are used by many countries in legislation governing permitted exposures and permitted discharges to air or water from nuclear plants. Regulations must of course take into account the

Table 12.6 Recommended allowable doses

These limits are recommended by the International Commission on Radiological Protection.
Doses to particular organs are the subject of more detailed recommendations.

Maximum cumulative dose in one year	**Millisieverts**
Occupational, to any individual	50
Population, to any individual	5
Population, average dose	1.7

particular danger from isotopes of elements such as strontium, iodine, carbon, etc. which can remain in the body for many years, and must try to allow for all the routes by which these can reach the population.

The magnitude of the problem of discharges – or of its inverse, containment – can be illustrated by a few figures for one fission product, iodine-131, a beta-emitter with a half-life of about eight days. Iodine is a solid at room temperature but vaporizes easily and therefore usually appears as a gas in the hot interior of the reactor. Discharged in the atmosphere, it reaches us either directly in the air we breathe or by a more circuitous route when deposits on grass are eaten by cows whose milk we then drink. I-131 is a health hazard because iodine concentrates in the thyroid gland, and concentration of any radioisotope means an increased dose to the surrounding tissue. Of course the short half-life means that long-term accumulation is not a problem: over nine-tenths of the activity will have disappeared in a month. Estimates of routine I-131 emissions vary, but usually range from under 1 MBq to about 10 MBq a day for a 1,000-MW plant. To convert these figures into expected health effects we would need data on weather, population distribution etc., and a more detailed analysis than is possible here; but it is fairly safe to conclude that this low level of discharge is unlikely to induce even one cancer over the life of a power-station. However, the *potential* for release is much greater: more than 100,000 million times greater, because the total I-131 content of a reactor can exceed an *exa*becquerel. In normal operation most of this, like other fission products, should be effectively encapsulated in the fuel cladding, but the potential for release is there if something should go wrong. It is estimated, for instance, that about an exabecquerel of radioactive iodine was released by the failure of fuel cladding in the Three Mile Island accident, and it speaks well for the containment that under 1 TBq – less than a millionth – escaped to the environment.

I-131 is of course only one of a number of radioisotopes emitted in routine discharges to the atmosphere, to ground water and to the oceans. And the discharge from the reactors themselves is in turn only one of a number of sources of environmental radioactivity associated with a nuclear power programme. Processes such as fuel fabrication and transportation contribute. Reprocessing, if carried out, is likely to produce greater emissions than the reactors themselves. Then there are the uranium mines. Lung cancer due to inhaling radioactive radon gas – inevitably released when uranium ore is broken up – is a classical example of occupationally-linked disease, and emissions from mine tailings have been responsible for some of the highest radiation levels to which population groups have been routinely exposed. Both these hazards can be reduced by proper management, but recent American figures, allowing for such reduction, still estimate the population exposure from mining and initial processing to be twice that from reactor operation. (This exposure will not of course appear in data for countries such as Britain that import their uranium, but this can hardly justify ignoring it.)

No one will be surprised to learn that data for routine exposure to radioactivity are the subject of a great deal of controversy, and the reader who has followed the discussion in this and the preceding section will be aware of at least some of the reasons. So, once again, the figures we give must be treated with great caution, as indicating little more than orders of magnitude. One main point to bear in mind is that whole-body population exposures are averages in two ways. They can hide large variations between groups of people and large variations in exposure of different organs. We take Britain as our example, which means that current data include contributions from reprocessing but not from mining. For the USA, with mines but no civil reprocessing, and again roughly 10 per cent of electric power from nuclear plants, the averages are not significantly different. In neither case are full data available for exposures due to nuclear weapons programmes – an obvious difficulty for epidemiological studies designed to assess the effect of *power* programmes. Estimates of total emissions and studies of critical pathways, records of levels near nuclear plants, and data from individual monitors worn by workers are all used in reaching the final figure, and that which is given here has been accepted by a number of recent analyses.

Subject, then, to all these reservations, the present annual level of whole-body exposure in Britain due to the nuclear power industry may be expressed in three equivalent ways:

- The total annual exposure is about 150 person-sieverts.
- The average annual exposure of each member of the population is about 3 μSv (microsieverts).
- The annual exposure per gigawatt of installed capacity is about 25 person-sieverts.

From these, and using data introduced earlier, we can draw some conclusions and make some comparisons. The number of future cancers attributable to the present nuclear power programme is about two per year of operation. The average radiation dose of 3 μSv is less than half a per cent of the natural background dose, and if all Britain's electric power were to come from nuclear reactors, would still be only a few per cent. These are the reasons behind the remark at the start of this section, that *routine* operation of *current* nuclear reactors does not constitute a major *present* hazard to health.

The carbon dioxide problem

As we saw in Chapters 6 and 7, the burning of fossil fuels inevitably produces carbon dioxide – a few tonnes for every tonne of fuel burnt. It follows that a world population consuming about 10 million MJ of fossil fuel energy each second throws into the atmosphere over 500 te a second of CO_2.

Now CO_2 is not toxic in low concentrations. It accounts for about one three-thousandth of the normal atmosphere, and the 20,000 Mte we add each year is a small fraction of the millions of megatonnes already there. Nevertheless it causes concern. Not because of any direct effect on health but because even a small increase in CO_2 concentration may change the world's climate. How much it will change it, or even which way, is unfortunately not clear.

We do know two things: that the amount of CO_2 in the atmosphere is rising, and that this is likely to affect the temperature at the Earth's surface. The amount has probably increased by a fifth in the past century and the present rate of rise is about one third of a per cent a year. The predicted effect on temperature comes from the fact that the CO_2 transmits most of the solar energy falling on the atmosphere but absorbs infra-red radiation travelling outwards from the Earth (the greenhouse effect mentioned in Chapter 10). If so, rising amounts of CO_2 will provide a better and better blanket to keep us warm.

Put this idea together with the assumption that within a century we'll be burning fossil fuels at 100 times our present rate and you can produce some startling figures. It was not uncommon a few years ago to find writers predicting temperature rises of more than 10°C (20°F), with the polar ice-caps melting and correspondingly dramatic changes in our environment.

Nowadays the suggested rises are more likely to lie between half a degree and five degrees. (Some even expect a *fall* in certain regions!) The generally lower figures follow in part the reduced estimates of fuel consumption, but the range reflects a much more cautious approach which has come with growing awareness that predicting climate is a very complicated matter. (Not too surprising, when we think of the problems with those short-term variations we call weather.) The Earth and its atmosphere form such a complex system that even if all the facts were known predictions would vary widely, and of course all the facts are *not* known.

It is known, for instance, that the annual increase in CO_2 calculated from measurements around the world amounts to only half that produced by burning fossil fuels. The rest must go somewhere, and the current view is that it dissolves in the oceans. Oceanographers claim that the figures just about come out right. However, it has been pointed out that we are changing the CO_2 balance in other ways. Living plants remove CO_2 from the atmosphere (see page 102), and the world's forests account for nine-tenths of all the carbon in living matter (about as much as in all the atmospheric CO_2). It is argued that the rate at which they are being cleared is effectively adding appreciable CO_2 to the atmosphere each year. How much, is a matter of controversy, with claims at one extreme that it is greater than from burning fossil fuels and at the other that it is negligible – or even negative!

This is just one of many problems to do with the facts. But even if these can be cleared up it is by no means certain that reliable predictions of the climatic effects can be achieved in a foreseeable time using foreseeable theory and foreseeable computers. Direct measurement is almost as difficult because climate is so variable over the span of a few human generations that it is not easy to link effects and causes.

Thus, we don't know exactly how much CO_2 is being produced at present, or what happens to it. We don't know how much the amount in the atmosphere would change if we were to burn more fuel. And we don't know how such a change would affect the climate. Nor do we know whether a one degree rise in temperature would be a good thing or a bad thing. It is hardly surprising that the final sentence of almost every account of this subject is somewhat as follows: 'The CO_2 problem is potentially serious and we must certainly bear it in mind.'

Probabilities and consequences

It is proposed to build a nuclear power-station near your home, and the chances are one in ten million per year, you are told, of an accident serious enough to cause ten immediate deaths, evacuation of the town for months and 10,000 ultimate deaths from cancer. How do you react?

A: 'That would be terrible. I'd rather live without electric power.'

B: 'So what. In 10 million years I'll be dead anyway.'

C: '£50,000 to a penny that it'll survive the first year!'

D: 'How do they know?'

Accidents are not of course confined to nuclear reactors. We've seen that they still cause some fifty deaths a year in Britain's mines. And North Sea oil rigs can collapse, liquefied natural-gas tanks explode and dams burst, killing hundreds in a single incident. Air disasters bring similar numbers of fatalities, and of course we kill ourselves in our ten thousands on the roads each year. In contrast, there has yet to be even one immediate fatality in Britain or the USA from a reactor accident in a public supply nuclear power-station. (The qualifications limit the statement to situations where we have reliable information.) Nonetheless it is the reactor accident which causes concern – which hits the headlines even when nobody at all is injured. In an earlier chapter we looked briefly at possible reasons for the unease, but now we must try to go further. We want to find out whether there is any rational justification for this particular concern with reactor safety. Or is it, as many in the nuclear industry seem to think, entirely due to a combination of ignorance and malevolence?

It is not the intention here to try to assess the probabilities or the consequences of particular accidents in energy systems. Only someone with detailed expert knowledge can do that. What we can do is to ask in general terms how they reach their figures. What sort of methods are used and what sorts of factors are taken into account? How *should* we react to statements like that at the start of the section?

Chances mean probabilities, and we begin with the obvious point that probabilities are not certainties. The chances are one in six that your dice will show a two. This does not mean that every sixth throw will be a two. Nor even that one throw in every six will be a two. It is perfectly possible that you get no twos in a dozen throws, or that you throw half a dozen in succession. In this sense our respondent B shows a poor appreciation of probabilities, while the betting man at least understands what they mean. We, however, shall adopt our usual role as expressed by D. How does anyone reach a figure for the probability of a certain type of accident?

The whole idea of expressing risks in this way is fairly new (to engineers, if not to gamblers), but when you do decide to look for numerical estimates one rather simple method is likely to suggest itself. Use past history. You can ask different sorts of question. 'What is the average life of this type of light bulb?' or, 'What proportion of these cars arrive without steering-wheels?' The general idea is the same, and it is obvious that you can plan on this basis, keeping an appropriate stock of spare bulbs, for instance, or steering-wheels.

The method is not foolproof of course. For one thing, history may not repeat itself. One minor reactor accident was complicated by switches which stuck, a fault traceable to the fact that a previously reliable supplier had moved his factory, taken on new, inexperienced workers, and failed to notice the quality deterioration. But on the whole the approach has proved sufficiently reliable to be used routinely as a tool by designers of many complex systems.

This sort of risk assessment seems to have moved from the design office to the public domain largely with rising concern about reactor accidents; and indeed it could be seen as a counter to the type of reaction expressed in the first of our four responses. As people were made aware of the appalling consequences of the worst conceivable accident it became necessary to point out (or to find out, according to your view) just how small the probability was that such an accident would occur. A rational person, it is argued, will reach decisions on a basis of *probability multiplied by consequences*. If I'll accept a situation where there's a one in five chance of two fatalities a year, then I should accept a one in 5,000 chance of 2,000 fatalities a year. And similar reasoning should determine my choice between alternatives.

Accidents in complex systems like power-stations are caused by combinations of circumstances, so we need combinations of probabilities to assess their likelihood. If one in every 500 jam doughnuts has no jam and one in every 2,000 meat pies has no meat, then the probability that your lunch will turn out to be a meatless pie *and* a jamless doughnut is one in a million. If we know the individual probabilities we can calculate the overall probabilities for complex sets of events – like a switch failing to open *and* a valve failing to close *and* . . . 'The reasoning can be run either way. You can say that, in order for the core to melt, the following *and* the following *and* the following must all happen, and then calculate the probability of this conjunction. Or you ask what happens if X fails, and in addition Y fails, and so on, and in this way find the probabilities and consequences of all possible accidents.

There are complications of course. If your meat pie and doughnut were made by the same baker, who was having a bad day, the coincidence of two missing fillings would be more likely than the calculation suggested. Your lunch would be the unfortunate result of a *common mode failure*. Allowing for common mode failures is essential in assessing risk, and eliminating them is an important aspect of design. The best-known example of what not to do is probably the 1975 BWR accident at Brown's Ferry in Alabama, where a single fire affecting hundreds of cables put out all the carefully-planned emergency cooling systems in one go. Cables are not now usually run in common channels.

Two other difficulties are 'unknowns' and people. With any new technology there will be parts of the system for which probabilities based on past history are not available. Risks must then be assessed by calculation, analogy with similar cases, and where possible experiment. The behaviour of the massive pressure vessel of a PWR under the stresses resulting from an accident, and the behaviour of the emergency cooling water when it hits the hot core, are examples where estimates rely heavily on theoretical predictions, and there is controversy over both.

People present similar problems. Brown's Ferry is only one of many cases where human ingenuity *prevented* serious accidents; but another well-known incident of recent years, at Three Mile Island, seems initially to have been the result of a determined effort on the part of all present to counter the action of the safety systems. (It is only fair to add that these people were in a control room with at one moment *over 100* alarms sounding and lights flashing. Perhaps they should be congratulated for staying at all.) Any estimate of risk which ignores the human factor will obviously be of little value, yet to give figures for the chances that people will behave in certain ways clearly introduces a whole new spectrum of uncertainties.

Despite all these difficulties, estimates are made. The most famous is the Rasmussen Report of 1975, the result of three years' work, at a cost of $3 million, and consisting of some dozen volumes of detailed analysis of probabilities and consequences of LWR accidents. (For details, see Further reading.) The Rasmussen conclusions have been criticized, not least for the degree of certainty they claim. Nevertheless it is a major work which has provided the basis for a number of subsequent analyses. There has been little chance to test its predictions since the number of serious reactor accidents has been small; but it did give a probability between one in 300 and one in 30,000 per reactor-year for an incident of roughly the TMI type, which in fact occurred after a total of 400 reactor years. (It is very odd to find in a recent major energy study the claim that therefore 'the range has proved to be correct'. They must lack experience with dice.)

The Rasmussen Report and its successors have been attacked not only on matters of detail but on the more general ground that most major accidents of any sort have been caused by totally unforeseen factors, rather than unlikely combinations of known possibilities. This may be so, but it can still be argued that the detailed step-wise assessment is useful. Not only is it an aid to designers; at the very least it offers lower limits for the likelihood of various happenings. (After all, if analysis showed a one-in-three chance of a hundred-fatality accident in the coming year we'd surely take notice of it.) So, with all the reservations, we'll nevertheless look at a few figures.

Table 12.7 shows estimates of probabilities and consequences for three increasingly improbable major accidents: a core meltdown with little or no release of radioactivity, a meltdown with an explosion violent enough to breach the containment, and the second of these again but with the reactor in a densely populated area and weather conditions which maximize the effect, exposing a total population of 10 million or so. The first column of figures reflects the range of physical situations and hence the amounts of radioisotopes released. The next three include the further uncertainties discussed earlier, about the resulting doses and their effects. (The Windscale fire of 1957, for instance, released some 1,000 TBq of I-131 but is claimed not to have caused a single death.) Then there are the estimates of probabilities, and it must be noted that even the wide ranges shown here are too narrow for many critics. Finally there are the numbers obtained by multiplying deaths by probabilities. With some 200 reactors in use world-wide, these can be interpreted as the predicted numbers of deaths in an average five-year period.

All these figures are of course only elements in a long and complex analysis which goes on to work out death, sickness and even property damage rates per 1,000 reactor-years from 'all possible' accidents under a variety of circumstances. We shall not follow them further. We've got an idea of the magnitudes involved, so let's see how these death rates compare with others.

The death rate for any population is well-known: it is one per person. Try again. If you are the average inhabitant of a Western country your chances of dying in the coming year are about one in a 100. About half a million people die each year in Britain, some

Table 12.7 Estimates of accident probabilities and consequences

See the main text for further explanation and comment on these figures.

Event	Activity released (curies*)	Total dose to population (person-sieverts)	Consequent number of deaths		Probable number of events per million reactor-years	Predicted number of deaths per thousand reactor-years
			Immediate	Delayed		
Meltdown without major breach of containment	0–10,000	0–1,000	0	0–10	10–100	0–1
Meltdown with breach of containment, under average conditions	10–100 million	0.1–1 million	1–10	1,000–10,000	0.1–10	0.1–100
Meltdown with breach of containment under worst conditions	10–100 million	1–10 million	1,000–10,000	10,000–100,000	0.001–0.01	0.001–1

* To obtain orders of magnitude in TBq, divide the numbers given by 20.

100,000 of them from cancer; and if the figures earlier in this chapter are correct, about 1,000 of these cancers are due to natural background radioactivity. Now let us suppose that nuclear power-stations are to provide almost all Britain's generating capacity. Say 50 GW, or roughly 100 reactors. Routine emissions (if the figures cited are reliable) would then account for about ten cancers a year. And if we take the 'worst case' from this section, accidents would add another ten – on average.

And there's the rub. In the last two words. When all the sums are done, how *do* we respond to probabilities? If the rational response is to compare probability-times-consequences for different alternatives, then as far as routine emissions and reactor accidents are concerned the figures – if even remotely reliable – make nuclear power at least as safe as other energy sources. But perhaps this response is not appropriate for very large-scale, low-probability catastrophes? It has been argued that, as individuals or societies, we normally don't bother at all about risks whose chances are less than perhaps one in a million per year. We do in practice adopt response B of page 259. On the other hand there are those who would argue for A: that if the consequences are sufficiently awful we shouldn't accept the risk at all, no matter how low the probability. To dismiss any of these three views as 'not rational' seems, to the present writer at least, neither justifiable nor particularly useful.

13 Prediction

Exponential growth

Despite their remarkable accuracy, predictions based on incantations over boiling cauldrons have fallen from favour in recent times, and today's presidents and prime ministers, whatever methods they actually use, expect their advisors to provide a more scientific justification. The attempt to develop better forecasting methods, and in particular to find a scientific basis for predicting *demand*, is one main subject of this chapter. The other is the question which has brought the word 'energy' out of the technical journals and into the daily headlines. How much oil is there still in the wells? Or more generally, how much do we know about *supplies*, not only of oil but of other fuels, and not only for the present but for the future? We'll look at these questions in the later parts of the chapter.

Suppose we begin at the beginning and ask how one might go about constructing a science of prediction. Any science needs rules – scientific laws – and it's always a good idea to try the simplest first. Let's start with a fairly obvious one:

- Things will be the same in the future as they were in the past.

In energy matters many of us seem to work on this assumption. We expect to need about the same energy next year as this, and we assume that we'll get it. Unfortunately we've failed to notice that we've actually demanded just a little more each year, and a glance back at Chapter 2 shows that the above theory, used thirty years ago, would have given very poor predictions indeed. So let us move up a step. We observe that there have been certain trends, and this suggests an improved theory:

- *Trends* will be the same in the future as they were in the past.

This too seems reasonable, and it is by no means a new idea, but its consequences can be startling. A number of writers during the past half-century have looked at the trend in world oil consumption and worked out what would happen if it continued. Their results were alarming, but most of us took very little notice – and consumption continued to rise as they had predicted.

Unfortunately for the scientific method, it wasn't the sudden dawn of understanding which at last made us pause. Wars and politics led to the shortages and price rises of the 1970s, and the lesson we learned was that the world's present oil production is determined by the world's present oil producers. However, it will be a pity if our current fascination with this obvious truth blinds us to the underlying situation. Whatever we may think of the producers, OPEC or non-OPEC, national or multi-national, the idea that if only 'they' would leave things alone there'd be cheap oil and gas for every one for ever must surely be a fantasy.

To see just how important it is not to treat the 1970s as an unfortunate interruption in a normal state of affairs which will continue peacefully once the fuss is over, it's worth looking at a few of those past trends and the future we face if we resume them. We'll start with world oil consumption because of its central importance. Fig. 13.1 uses data for the period from 1900 to 1980 in two ways. The first curve shows the rise of *annual* production and the second the *cumulative* production over the years. We see that we currently use about 3 billion tonnes a year, and that so far we have consumed some 60 billion tonnes in all. (We shall ignore the relatively small difference between production and consumption worldwide.)

Let's start in 1972, before the swoops and dives of recent years. How would we have predicted future consumption from the smooth curves up to that point? We can sketch various continuations but it is not easy to see that one is better than another, so we'll go to the

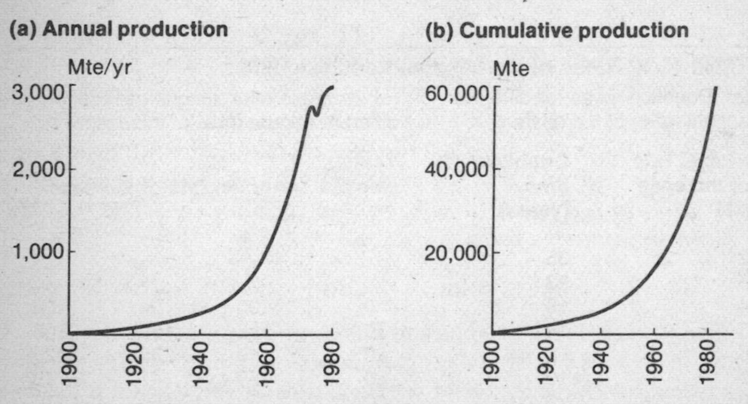

Fig. 13.1 World oil production 1900–80

numbers themselves. Table 13.1 shows annual consumption, and the pattern is obvious: it slightly more than doubles in each decade with the exception of the periods of the 1930s depression and the Second World War. For the rest, the increase is by equal *multiples* in equal times, a pattern called *exponential growth*. Further, if the doubling time is just over nine years, the increase in each year must be about 7½ per cent (Table 13.2).

Prediction on the basis of continuing trends thus means calculating the effect of this 7½ per cent rise for each of the required number of years into the future. But a graph is often more illuminating, if it is a suitable one. Having observed the pattern, we can re-draw the curve using a vertical scale where equal lengths represent equal multiples (rather than equal additions as in Fig. 13.1). If the growth is indeed exponential, this new graph will be a straight line, and predicting the future is a matter of continuing the line. Fig. 13.2

Table 13.1 World oil production 1900–70

Year	Annual production (million tonnes)
1900	19
1910	44
1920	95
1930	193
1940	294
1950	519
1960	1,050
1970	2,270

Table 13.2 Rates of increase and doubling times

(a) Doubling times for different annual rates of increase

Annual rate of increase (%)	Doubling time (years)
2	35
3	23
4	18
5	14
6	12
7	10
8	9

(b) Increase over various periods for different annual rates of increase

Years elapsed	Annual rate of increase		
	2%	5%	7½%
0	100	100	100
10	122	163	206
20	149	265	425
50	269	1,147	3,719
100	724	13,150	138,310

shows the usefulness of the method, because it reveals much more clearly than Fig. 13.1 that the exponential growth of the period up to 1930 did not continue through the war. However, it also shows the remarkable return to almost exactly the original growth rate.

What would happen if we were to treat the 1970s as a similar interruption and resume 'natural' growth after 1980? The dotted line shows the answer: 14 billion te a year by the year 2000, and ten times our present consumption by 2010, a mere thirty years away. However, the corresponding graph for cumulative consumption, where the shaded region shows the range of present estimates of *total* recoverable oil, tells us that supplies would have run out before then in any case.

As we have said, these calculations are not new. Neither are the responses, which reflect a wide range of views about society in general and economics and technology in particular. We'll return to some of these in Chapter 15, but here we are concerned with two key questions about Fig. 13.2(b) and similar data. The whole picture would be entirely different if *either* the predictions of future demand *or* the estimates of ultimate resources were seriously wrong. If, for instance, demand for oil stopped rising and continued at its present rate supplies would last for more than eighty years, and our problem would at least be less immediate. Or if there were 600 rather than 300 billion te of recoverable oil, we'd gain ten years or so even at the historic growth rate. So these will be our two topics:

- Is it possible to predict demand other than by simple extrapolation from the past?
- What do we know about the total resources which are potentially available?

First, however, a little case-study to show that although planning may be wise it's not always easy.

Predicting power

As one example of the difficulties and dangers of prediction we'll look at the efforts over the past 35 years of that much-abused body the CEGB. The Central Electricity Generating Board (or its predecessors) are responsible for power-stations in England and Wales, and like any electric power company must satisfy two demands from consumers. Naturally we expect them to supply the total ouput we want in any year – the total *gigawatt-hours*. Further, we expect them to supply it when we want it. Late in the afternoon

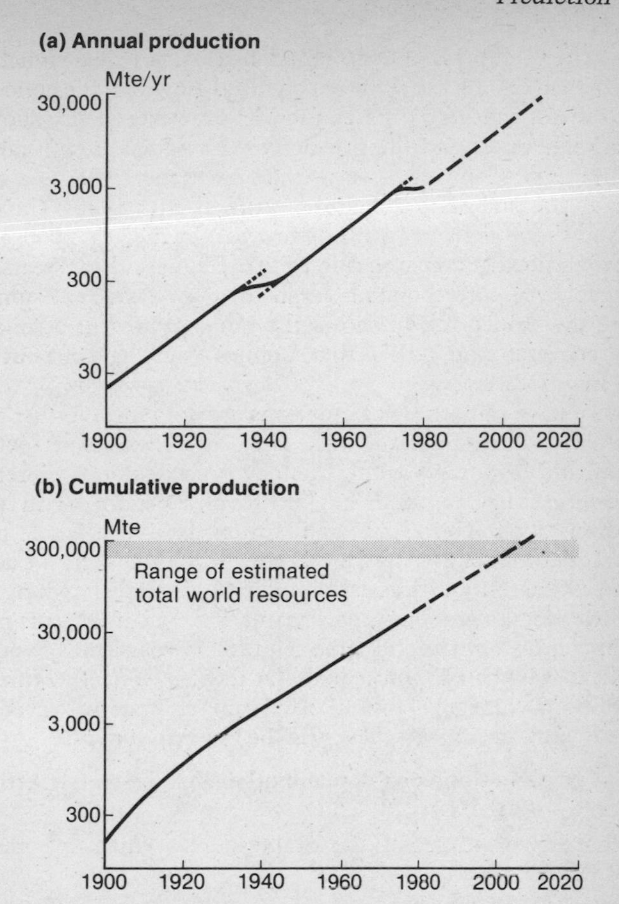

Fig. 13.2 World oil production, exponential growth 1900–2010

of the coldest day of the winter, demand will reach a peak. We'll be using power at almost twice the average rate for the year, and if the output capacity – the total *gigawatts* – is not enough, there'll be power cuts. So, until a way is found to store electrical energy in enormous quantities the system must have sufficient capacity for this peak. To be safe, it should have enough and to spare, partly because peak demand will depend on such unpredictables as the weather, but also to allow for power-station failure. The CEGB planning margin for contingencies has risen somewhat over the years and now stands at 28 per cent.

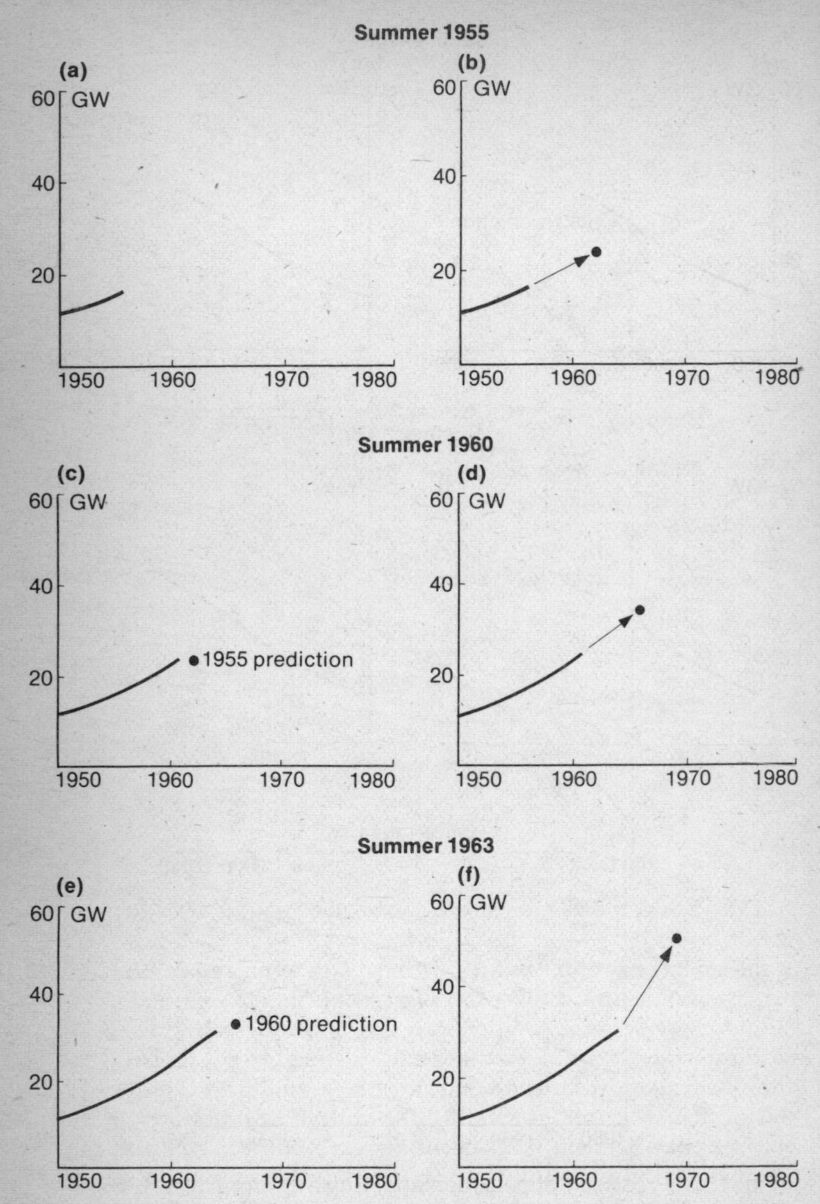
Summer 1955
(a)
60
GW
40
20
1950
1960
1970
1980
(b)
60
GW
40
20
1950
1960
1970
1980
Summer 1960
(c)
60
GW
40
20
1955 prediction
1950
1960
1970
1980
(d)
60
GW
40
20
1950
1960
1970
1980
Summer 1963
(e)
60
GW
40
20
1960 prediction
1950
1960
1970
1980
(f)
60
GW
40
20
1950
1960
1970
1980

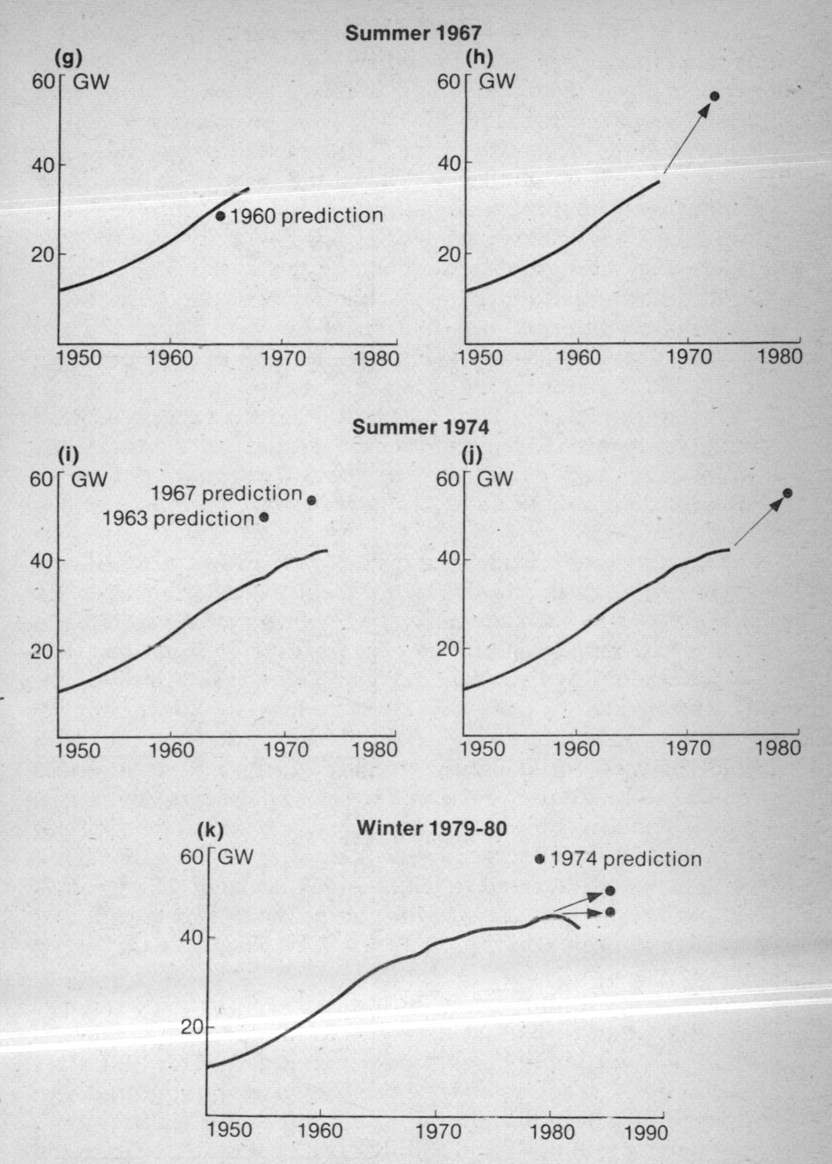

Fig. 13.3 Prediction of peak demand, England and Wales 1955–80

As long as total capacity will meet predicted demand (including planning margin) the actual numbers and types of power-stations can be chosen – in an ideal world, at least – for the best balance of cheapness and energy efficiency. Factors to be considered include the availability and price of fuel, construction times and costs, location and environmental questions, and the need to have some plant whose output can be altered quickly and efficiently to follow demand. Base load can be supplied by less flexible types if they are preferred on other grounds. So much for the ideal world.

The real world is different. In this brief study we can consider only one requirement: meeting peak demand. But we'll play fair by putting ourselves in the position of the planner, with only the past to go on. (Prediction after the event is remarkably easy.)

We start in mid-1955 (Fig. 13.3(a)) with the record of peak demand for the years since 1950, the aim being to predict peak demand in the winter of 1962, six and a half years ahead. You may care to try for yourself before moving on to (b) which shows the planner's prediction.

Unfortunately demand rose more rapidly than expected, with the predicted 1962 peak already reached (c) in 1960. You are of course aware of this in the summer of 1960, when you are required to predict the winter peak for 1966. (The real exercise is an annual one but we use roughly five-year intervals.) The next sketch shows the 1960 prediction.

Only a couple of years later (e), things are looking bad, with rapidly-rising demand and the freezing winter of 1962–3. At this point someone seems to have been struck by the idea of exponential growth. Our diagrams don't show it clearly, but for about a decade the rise had been 7 per cent a year: a ten-year doubling time. How would you then have predicted the 1969 demand? The planners seem to have been rather overwhelmed by thoughts of growth, as 7 per cent a year would have led only to 45 GW; but at least they are on the safe side (f).

And so to 1967 (g). It appears that the 1960 prediction was not so bad after all; but what has happened to the exponential growth? Where do you go from here? A prediction is needed for 1973. It's a little difficult to understand the one chosen (h), but the argument seems to have been that the falling-off in the middle sixties was a brief interruption in a trend and that the 7 per cent growth would soon be resumed.

The 1970s were of course a nightmare for all energy planners, with oil prices leaping wildly in 1973 and continuing to rise, and recession everywhere. It is difficult to argue strongly for one rather than another prediction from 1974 (i) for 1980. But (j)? The best we can say for it is that the rate is only 5 per cent per annum and not 7 per cent. Of course 1974 is already in the period where it takes ten years or more to construct a power-station, and those ordered in the late 1960s are too far advanced to be cancelled. Is it possible that predicted demand is being influenced by anticipated supply?

We finish with two predictions for 1985, one made just before and one just after the turn of the decade. The gloomy 1970s have had their effect at last. It should be said at once that there are many other examples of such downward revisions, and to be fair to the CEGB and the government departments advising them, they are not the only ones to learn that people have the irritating habit of behaving consistently for short periods and then switching to an entirely different set of rules. It would also be unfair to imply that Britain's energy planners do nothing but extrapolate from the past. Energy predictions are nowadays closely linked with economic planning in general, so our next task must be to look at the nature of these links.

Energy and economics

If prediction by extrapolation is unreliable, what are the alternatives? Should the OPEC producers or the Department of Energy carry out a sort of annual Gallup Poll, asking everyone how much oil, electric power etc. they'll need in twenty years' time? Not the most practical of schemes, and in any case the consumers would of course demand to know the *prices* in twenty years' time before answering. Whatever the method used, it is obvious that energy planning must be related to economic forecasting.

Leaving for the moment the rather separate question of prices and resources, we'll look here at three aspects of the relationship between economics and the demand for energy. There is the general question of the link between national energy consumption and national well-being measured by *gross domestic product*. There is the specific matter of the energy needed per unit of output for different products: their *energy intensities*. And there is the relationship between what you are asked to pay for energy and how much you'll buy: its *price elasticity*.

It should be noted that our study is confined to the *links* between energy and economics. We make no attempt to discuss economic forecasting itself. In Chapter 15 we'll look at a few proposed 'futures' to see what they mean in energy terms; but the methods used by economists in constructing their scenarios are beyond the scope of this book. (The interested reader will find more on this subject in some of the studies mentioned in Further reading.)

Energy and GDP

Diagrams similar to Fig. 13.4 have appeared in almost every energy book of the past twenty years. We have already met such data in the form of Table 1.2, and noted that the trend is what one might expect: wealthy countries use more energy per caput than poor countries. The question, however, is whether the reasoning works in reverse. Is a high per caput energy consumption *necessary* for wealth? More specifically, if the industrial countries want continued economic growth, does this mean continued growth in their use of energy? For a number of years the accepted view was that you sketched a line

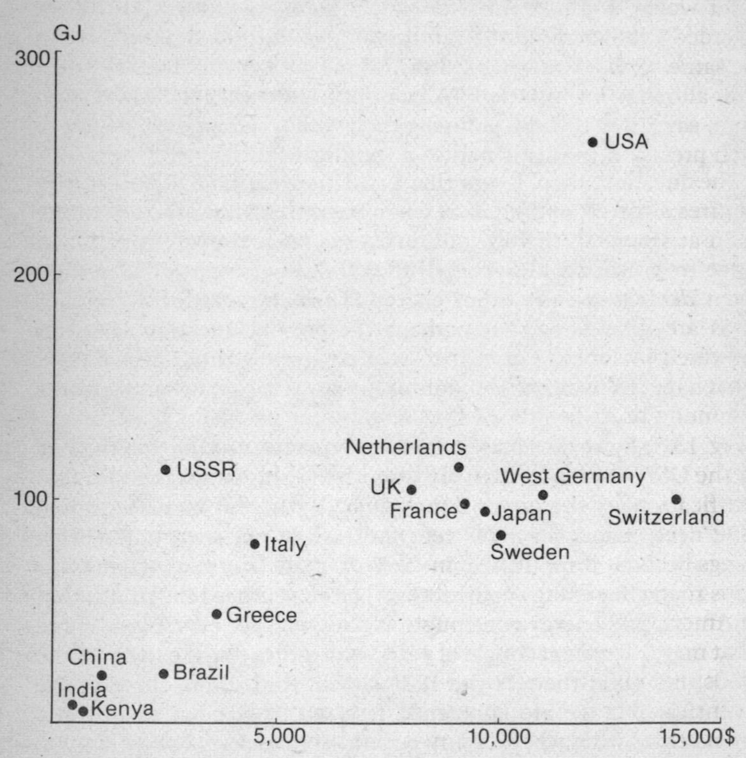

Fig. 13.4 International comparison of per caput annual energy consumption and per caput GDP

through the middle of the points of Fig. 13.4 and used it to predict that, for the per caput GDP to reach X dollars, the annual per caput energy must rise to Y megajoules.

Before looking more closely at this theory, we'd better be clear what we are talking about. The assessment of national energy consumption was discussed in some detail in Chapter 3, and the only point to mention here is that *net energy* rather than primary energy is used in Fig. 13.4, because it is probably a better indicator of consumer demand. The question of *exchange rates* was also mentioned earlier (Table 3.4, p. 44), and these obviously affect the pattern shown. However, the new feature is that we now want to talk about changes with time. Energy is not so much the problem because a joule is a joule is a thousandth of a BTU; but the 1982 pound is a very pale thing compared with its 1962 predecessor. In other words, we have inflation, and a doubling of the GDP is not regarded as an increase at all if the value of the pound has halved in the same period. So a 'change in GDP' means a change in *real terms*, after allowing for inflation. (A complication is that the conversion from, say, 1970 to 1980 pounds can vary, but our analysis is not of such precision that this makes a significant difference.)

The question, then, is whether a real increase in GDP necessarily requires a corresponding increase in energy consumption. Looking again at Fig. 13.4, the obvious answer is, 'Not if you emigrate.' A move from Britain to Switzerland doubles per caput GDP with a slight *decrease* in per caput energy. These international comparisons are interesting, but perhaps the present question is better answered by looking at year-to-year changes within one country. What happens to all those points as we move backwards or forwards in time?

Fig. 13.5 shows the answers for two countries. Taking first the line for the USA, we see a relatively smooth rise throughout the 1960s – justification for the opinion that each 10 per cent increase in real GDP needs about an 8 per cent increase in net energy. But then things become complicated. In 1969–70 there is a 'hop', when GDP stops rising but energy continues as before. (Perhaps the rule is that an American's energy consumption rises by 5 per cent a year come what may?) The next couple of years see a return to the pattern of the 1960s, but soon there is the 1973 shock. And so on, through the seventies. But we are concerned less with the detailed ups and downs than with whether any general conclusions can be drawn. There is obviously no simple proportionality between GDP and energy. The 1975 energy, for instance, is some 5 per cent lower than

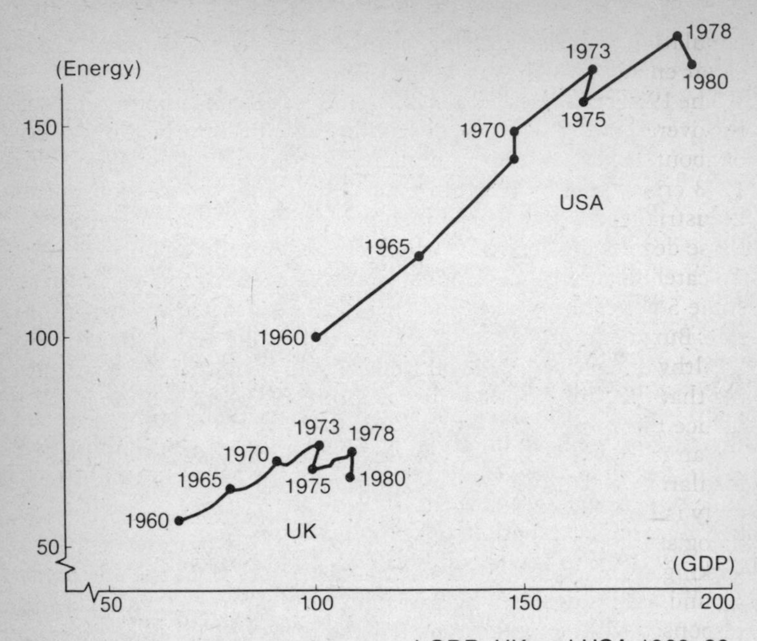

Fig. 13.5 Per caput net energy and GDP, UK and USA 1960–80
Both the per caput energy and the per caput GDP (in real terms) are given as percentages of the *USA 1960* values.

for a point with the same GDP on the line between 1970 and 1973. The rather drunken line near the bottom of Fig. 13.5, reflecting Britain's varied economic life during the past two decades, shows a similar fall in the energy/GDP ratio; and again we might ask why, if a lower energy is possible, was a higher route followed in the first place. Why use more energy than necessary? It may not answer the question completely, but certainly one difference between 1972 and 1975 was the *price* of energy, and we'll return to this later.

We shall not follow the changes in our other two countries in such detail, but the Indian situation shouldn't pass without comment. In the developing countries, because they are indeed developing – developing energy-intensive industries, and power systems where there were none before – the rate of growth of commercial energy consumption has tended to be considerably *greater* than the rate of growth of GDP. In India during the 1960s the GDP grew at about 3 per cent a year and energy consumption at nearly twice this. The population was also growing, at over 2 per cent a year, so the per caput energy growth was 4–5 per cent.

The 1973 price rises led to a fall in India's oil consumption and cut the overall rate of rise to a per caput average over the next five years of about $2\frac{1}{2}$ per cent. As a number of writers have pointed out, the 1973 crisis, seen by Western eyes as the cause of recession in the industrialized world, had a much more severe economic impact on those developing countries which depend heavily on imported oil. To catch up at all, a developing country needs economic growth at some 5 per cent or more, and energy growth at perhaps twice this rate. But energy growth at 10 per cent is possible only if you are very wealthy or have very large oil wells of your own, so it is not surprising that the Third World is just as interested as the West in how to reduce the energy/GDP ratio.

Can it be done? Ask a technologist and the answer will involve insulation, waste heat, and such matters. Ask an economist and the reply is likely to start with prices. We've tended to adopt the technologist's approach so far, pointing out the possibilities and then looking at their costs. Most energy forecasters work the other way around, assuming that if the pressures are great enough the technical responses will be found. Whether the pressures are to come from high prices or government intervention – or high prices due to government intervention – depends on your choice of economist, but we'll finish this discussion of national data with an outline of one approach.

Suppose that the aim is to predict the demand for energy in ten years' time. The method works as follows:

(1) *Energy/GDP ratios* for different years are calculated using data such as those in Fig. 13.5.
(2) Average energy *prices* for the same years are calculated.
(3) The data from (1) and (2) are used to find a relationship between energy/GDP ratio and price.
(4) Normal economic forecasting techniques are used to predict the price of energy in ten years' time and
(5) the GDP in ten years' time.
(6) The energy/GDP ratio in ten years' time is found from (4) and (3).
(7) Finally, the energy demand is calculated from (6) and (5).

You might just possibly be able to foresee a few difficulties – and you may care to try using Fig. 13.6 to carry out step (3).

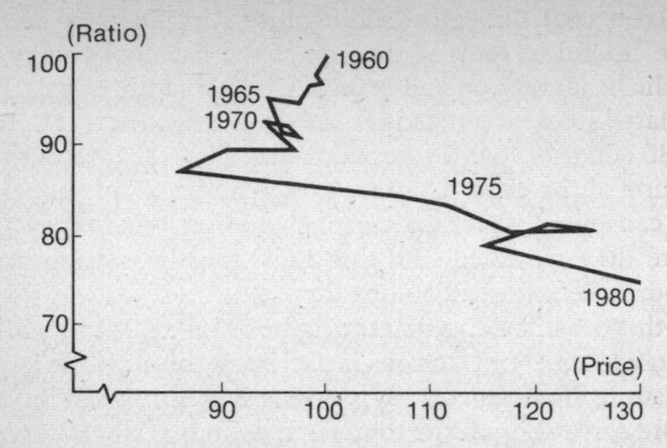

Fig. 13.6 Energy/GDP ratio and price of energy, UK 1960–80
Both the energy/GDP ratio and the average price of energy (in real terms) are expressed as percentages of the 1960 value.

Energy Intensities

One criticism of the simple energy/GDP approach to prediction is that it is too coarse-grained. The same GDP could result from very different mixtures of economic activity; different types of industry, different proportions of industry and agriculture, entirely different patterns of domestic consumption, etc. And the same energy consumption could of course be in different forms and for quite different purposes. Whether or not this explains the lack of success in the search for a simple relationship between energy and GDP, it does at least suggest an alternative. Why not separate our activities – motor car production, house-heating, paper-making or egg-laying – measure the energy needed per unit output, per household, etc., decide how much of each activity we'll have in ten years' time and work out the resulting total energy. If we expect to produce half a million more tonnes a year of crude steel or paper towels and we know that the energy input per tonne (for either) is about 20 GJ, then we know that we'll need an extra 10 PJ a year for this purpose. A bonus is that we also know whether we need it as 400,000 te of coal, 3 TWh of electric power, or some other mixture. And a further bonus is that the exercise may show up possibilities for conservation.

There are some decisions to be made, however, before the method will work. Consider tin cans. Some 10 billion a year are used in Britain. Suppose we expect to use 10 per cent more in five years' time. We look up tin cans and find that the energy needed to manufacture one billion is about 4 PJ – equivalent to roughly

150,000 te of coal. But before adding this to the predicted demand we must be careful to look at the rest of the national energy budget. Under the heading 'Iron and Steel' we may find the extra 100,000 te of tin-plated steel for our cans already entered, with its allowance of some 130,000 te of coal. So we need add only 20,000 tce for the actual production of the cans. Then we notice that a couple of per cent of the 'tin can energy' is for delivering them from where they are made to where they are filled, and this 3,000 tce already appears in the 'Transport' section. And so on.

Now this doesn't mean that the figure of 150,000 tce per billion for the *energy intensity* of tin cans is useless. If the question is whether we consume more energy by packing soft drinks in tin cans or cardboard cartons, then the *total* energy input is what we want, and that used in making steel or cardboard must certainly be included. But if the purpose is to estimate national demand we must divide our activities in such a way that each need is counted once and once only.

In practice the world is so full of a number of things that a very detailed *disaggregation* into 'tin can energy', 'cardboard carton energy', 'milk bottle energy', etc. would be quite unmanageable. From having too little information we have overshot to more than we can handle. Perhaps a more realistic target is to try for a division into somewhat broader categories: 'chemicals', 'paper-making', 'food', 'transport of goods', etc. An energy intensity in MJ/te or MJ/£ can then be assigned to each category. The 'food' count, for instance, would not include energy used for packaging, delivery or chemical additives. This is the present approach adopted in some national statistics.

The next question is how to measure energy intensities. It should be fairly simple. You stand at the factory gate and count the joules going in and the tonnes or £s coming out. Or more realistically, you ask the owner to do effectively this. Much information on energy intensities has in the past come in a rather indirect way from published financial accounts. The energy input is calculated from the amounts spent on different fuels, and the product output from the earnings, so that in a sense the financial accounts are turned into energy accounts. (A more wholesale version of this method associates an energy input with *every* expenditure, not just on fuel but on raw materials, bought-in components, labour, capital investment, and so on. Thus, a part of the input associated with wages is the energy needed to produce the cars in which people come to work. If the factory then happens to make glass for car windows, *its* energy per unit output contributes in turn to the energy needed to produce a car. The idea is to find a complete and consistent set of energy intensities for a whole economy. We shall not follow it further.)

In recent years, with everyone beginning to take an interest in energy costs, much more detailed information has become available. Some heavy users of energy such as steel and chemicals have made careful analyses, and there have been many studies of particular products – too lengthy, unfortunately, for inclusion here. A general impression of the British picture is given by Fig. 13.7, and the wide range of energy intensities is clear. However, these are *present* values. The forecaster needs future intensities. Shall we produce a tonne of steel, paper or polythene with less energy in the year 2000 than at present? We've seen that it is difficult to find any simple relationship between price and consumption on the national level. Can we do any better in the disaggregated world of individual types of consumer?

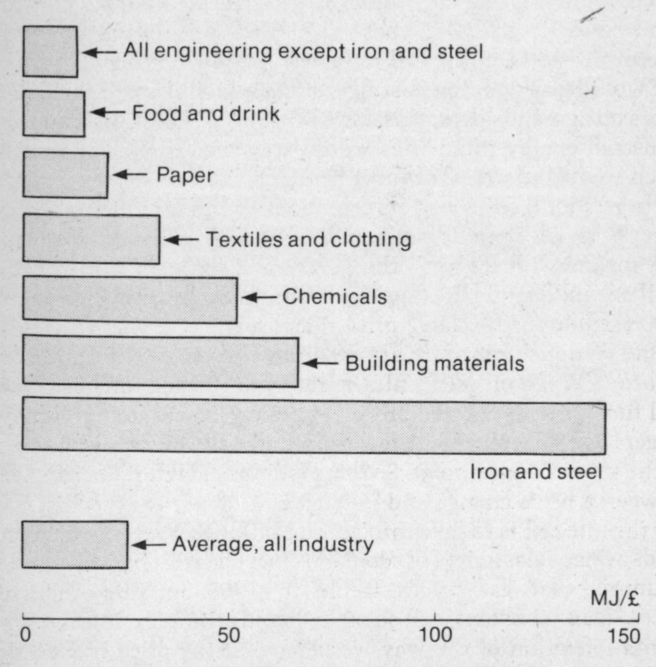

Fig. 13.7 Energy intensities, Britain

The bars show the relative intensities for different sections of industry. The scale shows the megajoules per 1980 pound.

Elasticities

We have already used the idea of an elasticity a number of times without giving it a name. Net energy consumed in the USA rose during the 1960s by about 8 per cent for every 10 per cent increase in GDP – an elasticity of 0.8. What is your *income elasticity of demand for energy*? If your income were to rise by 5 per cent in real terms, how much more energy would you consume? Two per cent? Then your income elasticity of demand is 0.4. (Of course you might invest in insulation and actually use 20 per cent *less* energy – an elasticity of minus 4!) Notice that your 'demand' is how much you'll buy in the circumstances. How much you would really like if only you had more money or it cost less is a different matter.

Price elasticities, rather than income elasticities, are our main concern, however. Consider your *price elasticity of demand for electric power*. How much less will you use if the price rises by 20 per cent? An elasticity of minus 0.5 means that you'll cut consumption by one tenth (0.5 × 20 per cent). Minus one means that you are determined to keep the bill the same come what may.

It would be a poor forecasting system which ignored the effect of prices on demand – which assumed that household consumption or industrial energy intensities would stay the same no matter how much we had to pay. Of course, if energy accounts for only a small fraction of total costs, or if its price doesn't change much, then price won't be a very obvious factor in determining demand. This was the case for many years until the 1970s, but certainly not since; and we'll see in Chapter 15 that even optimistic forecasts for the year 2000 assume further large price increases.

One complication in relating demand to price is *inter-fuel substitution*. When oil prices hit the roof you go back to your good old coal fire. Or rush out and buy a gas boiler. Except that you may no longer have a fireplace, and few people actually buy heating systems on the spur of the moment. So there is another factor: the *time delay* between a price change and its effect.

If the forecast is to take into account all these factors, the planner needs a price - elasticity - of - demand - for - electric - power - in - 1990 - assuming - that - gas - prices - treble - in - 1985 - and - all -other - fuel - prices - stay - constant, and so on. It sounds complex, but is no more than a reflection of the way we behave. If the price of electricity doubles, you are likely to turn down your storage heater and use the gas fire more. But not if the price of gas has trebled.

After all this you may be disappointed – but hardly surprised – to learn that there are no clear-cut, agreed figures for elasticities. It is certainly possible to put together a collection of numbers, prices and incomes and quantities of fuels purchased, analyse them and obtain price elasticities or income elasticities. As someone said in another context,* 'Given a formula, many people can calculate the unknowns with any desired precision (even with much undesired precision, if a computer is available).' You can certainly find published figures for the income elasticity of demand for household electric power. In fact you can find almost any figure you like for the income elasticity of demand for household electric power, from something greater than one down to zero – or even, as a recent writer says plaintively, 'of the wrong sign'. (Is it so obvious that people with higher incomes are bound to use more power?)

So we shall not finish with a nice table of elasticities. The justification for the discussion lies not in any numerical results but in the fact that almost any prediction of energy demand involves some assumptions about elasticities, and we should at least know enough to recognize this and look out for them – particularly when they are not spelled out. Nor are there any tidy conclusions on prediction in general; no 'best buy' in futurologies. Just the rather untidy conclusion that we need to look carefully at the data and assumptions behind any figures we are offered – and that we shouldn't be fooled into believing that any number of pages of mathematics can produce reliable predictions from uncertain data and doubtful assumptions.

Reserves and resources

There is only a limited amount of oil and natural gas in the world. This is obvious because the Earth is a finite object; but the question which concerns us in our short-sighted way is whether there is enough for twenty years or 200. The answer of course depends on demand, but the two figures indicate the range which appears in discussions of total conventional oil and gas resources. (New types of supply such as oil shales and tar sands are not included here, and we return to them later.)

If we start by asking why there should be such a degree of uncertainty, the first disagreement we find between the optimists and the pessimists is over the likelihood that there remain large unexplored fields. The general view is that present geological knowledge is good enough to rule out the possibility of further major surprises, but the optimists point to the recent example of Mexico, whose resources

* Editorial, IEE Transactions on Reliability. R 28 (1979).

are now claimed to rival those of Saudi Arabia. The next issue is the reliability of published estimates. The problem is that the most detailed data are obtained by oil companies (or their equivalent in Communist countries) and are not published in full. Various writers, holding that the figures are deliberately understated, claim that total recoverable oil under the North Sea, for instance, or in the USSR, is many times the current accepted estimates. Naturally this is denied by those who produced the figures in the first place. Finally, a related and equally important factor is that 'recoverable' quantity of oil from any field depends on the techniques used to recover it. Which means that it depends on the amount the producers find it worthwhile to spend, and ultimately therefore on the price the oil will fetch.

This question of 'economically recoverable' quantities is of course crucial. A reservoir of oil is not a large pond to be bailed out with a bucket until it's empty. There's no 'hole in the ground' – just a mesh of tiny channels in porous rock. Oil, gas and water migrate through these pores under the infiuence of local pressures until they are trapped up against caps or domes of impermeable material. Held there under pressure, they will rush out violently if the overlying material is punctured. In the first instance it is this natural pressure which propels the oil to the surface, and careful control is necessary in order to recover as much as possible before the pressure is lost. *Secondary recovery*, using reinjected gas (or water) to maintain the pressure, is important in increasing the fraction recovered, and there are also *tertiary* methods such as heating the oil to make it flow more easily – though these are not yet economic in most circumstances. Depending on the nature of the oil and the rock, and the recovery techniques used, the fraction recovered varies from less than a tenth to nearly three-quarters of the oil present.

So when the Minister informs Parliament that North Sea reserves are between 2 and 4 billion te, or the international expert talks of total world resources of 250 billion te, we need to know, at least in general terms, what sort of figures these are. Are they the amounts in existing fields where producers know pretty well what can be recovered at costs which make economic sense today, or estimates of totals which are potentially recoverable under some future conditions, but possibly not even explored yet? A detailed terminology is used to specify the different degrees of certainty in the case of oil (and gas). At one extreme are the *proved reserves*, identified and known to be economically recoverable with existing techniques. (The original idea was to protect the gullible from investing in companies offering untold mineral wealth!) Published proved reserves are almost always revised upwards over the years, and certainly do not indicate the total recoverable oil. At the other extreme, *reserves in place* or *total resources* include all there is, whether recoverable or not.

In order to have a simple way of specifying and comparing quantities – not only of oil and gas but of other fuels or energy sources – we shall use just two terms here and in subsequent discussions.

- *Reserves* will mean identified quantities known to be economically recoverable with existing techniques. In other words, the most cautious estimate, equivalent to the proved reserves of oil terminology.
- *Resources* will be used in a less restricted sense, to mean the total quantities thought to be present and which we might reasonably expect to recover. It covers a range from discovered resources, not yet counted as reserves because they aren't profitable at current prices, to undiscovered quantities in places where geological reasoning tells us we might expect them – undeveloped parts of the North Sea, for instance, or more speculatively, the unexplored Continental Shelf. The term does not mean 'all there is' because it takes into account costs and partial recovery.

The general significance of the two terms is fairly obvious. Reserves are naturally very important for the immediate future, and in the case of oil or gas they will have great national significance for the producing countries (including Britain, the USA and India). But the total recoverable resources are an entirely different matter. Their figures, if reliable, tell us how much there is before we run into serious problems which cannot simply be remedied by drilling more wells, or exploring more fields.

Needless to say, defining the terms doesn't remove all disagreement about the figures. The ranges in Table 13.3 cover the most generally accepted opinions, but as we've noted, extreme optimists would put the figures for both oil and gas several times higher. The 'energy' columns give the heat energy corresponding to the reserves and resources, to allow comparisons between fuels. (The uranium figure assumes conversion at the efficiency of the majority of present reactors.)

In the final four columns we've taken quantities in the middle of the ranges and asked how long they would last if used at present rates, or at a rate rising by 5 per cent a year. (For oil, these can be compared with the 7½ per cent growth of Fig. 13.2.) Naturally these

Table 13.3 World fuel reserves and resources

See the main text for explanations of the terms and comments on the figures.

Fuel	Reserves	Resources	Energy (EJ or quads)		Years at present production rate		Years at 5% growth rate	
			Reserves	Resources	Reserves	Resources	Reserves	Resources
Oil	600–750 billion bbl *or* 80–100 billion te	1,600–2,000 billion bbl *or* 220–300 billion te	3,000–4,000	9,000–12,000	30	80	18	30
Natural gas	2,200–2,500 thousand billion SCF *or* 65–70 thousand billion cu m	6,000–12,000 thousand billion SCF *or* 150–300 thousand billion cu m	2,200–2,500	6,000–12,000	50	150	25	40
Coal	600–800 billion te	6,000–12,000 billion te	15,000–18,000	150,000–300,000	160	2,000	45	90
Uranium	2 million te U_3O_8	8–12 million te U_3O_8	350	1,500–2,000	50	250	25	50

numbers don't mean that on 20 September of the year 2000, or even 2012, we shall suddenly find all the pumps empty. Even if the estimates are right, the more probable sequence would be rising prices and gradually falling consumption accompanied by a transition to alternative fuels. The significance of the numbers is that they show us roughly how long we've got to *find* these alternatives, and how this period depends on our patterns of consumption in the near future.

Future supplies

Having seen the uncertainties about known fuels we are hardly going to expect precise estimates for resources which are barely developed at all. Nevertheless, there are three which, although contributing jointly less than a thousandth of present world energy, are usually taken into account in estimates of future supplies. They are the tar sands and oil shales, and geothermal energy. One reason for their inclusion is that they are potentially large sources, but the first two also have a special value. If conventional oil resources are indeed sufficient for only a couple of decades (and particularly if tertiary recovery methods fail to work with the heavier oils) there'll be a desperate search for *liquid* fuel. The only substitute we've seen so far comes from coal, won't be cheap, and has major environmental penalties. So the importance of tar sands and oil shales is that they are the principal known alternative sources of oil.

Both are solids which contain hydrocarbons, and thus both mean 'mining for oil'; but they are very different in character. The tar sands, as the name implies, hold a very thick oil, or tar, which must be extracted by heating to about 80°C (180°F). Oil shale, on the other hand, is a rock containing a *solid* hydrocarbon called kerogen, and the crushed rock must be processed at 400°C or more (up to 1,000°F) to produce the required fluids. In either case, between one and five tonnes of material must be mined or displaced for every barrel of oil produced. Just a hundredth of present world oil production would mean extracting 2 million te *each day*.

Canada's Athabasca tar sands, occupying an area nearly large enough to accommodate Switzerland, are the most developed of these resources, but they have not proved easy. Northern Alberta has a fierce climate, so cold in winter that the tar becomes a hard glass. Extraction of the oil has meant solving problems of huge surface-mining machines which sank into boggy peat in summer and broke their massive steel teeth after a day's use in winter.

Having lost money for a decade or so, Canada's venture is said to have become profitable in the middle 1970s and production, now some 40 million barrels a year, is to be increased until, it is claimed, tar sands will supply half Canada's oil. The total Canadian resource is said to be over 1,000 billion barrels, but less than a tenth is accessible by surface mining, and techniques for dealing with the remainder are not yet developed. The USA has perhaps a third as much, and no other deposits comparable with the Canadian are known, so the total world resource seems to be about 2,000 billion barrels.

Environmentally, extraction is as horrifying as all surface mining, and it is not yet clear how far the land can be re-established. This huge tract of virgin tree-covered swamp, mosquito-ridden in summer and intolerably cold in winter, has its admirers, but they have not so far been vocal enough to impede development.

Not so with oil shales. The proposed development of the Colorado shales has brought many objections, to the mining itself and to its heavy use of scarce water. The oil shales of the northern states are thought to contain up to 1,000 billion barrels of recoverable oil, several times America's conventional resources. Throughout the world, shales containing more than 10 per cent oil are thought to represent a total potential several *hundred* times as great as the ultimately recoverable conventional oil resources. So the question of extraction is of undoubted interest. Estimated costs tend to be above current oil prices but below those of coal liquefaction; but environmental issues add a major uncertainty, and it's easy to see why. Every tonne of oil extracted means ten tonnes of residue; and moreover, the volume of this is *nearly a third more* than that of the original mined rock. So it can't be returned underground. One of the richest shale deposits is in the Piceance Basin, just west of the Rockies in Colorado. If all its shales were mined and retorted to extract the hundred billion barrels or so of oil, the entire area of a couple of hundred square miles would be left covered to a depth of several hundred feet by a layer of burnt-out rock.

An alternative would be to extract the oil underground, by shattering the rock and then using heat from its combustion to run what is effectively a huge subterranean retort. A few hundred million dollars have been invested in this *in situ* method, but it is by no means certain that it will work in a controllable way.

Oil shales have been little developed in other parts of the world. China is reported to produce a few million barrels a year, and Scotland had a shale industry for a century or so, producing at rates up to a million barrels a year. But this became uneconomic with the low oil prices of the 1950s and production ceased. Africa has a large potential but with even less oil per tonne of rock than in the USA. So, if the pessimists are right about conventional oil, and we don't find another way of travelling, it seems that the world will be studying Colorado and its neighbours with some interest. After all, in one visit, and no more than a day's journey, we can compare the attractions of oil shales on one side of the Great Divide with those of surface coal-mining and liquefaction plants on the other.

Our last resource is the warmth of the Earth itself. For a long time it has been known that the temperatures of rocks rise as you go down, by about 3°C per 100 m, and this has two implications. First, there must be an *outward heat flow* up through the layers of rock – just as there must be a flow through a wall with a temperature difference between its sides. The flow has been measured with some care, and the average is a little over a twentieth of a watt through each square metre of ground. Not exactly an intense source, and it's hardly surprising that we don't notice it at the surface, where any effect is swamped by the much larger heat flows associated with solar radiation. The total flow from within the Earth amounts to about three times our primary energy consumption. So geothermal energy is not going to solve the world's problems. But it does fulfil one requirement. The heat is energy released by naturally radioactive uranium, thorium and potassium (mainly) in the rocks, and the supply is therefore unlikely to decline in the next thousand million years or so.

The other way of looking at the temperature gradient is to ask how far down you must go to reach a useful temperature. Again the answer is not very encouraging. Sending your domestic water over a mile into the Earth (and back) to heat it doesn't seem very likely to be cost-effective. But if we drop this rather general approach and look around us, we might observe hot springs, boiling mud pools and even geysers shooting steam high into the air. What we've not considered so far is that although the average may be 3°C per 100 m, the gradient in many places is very much greater – and the water may be down there already.

Geothermal energy in the form of *hot water* has certainly been used for millenniums. Today it provides heat for people as far apart as Iceland and New Zealand, for vegetables in hundreds of acres of Hungarian greenhouses, and for carp in Japanese fish farms. Individual householders in Nevada run simple central heating systems including a 200-ft loop down into the ground in the circuit. Some eighty countries are now studying the potentialities of geothermal heating, and there is no doubt that its use will increase. The total resource is not known, because geothermal fields have been studied far less than oil fields, but it is important to realize that these concentrations of very hot water and steam *can* be exhausted. The local supply is not maintained by the weak heating from radioactivity (even less by heat flowing from the centre of the Earth!) but by infiltration of hot fluids from the neighbouring rock. So there is a limit to both the rate of extraction and the total resource. Present estimates put the total potentially recoverable heat in the USA at about 2,000 EJ, roughly twice the sum of oil and gas resources. But in view of great uncertainties about development the figure must be regarded as very tentative.

Power generation is, of course, a relatively recent use of geothermal energy, though the Larderello plant in Italy has operated for over seventy years. It is no coincidence that this was the first, because the source produces dry steam from depths of only 500 m or so. This is rare, and one problem with the development of geothermal power is that every field produces a different combination of steam, hot water, dissolved impurities and insoluble gases. The Imperial Valley in California, for instance, is known to have enormous underground heat resources, but unfortunately what comes up is often a strong brine, and decades of effort have not yet solved the corrosion problems.

Provided there is a good flow and it's reasonably clean, temperatures as low as 180°C are still considered for power generation. (The highest known are about 350°C.) Present world capacity is some 2 GW, of which Larderello provides over 400 MW and the Geysers field in northern California, started in 1960, about 700 MW. The most optimistic forecasts suggest a world capacity in the region of 100 GW early in the next century. Even if this were reached, which doesn't seem very likely, it would equal only a twentieth of *today's* world total generating capacity. Again, the cost is claimed to be competitive with other sources of electric power, but estimates are based on the rather healthy fields used at present. Unfortunately, geothermal power is not entirely unpolluting either. The water, if it contains high concentrations of minerals, must be treated or re-injected to protect ground water. And then there is H_2S, known to every school chemistry laboratory. In low concentrations hydrogen disulphide isn't toxic, but our noses are extremely sensitive to its characteristic smell of rotten eggs and people living near geothermal plant, quite understandably, often have reservations about increased development.

Extraction of heat from *dry rock* is also being studied. The idea is to inject water into one bore hole and get it back hot from another nearby, thus extracting some of the enormous amount of heat in the rock itself. As water won't travel through solid rock it may be necessary to create fracture regions between the entrance and exit points. Some work has been done on this technique in New Mexico, and a £6 million project is underway in Cornwall, involving drilling 2,000 m into the granite, using explosives to fracture the rock, and high pressure water to extend the fracture region. Present data suggest that a 600-MW power plant would need a couple of hundred holes drilled to 6,000 m – the maximum practical depth with present techniques. The estimated cost for drilling alone is over £1,000 million, and there are many uncertainties. (It's not that easy to drill two *parallel* holes four miles deep.) However, as the Department of Energy is currently investing more in this project than in any other British alternative except waves (for which their support seems to be falling) the experts presumably see it as potentially worthwhile – unless it's the uranium they are after!

14 Conservation

Apology for this chapter

If a national plan for energy began by working out all its predictions in careful detail, using the best forecasting techniques, and then said, 'Now let's knock off 15 per cent for conservation,' it could possibly be accused of not having done the right sums in the first place. Indeed there are those who would argue that to deal separately with conservation at all is a sign of the wrong approach. It shouldn't be something which you grab from the medicine chest when you feel the effects of over-indulgence. Conservation should be a way of life, not a remedy for a hangover.

All true, no doubt. But as we have already identified a number of examples of conspicuous waste, perhaps it is worth looking at possible remedies to see how effective they might be – and how cost-effective. The second point is important; you don't need expert knowledge of price elasticities to guess that conservation measures are more likely to be adopted if they save money as well as energy.

We'll begin by distinguishing three sorts of measure. The first are the improvements which require only *effort* – common sense and 'good housekeeping'. We'll start with a short review of some savings which they might bring. Then there are the measures which need *money*: investment in insulation or new plant. These are the main subjects of the chapter, although one topic (CHP) is on the borderline of the third category: changes which essentially need *time*. Of course it takes time to do anything, but the distinction intended here is that some improvements still need development time. There are technical problems still to be solved, or products not yet fully tested in use. Taking a rather cautious view we might put into this group three important possibilities discussed elsewhere in the book: more efficient internal combustion engines, solar water heating, and heat pumps. This is not to suggest that these are totally untried new concepts, but just that we don't yet know enough to quantify with any confidence their likely contributions to energy-saving in ten or twenty years' time.

Bad habits

Lights left on, heating running at full blast with the windows or even doors wide open, hot water at 65°C instead of 50°C; these are thought to account for a major part of unnecessary energy consumption in offices, shops and public buildings – the 'other' sector of the diagrams in Chapter 3. And it's not negligible in the industrial sector either: it is estimated that up to a sixth of Britain's industrial energy is used not for manufacturing processes but for space heating and hot water. How much of all this could be saved by better practices?

It is interesting to compare offices with houses. Of course the ways we use them differ but both are essentially buildings for people to live in, and they use energy for much the same purposes. Suppose we compare their consumptions on a basis of equal floor areas. The average home in Britain has a floor area of about 80 sq m and uses some 730 therms a year. Fig. 14.1(a) shows the 9 therms/sq m divided between its uses, and (b) shows the average for all 'other' buildings: about 18 therms/sq m per year (roughly this figure for shops, more for offices, and less for schools). The heating contribution, we see, is nearly *twice* that for a house, despite the fact that the normally very high levels of artificial light should be contributing appreciable extra 'free heat'. Now offices and houses are both heated to keep people warm. Let us assume that they are heated for about the same number of hours a year and that their structures are equally well – or badly – insulated. (Both assumptions could be said to overestimate the relative needs of offices.) We then have three possibilities: offices are over-heated, houses are under-heated, or offices waste more energy.

The comparison is of course very imprecise and leaves out many factors, but it does suggest that a closer look would be worthwhile.

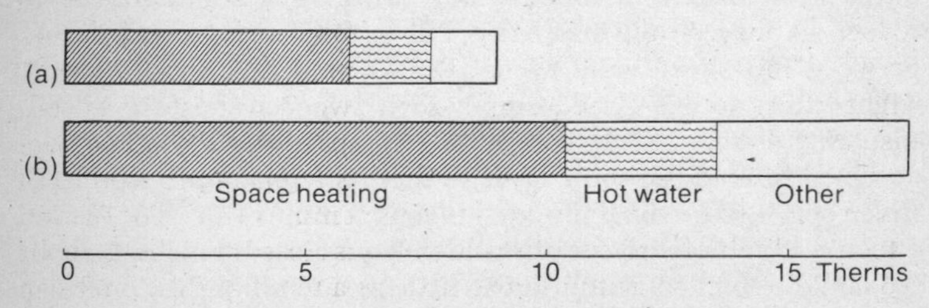

Fig. 14.1 Energy use in buildings

The bars show the annual energy consumption per square metre of floor area, for (a) dwellings and (b) offices, shops and public buildings.

We shall not attempt to reproduce the more detailed studies which have been made in both Britain and the USA, but the general conclusion is that *between a quarter and a third* of the energy consumed by this sector could be saved without any major change to structures or heating systems – just by improved habits. (What a pity that heat isn't bright red, so that we could see the energy pouring out!) Nationally, in either Britain or the USA, the estimated saving would mean a reduction of as much as a twentieth of net energy consumption – equivalent to between one and three times the total current contribution from nuclear power.

Insulation

So we can save ourselves a few hundred million pounds or thousand million dollars a year by effort alone. What if we were willing to spend a little money? Having closed the doors and turned down the thermostat, how about insulating the building? We'll talk mainly about houses, because in Chapter 3 we looked at their heat losses in some detail. We saw that a change from a present average British house to one with really good insulation could mean halving the annual heat loss. Our other two Western countries can provide an interesting contrast too.

Consider the older frame house common in many parts of the USA. An outer layer of overlapping half-inch boards, an inner layer of plasterboard, and nothing between. Single-glazed windows and perhaps a couple of inches of loft insulation. The result is an overall U value greater than two and very high ventilation losses. Fuel consumption for a household in the mid-west with American heating standards can be 4,000–5,000 therms a year. For contrast, take the Swiss mountain house – the cuckoo clock. Also wood, but an outer layer some four inches thick, with timbers interlocked to reduce air penetration. Inside this, two or three inches of insulation between foil sheets, and then an inner wall of half-inch wood. Tight-fitting double-glazed windows with wooden shutters. A well-insulated double ceiling below the roof space and a shallow roof designed to hold the snow layer so that its temperature won't fall much below zero even if the outside air is at minus 15°C. The overall effective U value (with curtains and shutters closed at night) is about 0.6, and the fuel consumption as little as a tenth of the American figure.

The comparison is of course unfair. The American house is larger (for the same family size) and hotter. We've not been careful to match the degree-days, and the solar gain during winter is greater in the Alps. Finally, these were extreme cases: an old American house and a Swiss one insulated to modern standards. As Table 3.3 (p. 43) reveals, Switzerland must have poorly-insulated houses and apartments too; whilst both Britain and the USA can offer examples, now, of houses matching our 'energy-saving' case of Table 3.9.

But unless you already live in one of these, the chances are that with better insulation and reduction of draughts you could reduce your heat losses to half their present value. Not the dramatic factor ten of our example, but a fuel saving of several hundred therms a year for the average house with central heating. Whether insulation to a high standard is *economically* worthwhile is less easy to say. There is a well-known rule of thumb, that people will invest in energy-saving measures only if convinced that they'll get their money back in two years. (This piece of folklore also appears in discussion of industrial and commercial energy saving.) It seems a rather short-sighted policy, but if true means that only householders with an extremely pessimistic view of fuel price rises will do more than scatter a few granules in the roof. The actual pay-back time will obviously depend on the cost of the conservation measures and the amount of fuel which they save, as well as trends in fuel prices. And another factor which enters is your *discount rate*, about which there is much argument.

If you owe £50 and are given the option of paying it now or in five years' time, which will you choose? Silly question. But suppose the choice is between £50 now and its equivalent after inflation in five years? The general view is that most of us would still prefer the latter. In other words, we are willing to pay more *in real terms* for the privilege of paying later. (As we may do with a deferred payment or hire purchase agreement, or a mortgage.) How much more you'll pay, for each year later, determines your discount rate, and clearly this enters when you balance your saving on fuel bills in 1990 against the sum you spend on insulation now.

Doing the sums isn't difficult. The problem lies in choosing figures for the discount rate and the future price of fuels, and estimating fuel saving for your particular house. Table 14.1 shows the widely varying results following from different assumptions. The sum you can afford to spend, for a pay-back time of ten years, varies from *six* to *twelve* times the amount you expect to save in the first year. Add to all this the fact that the cost of a given degree of insulation can vary greatly between houses, and we clearly have a situation where generalizations are not very useful. The nearest one

Table 14.1 Investing in insulation

	Case 1	**Case 2**
Assumptions		
Fuel price behaviour (real terms)	constant	+ 5% a year
Discount rate (real terms)	+ 10% a year	zero
Expected first-year saving	£100	£100
Spending on insulation now . . .		
for a ten-year pay-back time	£600	£1,200
for a five-year pay-back time	£400	£600

might come is to say that, in the British climate and a typical British house, and on assumptions somewhere between our two extremes, 'reasonably thick' roof insulation, some wall insulation and draught exclusion will probably pay for themselves in five years, while double glazing is more marginal but may save its cost if it reduces serious ventilation losses.

Another way of putting the question is to ask how much a nation should spend on conservation. Or perhaps how much a *government* should spend. We shall certainly not go into the question whether government spending is a Bad Thing or a Good Thing; but it may be worth seeing a few figures on a national rather than individual basis. The estimated total potential saving by reduction of heat losses from buildings, for instance, including the measures we talked about in the previous section, is about 10 per cent of national net energy consumption, for Britain or the USA. For Britain, this means a saving equivalent to a third of total annual consumption of natural gas – an enormous quantity worth, at final user prices, about £1 billion a year. But if the government were to pay for all the necessary work, the investment would be enormous too; certainly over £10 billion. For comparison, present annual government expenditure in Britain on all conservation in the domestic and 'other' sectors is about £100 million.

Waste heat

Each year we convert some 3×10^{20} J of chemical and nuclear energy into heat. We do it for two reasons: because we want the heat itself, or because we want mechanical energy. Looking back at Fig. 3.1 (p. 36) it seems – at a very rough estimate – that a little over half the world's primary energy is used in the first place to produce mechanical energy (all the transport consumption, all the input to electricity generation, and perhaps half the remaining industrial consumption). Now we know that the efficiency of this heat-to-mechanical conversion is only about 30 per cent and two-thirds of the input is bound to remain as heat. So if energy is in short supply, surely we should take care to use this 'free' contribution to our heat needs, rather than throwing it away and then paying in valuable resources to replace it. On the more local household scale, two-thirds of the price of your daily dozen kilowatt-hours goes in payment for heat donated to the English Channel, Pacific Ocean, Rhine, or other water, or to the atmosphere.

Now all this is well-known. Indeed, the number and variety of proposals for 'using waste heat' is quite bewildering, and even if we stay with existing, well-tried methods, a glossary is needed to untangle the terms. *Combined heat and power, co-generation, total energy systems, waste heat recovery*; even *urban waste disposal* and *district heating*. All these make their appearance, and the list is by no means exhaustive. We shall talk about one topic: how to use waste heat from power-stations. Some of the items on the above list have nothing to do with this, but it may be well to start with a few brief explanations to sort out the terminology.

To begin at the end, *district heating* is the provision of heat for houses, offices or other buildings from a single plant. Rather than distributing fuel or electric power, you distribute heat – usually in the form of hot water or perhaps steam. The name implies nothing about how the heat is produced. It could mean simply a central boilerhouse for a housing development, or it could refer to the use of heat from power-stations, refineries, or the local refuse incinerator. (The fact that a fifth of Sweden's households use district heating does *not* mean that a fifth are heated by 'waste heat'.)

Urban waste conversion is a rather general term. We've already seen in Chapter 2 that urban waste is a good energy source. Much of it is still dumped or burnt with no useful output, but this is changing. Wastes are already used as contributing fuels in power-stations, though not all the technical problems are yet solved. Where incinerating plant exists the heat output can be used for district heating, and other possibilities are the production of liquid or gaseous fuels, and the use of waste heat recovery systems.

Waste heat recovery normally means using something hot which is departing to warm something cold which is arriving. Allowing flue gases to pre-heat air for a boiler is an example. A special case, often called *power recovery*, is the generation of electric power using flue gases. Temperatures may be quite low, needing special turbines with fluids other than steam. Thermodynamic efficiencies are correspondingly low too, but to obtain even some power from a fuel which costs nothing may make the investment worthwhile.

The term 'total energy system' is not well-defined. It can mean anything from a massive installation producing hundreds of megawatts of electric power and thousands of therms an hour in the form of different sorts of steam, to a little engine running a 10-kW generator and producing some hot water. It is a rather fashionable term which is coming to be used for what would have been called either co-generation or combined heat and power.

These last two names started life in quite different circumstances, but seem to be converging. Both refer to the simultaneous production of electric power and useful heat in purpose-built plant. *Co-generation* was and is normally used for situations where heat is in any case being produced in large quantities for specific purposes – usually process heat for industry. It may then make sense to 'co-generate' electric power; but the heat requirements (so much steam at certain temperatures and pressures) tend to govern how the system is run, and the electrical output is topped up, if necessary, from the public supply.

Combined heat and power, on the other hand, refers in its original meaning to systems where electric power is the main product. Not exactly 'waste heat from power-stations' but power generation adapted so that the heat output is useful. Nowadays CHP is often used more broadly to mean almost any system designed to produce both electric power and useful heat, but we shall take the more restricted sense.

CHP

> *For Sale: Half a billion gallons a day of luke-warm water.*
> *Purchaser to pay delivery charges.*

The trouble with waste heat from power-stations is that it is too cold and too distant. This is a pity, because the total quantity of energy is enough to be very useful indeed. In Britain, it is equal to nearly two-thirds of the heat used for all space heating and hot water. All we need is to get this into our buildings, reduce heat losses by a third with improved insulation, and we won't need any fuel for heating at all! The American figures are slightly less favourable at just over half the heat needed, whilst Switzerland, with a very high proportion of hydroelectric power, has little waste heat from this source. We'll concentrate on Britain, where there have been a number of recent studies of CHP on the national and also a more detailed level. Is there, between the gloom and the euphoria of the two extremes outlined above, a realistic and economically viable possibility?

It's worth looking at the problem from both directions. What does CHP mean to the power-station operator, and to the householder? We'll start at the demand end. A typical British house with central heating is likely to have a boiler using some form of fossil fuel, and water circulating through 'radiators' to distribute the heat. (Like all objects they do radiate, but they should really be called convectors because that's how they give up most heat.) The rated output of the boiler will probably be in the range from 20,000 to 60,000 BTU an hour, and the water will leave it at a temperature between 135°F and 180°F (55–80°C). The return water should be about 20°F cooler. Whether the required heat is distributed by sending out very hot water for short periods or cooler water for longer periods is a matter of choice; but there will always be a limit, a lowest temperature, which depends on the rate of heat loss from the house and the number of radiators. If the water is too cool it won't supply enough heat even when circulating continuously, and there are many dwellings and offices in Britain which need water at close to 180°F to meet demand in the coldest conditions.

Now any alternative to the present system will be cheapest if it involves the minimum change, and most acceptable if it involves the minimum disturbance. So, from the consumer's viewpoint, if we are talking about *existing* dwellings, a central system will be best if it can supply water under roughly the above conditions. Huge quantities of water at 100°F (40°C) *could* heat our houses but would need enormous radiator areas – whole walls or ceilings. Hot air might use lower temperatures, but hot air needs ducts.

To conclude, waste heat is most likely to be useful if it arrives in controllable quantities of hot water (up to several gallons a minute) at a temperature of at least 150°F – and even then many buildings would need extra radiators.

Back to the power-station, where skilful engineering and 100 years' experience have led to a turbine efficiency of 45 per cent. One essential for this high figure is that the steam leaving the turbine is cooled to the lowest possible temperature. And now along comes some bright CHP enthusiast asking for waste heat at, let's say, to be on the safe side, and allowing for losses, 200°F. Put this figure into the calculations in Chapter 5 instead of the 80°F we used, and the turbine efficiency falls with a thud to well under 40 per cent, undoing the effect of several decades of research and development. You can understand why the committed power engineer is seldom a CHP fan.

The justification is that we would then make good use of the previously wasted energy – but would we? What happens when we all feel warm and close the valves on our supply pipes? Is the power engineer now to throw water at 200°F into the sea? Waste heat with a vengeance! Perhaps we'd better stop and think exactly what is proposed. The system implied by the figures above is the 'back-pressure' turbine: the steam leaves the turbine before it has given up the maximum possible mechanical energy. An alternative is the 'pass-out' turbine, where some steam is drawn off at an earlier stage. Choosing the best system involves a complex analysis. Flexibility is important: how far you can 'de-couple' the heat and power outputs in order to meet different relative demands. There are questions of heat transfer, and the distribution system. Should you have a lesser flow of hotter water or a greater flow of cooler water? Should there be one huge closed system with water from the power-station flowing through the domestic radiators, or would local heat-exchangers be better? How about the distances of large power-stations from people? Do we want them in the middle of cities? Nuclear power-stations? And finally, how will consumption be monitored? (You are going to have to pay 'heat bills' for all this.)

The present situation in Britain is that the CEGB has one small CHP district heating scheme operating (in Hereford) using two 7.5-MW diesel generators, while CHP and co-generation in various forms are widespread in industry, and several other European countries have quite large systems. Plant efficiencies can be as high as 70 per cent for power-plus-heat output. To summarize a number of recent reports in a few words: the view is that about a third of Britain's heating and hot water needs could be met eventually by CHP; that individual gas or coal-fired boilers are still marginally cheaper at current fuel prices; that the most competitive CHP district heating plant would have a power output of 200 MW or more, and could distribute heat over a radius of about five miles; that even the *planning* for a single-city project will need five years; and that a major network would take twenty or thirty years to develop.

Under most assumptions about future prices, the consensus seems to be that the utility *could* expect a reasonable return on its investment, at heat prices which consumers would find attractive compared with the direct burning of fuels. So if we are planning for the year 2000 or beyond, with increasingly scarce and expensive primary energy, surely we should be taking very seriously indeed this possibility for doubling the useful output from each tonne of coal or uranium entering the power-station. And we must not overlook another consequence. Doubling the useful energy per tonne means halving the pollutants or wastes associated with each joule we finally consume.

15 Alternative Futures

Introduction

Now that we know all the ingredients – the available resources, and the present and possible technologies for consuming and conserving energy – and have seen the recipes for preparing forecasts, it is time to look at some of the results. What do the experts in this particular branch of cookery offer for the year 2000? From literally hundreds of items we can choose only a few for closer study, and we start with three which are all of the meat-and-two-veg variety. They reflect the future as it appears – or appeared in the middle or late 1970s – to the sort of people who advise governments. To be precise, they are not predictions at all but scenarios, a distinction to which we return shortly. The other items in the sample are more exotic. Not all are worked out in the same detail, but they show the wide range of current ideas and will provide a background for some questions which we shall ask in the final chapter.

Scenarios

A scenario is not a prediction. Rather, it is an attempt to work out a consistent picture of a possible future state of affairs by asking, 'How would it be if . . . ?' So any scenario starts with a set of assumptions, some of which may be very well-based while others are much more open to question. It is pretty certain, for instance, that – short of an immense global catastrophe – the population of the world will rise by about 50 per cent in the coming twenty years. Knowledge of the age distribution of the present population, and the shortness of the time-span considered, make it unlikely that the figure will be very different from this. What will happen to the per caput income of this population is much less easy to say, and consequently we find different scenarios assuming a whole range of growth rates from zero to as much as 5 per cent a year. Then, when we come to energy itself, and need figures for elasticities of demand, or for available supplies at various prices, the uncertainties are very great indeed. Thus we shouldn't be surprised that there are enormous variations between scenarios in the patterns of consumption they show for any chosen future year.

On the other hand, there *are* rules. A scenario is not an 'anything goes' free-for-all. It is supposed to present a *consistent* picture in which the various parts fit together, without contradicting each other and in accordance with reasonable assumptions about economic behaviour and technological possibilities. A scenario in which the price of a certain fuel is less than the cost of extracting it, or one where the average household fuel bill is greater than the average household income, would at least have to explain how such a state of affairs is to be maintained. A five-fold increase in coal production by the year 2000 may be possible in certain countries, but a scenario would need to show that the capital cost of developing the mines and distribution system could be met within its other assumptions about the overall growth of the economy during the intervening period.

It would be nice, when we come to the different scenarios, to be able to study some from the past in order to see how well they have managed to represent the situation now. Unfortunately there are two reasons why we can't do this. Firstly, the construction of detailed energy scenarios is a relatively new art, developed almost entirely since the 1973 crisis. The maximum period over which we could test a scenario is thus well under ten years, and it is easy to see the source of the second difficulty when we look at the events of these years: recession in 1974, further massive oil price rises in 1979, and even deeper recession in 1981–2. Perhaps it is a valid criticism of the whole endeavour that, after the events of 1973–4, none of the mid-seventies scenarios allowed for the further doubling of oil prices before 1980 or the *falling* GDPs which we are seeing in the recession. To be fair, the authors always point out that no scenario can deal with surprises. It stands to reason that it can't: unexpected political events or scientific discoveries are by definition those we don't yet know about. So the best we can hope for from any scenario is that it will show how things might work out.

The following paragraphs contain brief accounts of just three of the very many recent studies; two dealing with the USA and with Britain and the third with the entire world. The choice has been governed in part by the fact that reasonably detailed accounts of all three appear in books written in fairly straightforward language, and are therefore accessible to any reader who would like to follow the arguments more closely. The relevant publications are: *Energy in Transition, 1985–2010* (known as the CONAES report), *Energy Poli-*

cy, a consultative Document (the 1978 Green Paper), and *Energy in a Finite World* (the IIASA study). Details of these and other useful books appear in the list of Further Reading.

USA, 2010

The method used in the CONAES study starts by dividing consumers into three categories: *industry, buildings* (residential, commercial and public), and *transportation*. Future demand in each category is calculated for a range of different assumptions about economic growth and energy prices. Then the three sets of figures are brought together to construct consistent patterns of consumption for the entire nation. (The method used is rather like that mentioned in passing on p. 279.) Finally, these patterns of demand are matched with estimates of available supplies – which may involve shifts from one energy type to another; from gas to electrical heating, for instance. Ideally, the supplies would be those available under the conditions used in the demand projections, but it is interesting to note that the panel dealing with supply did not in fact produce the required lists of amounts available at different prices in future years, arguing that production is influenced by so many political and other factors that the figures would be meaninglesss. (An endearing feature of the entire CONAES report is that it makes no secret of the wide disagreement between its experts. Indeed, a good starting-point for the reader is the appendix consisting of some thirty pages of dissenting views!) Rather than detailed costs, then, the supply situation is characterized by the effort needed to reach the output required in 2010: from 'business as usual', carrying on with the existing system, to a 'national commitment' involving major investment in synthetic fuels, a possible relaxation of environmental controls, and incentives to develop renewable resources.

Most of the ten scenarios studied assume annual growth rates of either 2 per cent or 3 per cent. Energy prices are constant (in real terms), rising at 2 per cent, or rising at 4 per cent a year (doubling or quadrupling between 1975 and 2010). The two shown here lie slightly towards the 'growth' end of the range (a), and more towards the 'conservation' end (b). Both assume a slight rise in population and no major changes in technology or in life-styles. (Neither includes breeder reactors, for instance, and neither assumes a great reduction in the number of road vehicles – though the engines may be more efficient.) The main differences in demand come from two assumptions:

Scenario (a) assumes that GDP increases by an average 3 per cent a year while energy prices rise by about 2 per cent a year.

Scenario (b) assumes that GDP increases by only 2 per cent a year but that energy prices rise by 4 per cent a year.

On the supply side, both assume a moderate emphasis on renewable resources and (a) requires major investment in synfuels.

Fig. 15.1(a and b) displays the main results, in the same style as Fig. 3.5 (p. 41) which showed present US consumption. Comparing the three diagrams, we see at once that total primary energy doubles according to (a) but increases only slightly in (b). In both cases coal production rises (doubling or quadrupling) to replace oil and gas, mainly in power production but also as synthetic fuels. Indeed (a) requires enough liquefication and gasification capacity to process several million tonnes of coal a day, producing the equivalent of nearly 10 million barrels a day of synfuels. Nuclear capacity, 40 GW

(a) 'Medium energy'

National consumption

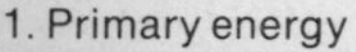

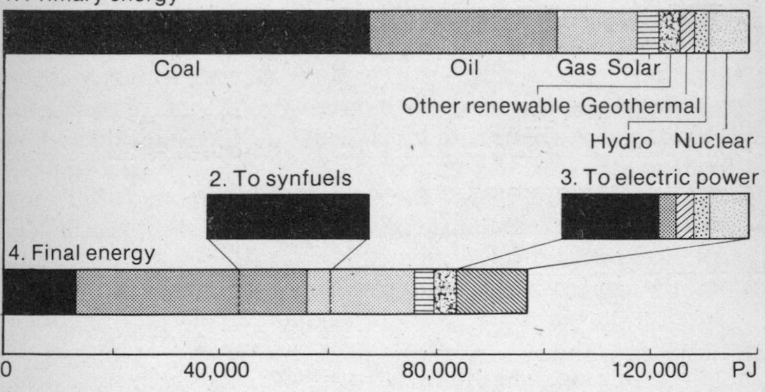

Consumption by sectors

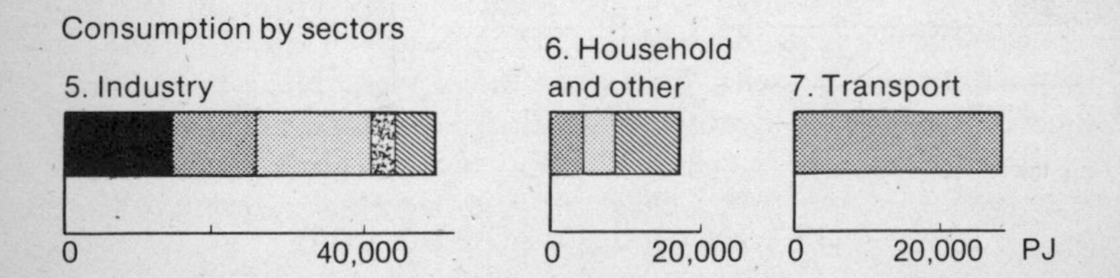

in 1975, rises to 350 GW in case (a) – effectively nine new power-stations a year over the 35-year period. Both scenarios include contributions equivalent to about a million barrels of oil a day from wastes and from geothermal energy, and enough solar energy to supply about a quarter of all residential heating.

Despite all these additional sources, neither scenario frees the USA from dependence on imported oil. Total oil consumption stays at roughly the present level in (a) and is halved in (b), and in both cases shales contribute a few per cent (up to a million barrels a day). Nevertheless, even with enhanced recovery, domestic oil is sufficient to meet only half the demand in each scenario, and imports are thus at about the present level for (a) or half this for (b).

Britain, 2000

The method used by Britain's Department of Energy in constructing their scenarios is not too different from that of the CONAES study.

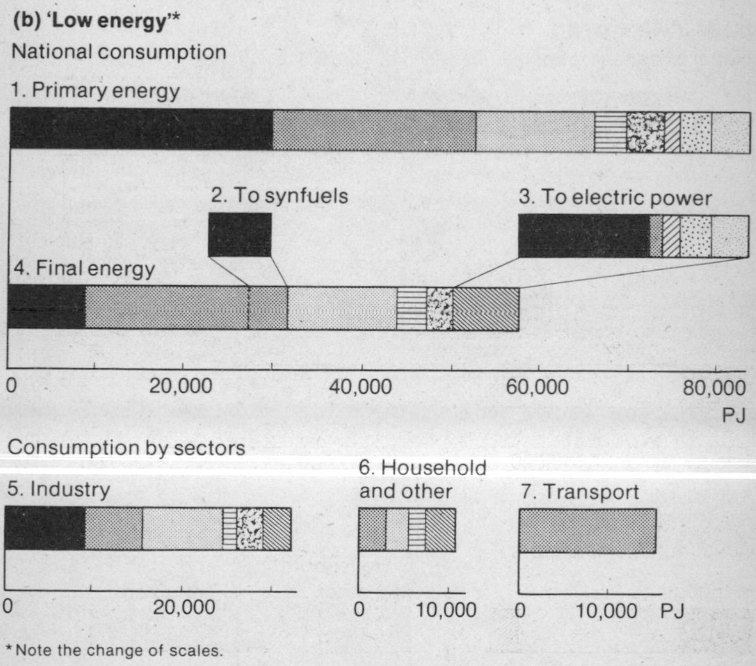

(b) 'Low energy'*

National consumption

1. Primary energy

2. To synfuels

3. To electric power

4. Final energy

0 20,000 40,000 60,000 80,000 PJ

Consumption by sectors

5. Industry

6. Household and other

7. Transport

0 20,000

0 10,000

0 10,000 PJ

*Note the change of scales.

Fig. 15.1 Scenarios for AD 2010, USA

Estimates of future demand under the assumptions of the scenarios are made for five main sectors of the economy: *iron and steel, other industry, domestic, commercial and public,* and *transport*. How this demand will be met by different fuels is then determined, taking into account availability and the choices consumers are likely to make in the light of future prices. The picture presented here is for the 'basic reference case' scenario, which assumes 3 per cent annual growth in GDP and a doubling of world energy prices between 1975 and the year 2000 (about 3 per cent annual increase in real terms). It is thus rather like scenario (a) of the US study, but with a shorter time-span. Its assumptions about technological and other energy-saving improvements are also similar; equivalent, it is claimed, to a 23 per cent reduction below present trends over the 25-year period.

Fig. 15.2 shows the outcome, and may be compared with the present pattern shown in Fig. 3.3. Primary energy consumption (which in both cases excludes non-energy uses) rises by a little under 50 per cent in the 25 years, during which GDP doubles. The first

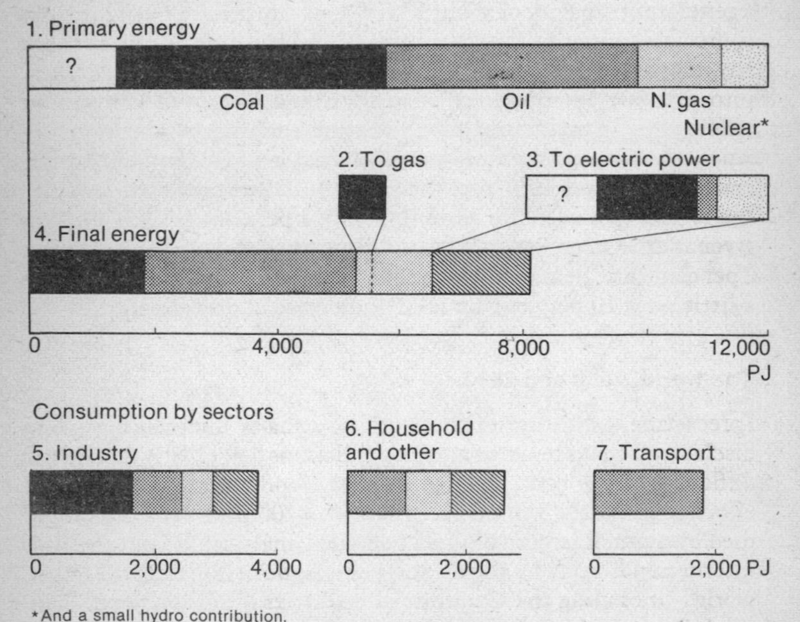

Fig. 15.2 Scenario for AD 2000, Britain

noteworthy feature in the diagram is the blank segment, representing a shortfall, and the Green Paper discusses how this might be filled. One possibility is that oil or gas reserves – or both – prove to be greater than estimated. (The gas production shown for 2000 is a little below the corresponding 1980 figure and oil about 10 per cent higher.) An alternative would be an even greater rise in coal production, but this is thought unlikely as it would mean an increase of more than 30 per cent above the present 130 Mte/yr within twenty years. A third suggestion is for Britain again to become an importer of fossil fuel.

The shortage of gas presents another problem, because the original demand projections for this scenario show it as a preferred fuel in the domestic and commercial sectors. In view of estimates of supply, the original consumption patterns for both these sectors are altered by substituting oil for about one-third of their gas demand. The predicted shortage also explains the only new technology in this scenario for the year 2000: the production of gas from some 100,000 tc a day of coal.

Nuclear energy contributes over a third of the electric power, with a generating capacity of about 35 GW – six times the 1980 figure. No appreciable input is expected from renewable sources, nor is there any major saving from CHP. The main conservation measures taken into account are improved insulation and improved conversion efficiencies in industrial processes and internal combustion engines, to be brought about by a combination of government action and increasing prices. To see the effect of GDP on predicted demand, the results of a scenario assuming only 2 per cent growth are also given in the Green Paper. With the same price assumptions as in the 3 per cent case, demand in the year 2000 becomes 10 EJ: a rise of only a little over 10 per cent for a GDP increase of 50 per cent.

The world, 2000 and 2030

Forecasting the future of the world is a major undertaking – and disclaimers that scenarios are not predictions don't really reduce the difficulty of the task. The IIASA study represents a monumental effort, seven years' work condensed into a 200-page book (accompanied by a much larger volume of detailed analysis). It isn't the first international study by any means, but it is the first to treat the entire world – including the Communist countries – in such detail. The method adopted for the study divides the world initially into seven regions: two for the non-communist developed countries (North

America, and the rest), two for the planned economies (the USSR with Eastern Europe, and China with the Asian countries), and three for the developing countries (the Middle East and North African oil producers, Latin America, and the rest). Demand predictions for each region are prepared, using a quite detailed breakdown of the economy into sectors and with two main sets of assumptions leading to *high-growth* and *low-growth* scenarios.

The first estimates of future demand to come from this exercise were so high, even with modest growth rates, that the original programme, designed to show how the world might manage the transition from fossil fuels to renewable resources, was drastically amended. The new question was whether, on any realistic assumptions at all, predicted demand could be brought down and predicted supply raised until they were at least within touching distance. However, before we move on to the final conclusions, it might be worth looking at one strand of the argument – in a very much simplified version – to understand the problem.

We start by noting that the present world annual energy consumption, some 300 EJ, can be split into three parts: the developed countries (a fifth of the world's population) account for rather more than 150 EJ, the Communist world for rather less than 100 EJ, and the developing countries, with half the total population, for 50 EJ, non-commercial fuels included. Per caput income in the developing world is about a tenth of that in the industrial countries, and we adopt the modest assumption that it will rise by 3 per cent a year, to reach in 2030 just under half of the *1980* figure for the wealthy fifth of the world. We further assume that the energy/GDP ratio needed for this industrial growth must at least equal that which the developed world now enjoys. Together, these assumptions imply almost a six-fold increase in per caput energy for the developing countries. But their population will more than double over the fifty years, so the total energy increase will be twelve-fold: from 50 EJ to 600 EJ. Add to this the 170 per cent increase resulting from a mere 2 per cent annual growth in the rest of the world and the total demand in 2030 becomes just under 1,300 EJ: over four times present consumption.

The broad results of the two main IIASA studies are shown in Fig. 15.3, for commercial fuels only. The high-growth total is appreciably below the figure from our rough calculation, while the low-growth version, holding the increase in per caput income below 2 per cent for almost all regions over the fifty-year period, has a 2030 demand which is about twice the 1980 consumption. The authors of

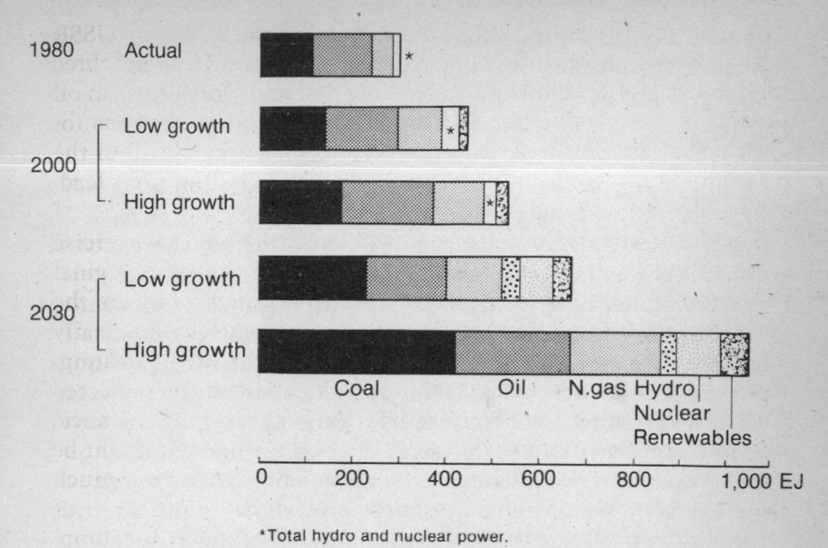

Fig. 15.3 Scenarios for AD 2000 and 2030, world primary energy

the study claim that supplies must be stretched to their limits to satisfy these demands. On thc assumptions that the Communist world will not need to trade in oil with other regions, and that the Middle East and North Africa will provide a steady supply of just over 30 million barrels a day, a major contribution is required from enhanced recovery of oil, tar sands and oil shales, and synfuels from coal. Together, these account for over two-thirds of liquid fuels in the 2030 high-growth scenario: 80 million barrels a day including over 30 from the conversion of more than a quarter of all coal production.

The huge coal total, nearly four times the amount we use today, raises the question of the effects of carbon dioxide; and the production of 10 million or more barrels a day from shales is not very inviting either. Then there is the nuclear input: a world generating capacity of nearly 5,000 GW, half of it breeder or near-breeder reactors. The figures are of course all scaled down in the low-growth scenario, but the total saving would do no more than reduce, say, the synfuels and the nuclear contributions by a half. It is argued that the 10 per cent or less supplied from renewable sources (including hydroelectric power) is the maximum which can realistically be expected within fifty years given the present state of their technologies;

though the long-term future is still seen to lie with the 'sustainable resources'.

Radical alternatives

Looking at the patterns of primary energy consumption in Figs. 15.1, 15.2 and 15.3 we must be struck by their remarkable similarity. The trends are really very much the same in all the scenarios studied. There is the thirst for oil as the only form of energy to satisfy the continuing needs of transportation, with a growing contribution from unconventional sources and coal liquefaction as conventional oil resources are exhausted; and there is an increasing demand for electric power from coal and nuclear energy, due in part to the shift of oil away from non-premium uses such as heating. It could of course be claimed that such resemblance emerging from three quite independent studies is evidence for the validity of their results, but not everyone would agree. An alternative view, noting that these and similar studies have all used the same type of analysis, drawn on much the same data, and – most important of all – adopted very much the same assumptions about economic growth, technical developments, and the ways people will behave, finds it not surprising that they have all reached much the same conclusions.

There's no shortage of more radical suggestions for the future. They cover a wide spectrum of opinion, but they do have one common feature. Where the conventional scenarios are telling us that we'll use more of the same (except when we haven't got it) the alternatives are looking towards something new, or at least a major change of emphasis, in the ways we use energy. Just a few of a very large number of ideas are discussed in the following sections, a somewhat arbitrary choice but one which includes examples of quite different views and also brings in one or two energy systems not treated elsewhere in the book. The results of the conservation scenario, one of the most detailed studies of an alternative to the conventional approach, are presented in the form used for earlier scenarios; but for the rest we concentrate on the essential features and critical points of the proposals rather than their detailed implementation.

The all-electric economy

The name is self-explanatory, and the proposal takes as its starting-point the fact that electricity provides by far the cleanest, and in many ways the most convenient, form of energy to the consumer. It can be produced from almost any energy source and used to satisfy almost any energy need. In the immediate future the primary sources would be mainly coal and nuclear fission reactors, while possible developments for the medium term are breeder reactors and improved combustion of coal – both options with no resource problems for many decades. For the more distant future, nuclear fusion will provide an almost inexhaustible source, but if it should not prove feasible, electric power is in any case the most likely – if not the only – way to use the renewable sources: the sun, tides, waves, wind, or OTEC.

The industrialized countries already possess distribution systems, and the developing countries are almost certain to extend theirs in any case, so the available financial resources can be concentrated rather than dissipated over many energy systems. Electric motors and heaters already exist in wide variety, which makes the move to an all-electric economy easier for the consumers (household or industry) and the equipment producers than any other alternative future.

Transportation is the main problem for the all-electric economy: the electric car, or rather its absence. If it existed it would fit very well into the plan, charging its batteries overnight when the demand for power is otherwise low. And electric motors can be amongst the most efficient of energy converters, while modern electronic controls are rapidly reaching the stage where there is no reason why the motor shouldn't maintain this high efficiency over a range of speed and power output. Then of course electric vehicles are virtually emission-free. The picture of a city or a motorway with every vehicle purring quietly and cleanly on its journey is indeed attractive, and the only snag is that the journey is likely to be a rather short one.

Let's recapitulate a few figures. The medium-sized, not very well designed car of Chapter 6 needed an engine output of some 80 MJ for its 100-mile journey: a little over 20 kWh. The best storage capacity which has so far been achieved with relatively cheap rechargeable batteries (variants of the lead–acid car battery) is a storage capacity of a little under twenty watt-hours per pound of battery weight. So in its 100-mile journey our car would use all the energy stored in about half a tonne of batteries! And that assumes 100 per cent efficiency and no losses in the control circuits. Rather different from the internal combustion engine which, even with its miserable 18 per cent efficiency, still uses only a little over two gallons, or about 25 lb, of petrol for the journey.

Electric vehicles do already exist, of course, and have done for a long time. Apart from industrial trucks and delivery vehicles there are several models of electric car. Britain's Enfield, looking like a little mini, has been on the road for nearly twenty years; but even with its light body, and a bonnet and boot full of batteries, its top speed is barely more than 50 mph and its range at best seventy miles. A somewhat larger and more powerful car, the ETV-1, has been developed with the aid of government funding in the USA. Using eighteen 6-V batteries (nearly twice the Enfield's power supply) it achieves marginally better performance with appreciably more usable space.

At this stage it is difficult to compare costs with those of conventional cars. If we take half a kilowatt-hour as a reasonable figure, the 'fuel' cost per mile in Britain, using off-peak power, would be about 1p – a quarter the cost of petrol. But to this must be added the cost of replacing the complete battery set every 10,000 miles or so. At some £1,000 a time, this increases the effective cost per mile to roughly 10p, and a similar figure is quoted for the ETV-1.

Everyone is aware of the potential market for a competitive electric car (and indeed for cheap storage of electrical energy in general) and a great deal of research effort is being devoted to batteries. Essentially they are chemical stores, releasing energy in the reactions which occur as a current flows; but no one expects a battery ever to equal the storage capacity of petrol. Much higher energy densities than with lead–acid batteries have already been achieved, but the problem is to find a method which is reliable over many chargings and dischargings, does not need expensive materials, and is safe and simple to use. Until it appears, it seems unlikely that the direct use of electric power will play a major role in the transportation sector of the energy economy.

The hydrogen economy

Hydrogen is in many respects the ideal fuel. The energy released per tonne burned is more than twice that of any hydrocarbon, and the sole combustion product is water vapour – no carbon dioxide because no carbon is involved. The difficulty, however, is that the amount of free hydrogen gas existing naturally is very small, so if it is to be a major fuel it must be produced from something else. There are well-known techniques for doing this, and the chemical industries of the world at present use tens of millions of tonnes a year. Unfortunately most of this is made from methane (and some from coal), so as a way of reducing dependence on fossil fuels and avoiding the CO_2 problem it is hardly a useful idea. But the advocates of the hydrogen economy see a much more plentiful source than any hydrocarbon: hydrogen accounts for one-ninth of the mass of all *water* in the world.

The hydrogen economy, then, would work as follows. Hydrogen would be extracted from water, distributed as we now distribute natural gas, and burned as a heating fuel or used in internal combustion engines for motive power, in both cases combining with oxygen from the air to produce water vapour. It doesn't require deep thought to spot the catch in this. You can't start with water, carry out a series of processes finishing with the same amount of water, and produce nothing but energy! (We rejected a similar idea for methane production on just these grounds in Chapter 7.) The point is that the hydrogen economy is a way of dealing with energy, not a way of producing it. You *use* energy to extract hydrogen from water, so the hydrogen is a store and a means of distribution. We might regard the idea as an alternative to the all-electric economy, and indeed the most likely method for producing the hydrogen uses electric power as its input.

In principle the electrolytic cell is simple – though the complexities in practice are considerable. A voltage generated by some external source is applied between two metal plates dipping into the water – or to be precise, a suitable solution – and this pulls the positive and negative ions in opposite directions. As they reach the plates, hydrogen and oxygen gases are formed and bubble up. The salt used for the solution, usually potassium hydroxide, is not consumed, and the sole effect (in the ideal case) is the breaking up of H_2O into hydrogen and oxygen. Moreover, there is even a slight energy gain in principle, because the cell takes in not only electrical energy but some heat as well. Theoretically, the fuel energy of the hydrogen which is produced should be about a fifth more than the electrical energy input. In practice it is about equal at best.

But why have a hydrogen economy at all? If you have to produce electric power in the first place, why not stay with the electric economy? The answer lies in two major advantages of hydrogen: it can be stored and it can be used in internal combustion engines (Table 15.1). All the technologies needed for its use are known: large electrolytic plants exist; many large storage tanks (each equivalent to about ten Dinorwics) and at least one 100-mile pipeline (in the Ruhr) have been in use for decades; vehicles have run on hydrogen, and we used town's gas (50 per cent or more hydrogen) for nearly a

Table 15.1 All-electric and hydrogen economies: a comparison

	The all-electric economy	The hydrogen economy
Supply	Sources: primary energy in almost any form. Present: fossil fuels, thermal reactors, hydro. Medium term: coal and breeder reactors (plentiful supplies), hydro. Long-term: fusion reactors, sun, wind, tides, waves (inexhaustible or renewable), hydro. Production costs: about ten times those of oil or natural gas (which vary over a wide range). Energy cost: output is 30–35% of primary input (except hydro), so two-thirds of energy is wasted.	Sources: as for electric power. Long-term additional possibilities: semi-direct use of solar energy; thermochemical production using the heat from nuclear reactors directly. Production costs: estimated to be about 10% more than the corresponding electric power. Energy cost: as for electric power.
Storage	Difficult. Pumped storage is the only large-scale method available.	As liquid hydrogen (maintained at very low temperature) in tanks holding a few thousand gallons.
Transmission and distribution	Mainly high-voltage overhead lines for long distances. Systems already exist. Costs: account for about 20% of final cost to consumer (in Britain). Energy cost: losses are about 8% of generated energy (in Britain).	Pipelines. Possible (but not certain) use of existing gas pipelines and distribution network. Costs: should be about half that of the corresponding electric power. (But possible initial cost for pipes.) Energy cost: should be about one-fifth of the corresponding electrical loss.

Final use	High efficiency. Technologies exist for heating and stationary motive power. Problems in uses for transportation.	Wide range of efficiency depending on use. Technologies exist for heating. Can be used in internal combustion engines – conversion necessary but no major technical problems expected.
	Cost to consumer: per unit of delivered energy, about the same as oil or petrol and about four times natural gas.	Cost to consumer: many unknowns. Estimates from about 20% below to 20% above electric power.
Environmental issues	Those associated with the primary sources. Waste heat. Transmission lines. Final use is very clean.	In production mainly as for electric power. Transmission by underground pipeline. Probably clean in final use (some NO_x but less than with oil or petrol).
Safety issues	Those associated with primary sources. Known final-use hazards of electricity.	In production mainly as for electric power. Transmission and storage possibly more hazardous than for natural gas. Final-use hazards associated with natural gas or town gas. Further uncertainty with hydrogen in vehicles.

century. Of course, the scale is far too small at present. A full hydrogen economy would need several hundred times the world's existing production capacity, with about a five-fold increase in generating capacity for the electrical input. As Table 15.1 suggests, there are still many uncertainties, and the concept can at best be regarded as an option for the mid-twenty-first century. Once the uncertainties are resolved the choice between electric power and hydrogen seems likely to be determined by weighing the extra capital cost and possibly greater hazards of the latter against the two advantages mentioned above. But as we shall now see, these are by no means the only options.

A conservation scenario

Figs. 15.2 and 15.4 provide an interesting contrast. Both show the results of scenarios for Britain in the year 2000. They start with the same data from the past, are based on similar disaggregations of the economy, and both assume a growth rate of about 3 per cent a year in real terms. The difference which leads to the one-third reduction in

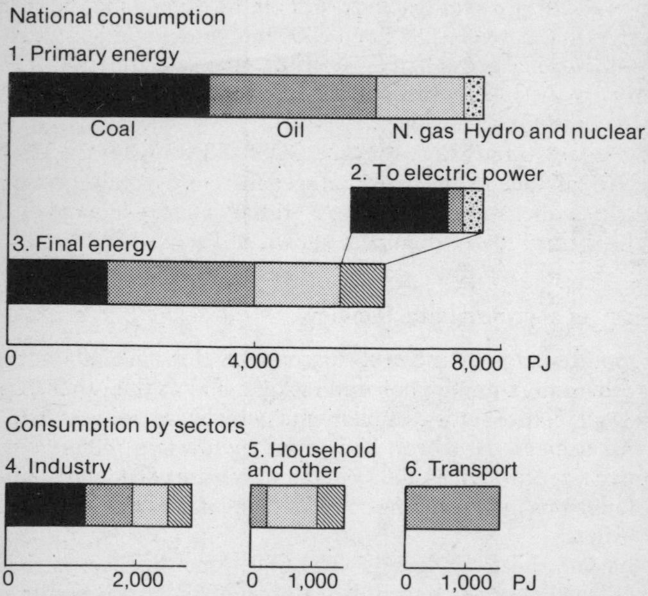

Fig. 15.4 Conservation scenario for AD 2000, Britain

demand is that the second scenario takes into account the savings which might follow a really serious conservation effort. The data for Fig. 15.4 come from a study whose main results were published in 1979 as *A Low Energy Strategy for the United Kingdom*. (See Further reading for details.) Space allows only a brief treatment of this very detailed analysis, and the interested reader is recommended to study the report. However, almost all the specific measures needed to reach the result shown have been treated in earlier chapters, and indeed the striking reductions in household and commercial consumption are very similar to those discussed in Chapter 14.

These energy savings are not based on revolutionary new technology or changes in life-style. Combined heat and power contributes only a thirtieth of household and commercial consumption, and solar energy and heat from the combustion of wastes roughly the same. Waves and wind together supply about a sixtieth of electric power, suggesting perhaps 250 MW of wind generating capacity and three times this for waves. From the viewpoint of 1982 the latter looks very optimistic, but its complete elimination wouldn't change the situation in a major way. The same could be said of the authors' assumptions about the rate at which we'll adopt heat pumps and waste heat recovery; but on the other hand the scenario assumes that between 1975 and 2000 fuel prices will double. As we have already had about half this rise in a quarter of the time, it would seem that the incentive to invest in conservation should be even greater than the study assumes.

It is interesting to note that the CONAES study for the USA (see page 303) included a conservation scenario corresponding roughly to the British one, and that it led to a primary energy demand of about 60 EJ – a third lower than that shown in Fig. 15.1(b).

Biogas: an appropriate technology

All *good* technology must be appropriate. If it isn't, it's not good. Sledgehammers make poor nutcrackers, and at the other extreme we've all come across implements which are so versatile that they're useless. However, the term *appropriate technology* has come to refer to devices and systems which are particularly suitable for conditions in developing countries, and it is in this sense that we use it here.

What are these conditions? The first important fact is that few developing countries have full-scale, reliable distribution systems for water, electric power or gas. Electric power reaches less than half the villages in India, for instance, and even in the electrified villages only one in seven households is connected to the supply. The absence of piped water means that pumps are needed for irrigation, household supply or local industry. Like the pumps which remained in use until quite recent years in many villages and remote farms in Europe, these have traditionally been operated by person-power, or perhaps animal power, and the same is true of other motive-power needs, including of course transporting people and goods. One way – perhaps the only way – of achieving the increased output necessary for economic growth is to call on the power of machines, and the evidence is that people throughout the world will adopt whenever they can one particular piece of appropriate technology: the internal combustion engine.

This is not what the advocates of appropriate technology usually have in mind of course, but there is no reason why the three-quarters of the world's population who are only beginning to have access to this versatile source of mobile power should be more immune to its charms than the rest of us. Unfortunately, however, it runs on oil products, and the price increases of recent years have been particularly catastrophic for people whose *only* modern aids are oil-based. The problem is further compounded by the other main uses of energy, for heat and light. With no electricity or gas, the only present alternative to ever-diminishing traditional fuels is kerosene – in British terminology, the paraffin stove and lamp. So it is not surprising that there is now a desperate search for other sources of energy. With no sign of a viable alternative to the internal combustion engine for *mobile* power, the question is whether a different fuel can be found to run it. For stationary power, and heat and light, the range of options may be wider.

Internal combustion engines will run on a variety of liquid and even gaseous fuels. Methanol and ethanol (methyl and ethyl alcohol) are both possible, and various mixtures of gases produced by partial combustion of coal, charcoal or wood have all been used. The largest existing petrol replacement scheme is Brazil's alcohol programme, processing sugar cane wastes and cassava to give some 50,000 barrels a day of ethanol. Cars designed for petrol run on mixtures with up to one-fifth ethanol, and it can also replace fuel oil for electric power generation. With millions of acres planted, and large distilleries, this is hardly small-scale, local technology. Nor is the energy balance entirely clear. As with other intensive bioconversion schemes, there is a considerable fuel *input* – for agricultural machines, fertilizers, transport, etc. – so not all the 50,000 barrels are net gain. (In the USA, the Department of Energy developed doubts during the 1970s and now refuses to fund any scheme for alternative liquid fuels unless it can be established that the net output will be positive!)

A fuel based on cultivated crops is unfortunately not very useful to countries like India whose people need every bit of good agricultural land for food. One alternative which has been suggested is methanol produced from wood. Given the increasing shortage of wood for fuel this may seem strange, but the argument is that the need to reduce dependence on costly oil products should take priority, and that burning wood on open stoves is in any case a very inefficient use of a valuable resource. If all the estimated 120–130 million te used annually for cooking in India were to be converted, the methanol produced would be sufficient, it is claimed, to run all the diesel engines needed for road transport and pumping water.

The question how people are then to cook their food is answered by what seems to be everybody's favourite appropriate technology: anaerobic fermentation. This is the process of decay of organic matter in the absence of air, and its product, called *biogas*, is a mixture with over 50 per cent methane. (The same as the marsh gas which bubbles up through stagnant water and, under suitable conditions, burns with a will-o'-the-wisp flame.) The idea is to run the digester (Fig. 15.5) on animal wastes; hence the Indian name *gobar* plant, from the word for cattle dung. In many ways it is the ideal energy converter. Biogas burns much more efficiently than dung cakes and the residue sludge is a better fertilizer than raw dung, so the output provides over twice as much heat and nearly three times as much useful fertilizer as would have been obtained from the original dung in its two traditional roles. Human wastes and some plant waste can also be processed, and as the treatment eliminates many pests and disease-carrying organisms, it also improves rural hygiene. (Both the gas and the residue are said to be clean and odourless.)

With about three cattle to every person, India has the potential to supply all cooking needs from biogas, with enough surplus in some regions to run diesel generators or pumps. (These are expensive, needing the dung from about fifty cattle for one five-horse-power motor for eight hours a day. And biogas, produced at low pressure, is not suitable for distribution over long distances, or for vehicles.) India's present 100,000 or so plants represent less than one per cent of the total potential, which could be of the order of an exajoule of useful heat a year. Capacity has expanded during the 1970s, though

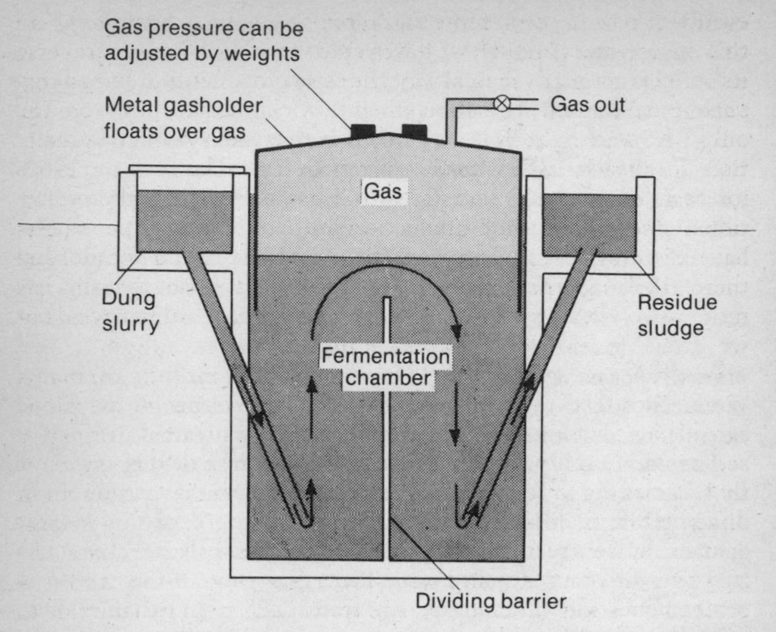

Fig. 15.5 Anaerobic fermentation plant

not always without problems. The biogas digester may be simple and robust, but it is no less demanding in its operating conditions than any other chemical plant. Incorrect feed, the wrong temperature, insufficient mixing, can all bring production to a halt, and like any failure of new technology, confirm the sceptic in his view that all innovation is a bad thing. And even this benign technology can cause unintended harm, when the hitherto free cow dung becomes the monopoly possession of the wealthy man who owns the digester.

Small is beautiful

Well before the first energy crisis of the 1970s a steady flow of books and articles had begun to appear whose theme was the need for a radical reappraisal of our attitudes to energy production and consumption. The authors deplored the accepted view that ever-rising demand was natural and inevitable, and in particular repudiated the fashionable belief in economies of scale – that bigger must be more efficient or more cost-effective, and therefore better. With the events of recent years, more and more people have come to share this unease, and although we have yet to see a whole country reverse its policies in a really radical way, the early proponents of *soft energy paths* have some justification when they claim that they were the ones who were right. It is certainly true that today's official predictions for the year 2000 often bear far more resemblance to the 1960s forecasts of the 'naïve and unrealistic' radicals than to the projections which those same officials were then offering. The experts have been revising their figures downwards for over a decade, and there's little sign that the process has stopped. It may be that this has more to do with economic recession than radical alternatives, but we should at least look at a few of the ideas.

What are soft energy paths? One thing is certain: they are many. We are not talking here about a single group of people with one agreed plan for the future. There *have* been a few detailed alternative scenarios, but it is in a sense a natural feature of small-scale systems that patterns of production and consumption should be different in different places, depending on local supplies and responding to local demand. So if there is a common view, it is probably better characterized by a set of aims than by one particular programme. And it is partly a question of emphasis. The transitions from exhaustible to renewable resources, from polluting to non-polluting technologies, and from energy waste to energy conservation should surely be welcomed by any disinterested observer of the energy scene. The difference comes in the interpretation of these phrases and the assignment of priorities. Which doesn't make it less of a difference, of course. However, we have already discussed all of these – or at least their technological implications – at some length, so perhaps it is more interesting to concentrate on the other plank in the radical platform, the one which gives this section its title: the belief that the *scale* of present energy systems is wrong.

The electric power industry is one major target for this criticism. (Oil is another, but the spread of do-it-yourself oil-wells is not exactly the desired end, so the emphasis is less on small-scale systems than on conservation: replacement of oil by other sources for heating, improvements in vehicle efficiency, and life-style changes such as increased use of public transport, bicycles, feet – or staying put.) Power production is crucial because it offers the greatest contrast between the systems we have and those we might have. The typical modern power-station in an industrialized country has an output in the gigawatt range, sufficient for about a million households. Its fuel is coal or uranium and (for either environmental or economic reasons) it is not likely to be sited in the middle of a town or city. It is large, it is distant, and it is seen as being controlled by a remote and inaccessible body.

The soft energy version would be very different. The demand for electric power would be lower, because resources now spent on devising ever more ways of using energy and persuading consumers to buy as much as possible would be devoted to developing energy-saving houses and energy-efficient equipment. Prices would encourage conservation, too, with the price per kilowatt-hour *increasing* rather than decreasing with the amount consumed. (This pricing policy is already adopted by some utility companies – in California, for instance.) The reduced demand resulting from all these measures would be as far as possible satisfied by renewable resources. Urban wastes; co-generation in industry; windmills, small hydro and geothermal schemes, or photovoltaics, where appropriate; all these, it is argued, could contribute much more power if given the financial support which existing fuels (nuclear power in particular) have received. Thermal power-stations (coal-fired plant with flue gas cleaning) should be in the megawatt rather than gigawatt range and built in or near towns and cities so that waste heat could be used in CHP schemes. Instead of the present competition between giant organizations each committed to one form of energy, local control of the full range of supplies would promote the best use of *all* resources.

This picture represents what might be called the conventional end of the range of soft energy paths. It has much in common with the conservation scenario already discussed and has indeed been worked out in detail in some cases. (Even by the US Department of Energy, in a 1978 study called *Distributed Energy Systems in California's Future*.) The systems discussed above, though smaller than today's norm for any of our three industrial countries, are by no means untried. There are already towns in the USA generating half their power from the wind; Switzerland has many small hydroelectric plants and an appreciable contribution from urban wastes; while Britain has a number of experiments, including, for instance, drilling for geothermal heat in the middle of Southampton. If we go to other countries, especially in Scandinavia, we can see integrated local systems in operation, and in particular, extensive use of CHP for district heating. But what about the really radical alternative, the softest possible path, the smallest scale of all?

The autonomous household is independent of all man-made energy supplies. The building is designed to capture as much solar energy as possible, and has some form of heat store: a large volume of rock or water, or the masonry itself. Electric power is supplied by solar cells or wind generators, with batteries providing enough storage capacity to run lights and other equipment when there is no input. Other sources of heat are biofuels, usually including wood, and possibly biogas produced by anaerobic fermentation of wastes. Autonomous houses do exist, and are proof that a return to medieval energy inputs need not mean a return to medieval standards of living. But the test is of course incomplete, because the present experiments all use the resources and technologies of the surrounding industrial society. (Photovoltaic cells *do* grow on trees, but they are called leaves, and they don't make very good power supplies.) To be fair, most of the enthusiasts for this life-style recognize that in its extreme form it is unlikely to appeal to everyone. Nevertheless, they make one point which ought to strike a sympathetic chord with just those supporters of modern technology who tend to reject their entire philosophy out-of-hand. The overwhelming majority of people in the developed world are totally dependent for their way of life on technical devices and systems which they do not begin to comprehend. Centuries of scientific advance have resulted in a situation where the chain of events between the coal-mine and the electric light-bulb is as much a mystery as thunderbolts from heaven were to our ancestors. Surely any serious technologist should welcome the idea of a world where people are not alienated from the mechanical objects around them, where an elegant solution to a technical problem might earn the same informed applause as a work of art or a display of sporting skill.

16 Future Alternatives

Choices

Every day, decisions are made about future energy supplies and future patterns of consumption. India accepts a tender for a new coal-fired power-station. The British government announces reduced support for wave-power. An American oil company abandons the development of oil shales. Swiss villagers adopt a proposal for combined waste-burning and hydroelectric power. Not headline news – just a collection of short items from the inner pages of a few days' papers. Such decisions are rarely considered important enough for detailed discussion even in the serious newspapers, yet it could be argued that it is the cumulative effect of these 'minor' choices which is determining the energy future of the world.

One purpose of the questions which follow is to make the connection, to draw attention to the fact that the sum of many little local decisions is a large-scale policy. And the converse: that any grand plan has small-scale implications which cannot be ignored. Large-scale or small-scale, the importance of quantitative reasoning is obvious. Our data may not be all that we would wish and we must be careful not to regard as certain that which isn't, but without some idea of magnitudes any proposal is useless. The questions posed here are clearly over-simplified, with heightened contrasts to bring out the issues, but they do preserve one important feature of the real world: it should be obvious that no matter how simple the questions there are no simple answers.

Resources

Arguably the single most important question about primary resources is how to make the best use of the remaining oil. Any oil producer, country or company, must have a depletion policy. Decisions must be made on how fast to extract the known reserves, how much to spend on enhanced recovery, when to invest in further exploration, and how soon to develop new discoveries. Unfortunately, even the most altruistic producer, with no concern but the present and future welfare of mankind, is faced with a dilemma.

Oil supplies are finite, so surely the right approach must be to limit the rate of production and use this valuable fuel only where there is no present alternative – which means mainly for transportation. Moreover, limiting the supply would encourage the development of more efficient vehicles and alternatives to petroleum, so that the world would be better prepared for the day when the wells do run dry. All very fine, but recent experience has shown only too clearly the effect of limiting the supply. Is that what we want – another doubling of oil prices, with all its consequences?

The alternative is to keep the oil flowing and not to let prices rise. Optimists point out that the growth in consumption has slowed or even reversed in the past few years, so perhaps the situation is not so serious. (Half a gallon to spare, and an 'oil-glut' appears in the headlines.) Whether there really is saturation in the developed world or merely a temporary effect of recession is a debated question, but we mustn't forget the developing countries, who would like very much to build their economic growth on cheap oil. Unless the total quantity turns out to be much greater than present estimates, a helpful policy of oil for all will certainly mean some hard thinking about alternatives, because they'll be needed within a few decades. This may sound a reasonable period – a whole generation – but history tells us that every new energy source so far has taken a good half-century of development to make an appreciable contribution to the world total.

So there's the question. If you were the world's energy manager and controlled its oil resources, would you limit supplies and accept the economic consequences, or follow the second course? Would your answer be different if you were managing the energy economy of the USA or Britain, or Saudi Arabia? And what would be your policy on alternative sources? Would you put your money into tar sands or oil shales, or coal liquefaction? How many of your billions would you invest in drilling deep into the Earth's crust to find the vast reservoir of natural gas which may or may not exist but would solve all our problems if it was there?

Conservation, prices and prediction

The questions about conservation are the obvious ones. We've looked at the technical possibilities: more efficient vehicles, better-insulated buildings, combined heat and power, and others. They'll all cost money now and save energy later. How much more is it worth paying for a 100-mpg car? Should something be done to reduce heat losses from buildings? Do we need legislation or more government spending, or should it be left to the individual?

This leads at once to the obvious question about prices: Is energy too cheap? The question is not frivolous. If it isn't worth our while to stop those draughts; if we don't think twice before using the car; if new 'prestige' office blocks have more-than-adequate heating and air-conditioning, but thin, metal-framed windows; if it's not worth investing in more efficient motors and pumps, or installing a heat exchanger to catch some of the hundreds of therms a day wafting out of the factory ventilator; in short, if conservation is just not economic, then perhaps energy *is* too cheap. Representatives of industry in Britain pleaded recently with government for lower energy prices. It would be nice to know that they had first asked for help in becoming more energy-efficient. Perhaps they had, but the issue didn't feature very largely in their submissions.

We certainly do use energy more efficiently now than 100 years ago. We heat our houses much better with very little more energy. We generate ten times as much electrical energy for each tonne of coal. Energy intensities and energy/GDP ratios are falling in all industrial countries. But is it enough? What would you do about prices if they were under your control? Raise them until a serious conservation effort *was* worth while? Is that the only sensible policy or a recipe for economic ruin? And would it not hit hardest those who can't even afford adequate heating at present prices?

Policies on conservation and prices are of course only the start. Your next task is to forecast demand in ten or twenty years' time. Some of the complexities and problems have been discussed in the past few chapters, and you'll find the arguments set out in detail in some of the books mentioned in the list of Further reading. That done, you'll need to look at the technical means for meeting the predicted demand.

Technology

The central issues are nearly all to do with electric power. How much, for instance? Should it provide more of our final energy than today or less? What about scale? Larger or smaller power-stations? Huge power-parks or individual household windmills? One immediate issue is the choice of primary energy. Should it be coal, nuclear power or one or more of the renewable sources? We've seen the potentialities and the penalties, and can recite the arguments.

The case for coal, whether its real supporters like it or not, is for many people the case *against* nuclear power. Coal is seen not as an ideal solution but as the lesser of two evils. Coal is dirty; mining it ruins the environment and is dangerous; and the carbon dioxide generated by combustion on a very large scale may have catastrophic effects. But a coal-fired power-station doesn't leave intensely radioactive wastes. It couldn't ever scatter exabecquerels of radioisotopes over the surrounding countryside. And world-wide growth in electric power from coal wouldn't have as an essential consequence the world-wide distribution of plutonium.

The trouble with the renewable sources is that although they are there, all about us, we haven't yet got the means to use them. There's no doubt that Britain or Switzerland, India or the USA, or the whole world, could satisfy *all* their needs from renewable supplies. But at what cost? Apart from the environmental questions, it's difficult to find agreement even on the order of magnitude of the necessary investment. Official studies all conclude that increasing the contribution of renewables to a half or more of national energy consumption would take at least half a century. We can't afford a faster growth rate; it would cost too much.

Once again you make your choice. If you were Prime Minister of Great Britain, would you put your faith in the assurances of the Atomic Energy Authority, the goodwill of the National Union of Mineworkers, or the winds and waves of the North Atlantic? If you were a Swiss mountain farmer, would you prefer rows of wind turbines along the skyline and dry rocky beds where streams have been diverted into pipes, or a dozen large nuclear power-stations in the lowlands? If you lived in the Third World, would you rather starve for lack of cooking fuel, or have your government and the one next door busily using their spare plutonium for all sorts of interesting experiments? Where ever you live, would you rather have the USA and the USSR developing their breeder reactors or competing for diminishing Middle East oil supplies?

A little bit over-dramatic? Too much gloom and doom? After all, modern technology does bring very great benefits, and if our ancestors had spent all their time reflecting on the possible ill-effects of change, we'd still be in the Stone Age. Our present energy systems do work, and their hazards are negligible compared with other natural and man-made risks which we happily accept. There have always been pessimists ready to forecast disaster within a decade, and history shows that they've usually been wrong. Coal and nuclear reactors both offer power from resources which could last for centuries. We already have long experience with the technologies, and the more efficient systems such as fluidized-bed boilers and breeder reactors are in a much more advanced state of development than any renewable source except hydroelectric power, whose contribution is limited. There is a reasonable chance of fusion power by the middle of the next century, and if coal and nuclear power together can meet world needs until then, why not use them?

Or is there an altogether different alternative? How about a worldwide electric grid? It could use underground and ocean-floor superconducting cables, and the power would come from solar farms in the world's major deserts, OTEC plants in tropical waters, and wave power-stations and wind turbine arrays in remote regions. No atmospheric pollution, no radioactive wastes, no use of valuable agricultural land or precious fresh water. Would it work? Estimates of world annual energy demand in fifty years' time lie between 600 and 1,000 EJ. Using the data from earlier chapters, it isn't difficult to find the size of installation for any selected contribution from each type of power plant. There are probably no insuperable *technical* problems. There is just one question. How do we get there from here?

Appendix A Orders of magnitude

The problem is how to deal with very large numbers without writing (or even worse, trying to recite) large numbers of noughts. There are two solutions: use an abbreviated form of arithmetic, or use special names.

Powers of ten

Two million is two times a million, and a million is ten times ten times ten times ten times ten times ten – six of them in all. This remark can be expressed mathematically:

$$2{,}000{,}000 = 2 \times 10^6.$$

The quantity 10^6 is called *ten to the power of six* (or *ten to the six*, for short), and the advantage of using this power-of-ten form is particularly obvious for *very* large numbers. For instance, Britain's annual primary energy consumption is a little under 9,000,000,000,000,000,000 joules, and it is certainly easier to write 9×10^{18} (or to say 'nine times ten to the eighteen') than to spell out all the noughts. And you can even observe that 18 is 3 times 6, so the figure is 9 million, million, million.

Prefixes

The special names used for large numbers are based on powers of ten, most commonly in multiples of a thousand (10^3, 10^6, 10^9, etc.). The table shows these, with some examples.

Prefix and abbreviation	Multiply by	Examples
kilo- (k)	10^3 (a thousand)	One kilowatt-hour is just over three and a half megajoules.
mega- (M)	10^6 (a million)	

(*Table cont. over page*)

Prefix and abbreviation	Multiply by	Examples
giga- (G)	10^9 (a billion)*	The output of a large power-station is about a gigawatt.
tera- (T)	10^{12}	The world consumes primary energy at an average rate of about ten terawatts . . .
peta- (P)	10^{15}	or about thirty-six petajoules an hour . . .
exa- (E)	10^{18}	or about three hundred exajoules a year.

* The now customary 'American' billion is used throughout the book.

Sub-multiples also have special prefixes. One *milli*metre is a thousandth of a metre – and a *micro*computer is technically a millionth of a computer.

Appendix B · Units and conversions

This appendix gives some of the figures you need for 'energy arithmetic'. Other data appear in the main text and can be found by using the index.

Energy

Basic energy units

Units based on watts × time

1 kWh = 3.6 MJ   1 MW–dy = 86.4 GJ   1 TW–yr = 31.5 EJ

Units based on the BTU

1 BTU = 1,055 J   1 therm = 105.5 MJ   1 quad = 1.055 EJ

Relating the two: 1 therm = 29.3 kWh.

Energy equivalents

Fuel	Quantity	Megajoules	Therms
Coal	1 tonne (1 tce)	29,000	275
Oil	1 tonne (1 toe)	42,000	400
	1 barrel (1 boe)	5,600	55
	1 gallon	160	1.5
Natural gas	1 tonne	55,000	520
	1 SCF	1.05	1/100 (1,000 BTU)
Uranium (natural)	1 kg	580,000	5,500
	1 lb	260,000	2,500
Wood	1 tonne	15,000	140
	1 cubic metre	9,000	85

(See the discussion in Chapter 1 on the variations in some of these quantities.)

Rate of consumption

Basic unit of power

1 watt = 1 joule per second

Some equivalent rates

1 kW is 86 MJ	*or* 0.8 therm	*or* 3 kg coal equiv.	*or* ½ gal oil equiv.	**per day**
1 MW is 86 GJ	*or* 820 therms	*or* 3 tce	*or* 15 boe	**per day**
1 TW is 31.5 EJ	*or* 30 quads	*or* 1,090 Mtce	*or* 750 Mtoe	**per year**

Note that 1 Mboe/day (1 Mbd) is 50 Mtoe/year.

Other quantities

Mass	1 tonne (1 te) = 1,000 kg = 2,205 lb
	(1 ton = 2,240 lb, 1 short ton = 2,000 lb)
Time	1 year = 8,760 hours = 31.5×10^6 seconds
Length	1 metre = 3 ft 3 in 1 mile = 1.6 km
Area	1 sq metre = 10.8 sq ft 1 acre = 4,047 sq metres
Volume	1 cubic metre = 220 gal 1 barrel = 35 gal
Speed	1 mph = 0.45 metres per second

Further reading

There is a *very* large number of books on energy, from popular paperbacks with pictures to massive tomes on technical topics, and it isn't easy to select just a few in such a wide field. Most of those mentioned here are chosen because they are well-written and reliable, and because they are indeed 'further reading': more detailed accounts of some of the many topics which are dealt with only briefly in this book. The titles are arranged in two groups, dealing first with the technologies and then with more general questions.

Technology

Unfortunately, the number of books which give enough technical detail to be interesting without drowning you in mathematics – or assuming that you already hold an engineering degree – is all too small. The following both come into this rare category in their very different ways, although it's a pity that Nero says little about Britain's AGR (perhaps someone will fill the gap?), and it should be mentioned that Park concentrates on small-scale systems.

Anthony V. Nero, *A Guidebook to Nuclear Reactors*. University of California Press, 1979.

Jack Park, *The Wind Power Book*. Cheshire Books, 1981.

The remainder tend to be a bit mathematical in parts, but have a lot of good accessible information between the algebra. Penner covers a wide selection of energy topics at a somewhat varying technical level, while Romer treats the basic science systematically in some detail. Both include masses of useful data. The rest are more specialized, but very informative and generally well written.

S. S. Penner and L. Icerman, *Energy* (3 vols.). Addison-Wesley, 1974–77.

R. H. Romer, *Energy, an Introduction to Physics*. W. H. Freeman, 1976.

F. S. Aschner, *Planning Fundamentals of Thermal Power Plants*. Wiley, 1978.

D. R. Blackmore and A. Thomas (eds.), *Fuel Economy of the Gasoline Engine*. Macmillan, 1978.

A. B. Meinel and M. P. Meinel, *Applied Solar Energy*. Addison-Wesley, 1977.

Other useful sources are the *Energy Papers* produced by the UK Department of Energy, which tend to lie in the region between technology and its wider implications. There have been studies of energy from waves, wind and sun, of CHP, the Severn barrage and many other topics. A list of current titles appears, for instance, in *British Books in Print*, under *Energy, Department of*.

In many ways the more popular scientific journals are the best source for the non-specialist reader, and first among these must of course be the *Scientific American*, with its excellent and authoritative articles. For topics of current interest, news items, and occasional longer articles on energy matters, there are *Science* and the *New Scientist* and a number of specialist journals, such as *Physics Today* and *Engineering News Review*, which aim to be accessible to outsiders. Then the individual energy industries produce masses of material, including journals such as the monthly *Atom* of the UKAEA.

Nuclear power

The central role of the nuclear debate in any discussion of the future, and the controversy surrounding nuclear power, are the main reasons for this separate section. (It must be emphasized that we are concerned here with nuclear *power*, not nuclear weapons.)

First there is literally a debate – the report of a Forum held at the Royal Institution. This impresses as one of the few publications in which those for and against nuclear power are trying to talk *with* rather than *at* each other.

G. Foley and A. van Buren (eds.), *Nuclear or Not?* Heinemann, 1978.

Next come four studies of some of the problems:

Royal Commission on Environmental Pollution, 6th Report: *Nuclear Power and the Environment*. HMSO, 1976.

Select Committee on Energy, 1st Report: *The Government's Statement on the New Nuclear Power Programme*. HMSO, 1981.

F. C. Williams and D. A. Deese, *Nuclear Non-Proliferation: The Spent Fuel Problem*. Pergamon, 1979.

Alan Cotterell, *How Safe is Nuclear Energy?* Heinemann, 1981.

These are obviously on the boundary between technology and the wider issues. The first is the well-known Flowers Report, and the second includes a fascinating and at times hilarious account of the committee's efforts to determine the cost of a PWR.

For anyone feeling like struggling with the mass of data, the enormous Rasmussen Report mentioned in Chapter 13 is:

NRC, Reactor Safety Study: An Assessment of Accident Risks in US Commercial Nuclear Power Plants. US Nuclear Regulatory Commission, 1975. (Also known as WASH 1400.)

And finally, two partisan paperbacks – perhaps best read with one in each hand.

Fred Hoyle, *Energy or Extinction? The Case for Nuclear Energy.* Heinemann, 1979.

Walter Patterson, *Nuclear Power.* Penguin Books, 1980.

Supply, demand, futures

Here the choice is very wide, and the following is just a small selection from literally hundreds of studies of the economic and environmental aspects of energy. We start with the four discussed in Chapter 15:

The National Research Council, *Energy in Transition 1985–2010.* W. H. Freeman, 1980. (The CONAES Report.)

Department of Energy, *Energy Policy, a Consultative Document.* HMSO, 1978. (The Green Paper.)

International Institute for Applied Systems Analysis, *Energy in a Finite World.* (Ballinger, 1981. (The IIASA Report.)

Gerald Leach, *A Low Energy Strategy for the United Kingdom.* Science Reviews Ltd., 1979.

Then a little book which argues strongly that there is far more oil than official figures suggest:

Peter Odell, *Energy: Needs and Resources.* Macmillan, 1977.

And a major study which sees fewer dangers than most, in a world future based on coal:

World Coal Study, *Coal, Bridge to the Future.* Ballinger, 1980.

A difficulty for any writer on energy is the speed with which the figures become out of date. Nevertheless, two books using data from the early 1970s should be mentioned:

J. Darmstadter et al., *How Industrial Societies Use Energy*. Resources for the Future, 1977.

P. D. Henderson, *India: The Energy Sector*. Oxford University Press, 1975.

One chapter from the second of these is reprinted, together with a number of other interesting papers, in

V. Smil and W. E. Knowland (eds.), *Energy in the Developing World*. Oxford University Press, 1980.

For comparison with the CONAES study there is,

Resources for the Future, *Energy: The Next Twenty Years*. Ballinger, 1979.

And for a survey of British resources, with a usefully long list of references at the end of each chapter,

John Fernie, *A Geography of Energy in the United Kingdom*. Longman, 1980.

A good recent critical study of epidemiological data and their problems is,

R. Wilson et al., *Health Effects of Fossil Fuel Burning: Assessment and Mitigation*. Ballinger, 1980.

And for a straightforward account of another health effect,

J. E. Coggle and G. R. Noakes, *Biological Effects of Radiation*. Wykeham, 1972.

Sources of data

Energy is important these days, and the *National Statistics* (statistical yearbooks or handbooks, digests of statistics, etc.) produced annually by almost every country include more information on the subject each year. For a 'bare-bones' treatment of all countries, the *United Nations Statistical Yearbook* and *Monthly Bulletin of Statistics* are both useful, while the volumes called *World Energy Supplies (UN Statistical Papers, Series J)*, which cover five-year periods, are invaluable. For the countries of the OECD the *Interna-*

tional Energy Agency publish detailed energy balance sheets and many other sets of data.

Apart from their annual reviews and papers on specific topics, the Departments of Energy of the UK and the USA both publish short monthly surveys:

UK DOE, Energy Trends (from the Central Office of Information).

USA DOE, Monthly Energy Review (from the National Technical Information Service).

And the UK DOE produces annually the smallest 'reference book' of all: a handy little card to fold up and put in your pocket, called *UK Energy Statistics*.

Index